THE PHYSICAL FOUNDATIONS OF
HYDRODYNAMIC PROCESSES

Macroscopic and Kinetic Approaches

THE PHYSICAL FOUNDATIONS OF
HYDRODYNAMIC PROCESSES
Macroscopic and Kinetic Approaches

Nikolay Nikolaevich Fimin
Valery Mihailovich Chechetkin

Keldysh Institute of Applied Mathematics of the
Russian Academy of Sciences, Moscow, Russia

 **World Scientific**

NEW JERSEY · LONDON · SINGAPORE · BEIJING · SHANGHAI · HONG KONG · TAIPEI · CHENNAI · TOKYO

Published by

World Scientific Publishing Co. Pte. Ltd.

5 Toh Tuck Link, Singapore 596224

USA office: 27 Warren Street, Suite 401-402, Hackensack, NJ 07601

UK office: 57 Shelton Street, Covent Garden, London WC2H 9HE

British Library Cataloguing-in-Publication Data
A catalogue record for this book is available from the British Library.

ISBN 978-981-121-115-7 (hardcover)
ISBN 978-981-121-116-4 (ebook for institutions)
ISBN 978-981-121-117-1 (ebook for individuals)

For any available supplementary material, please visit
https://www.worldscientific.com/worldscibooks/10.1142/11576#t=suppl

Desk Editor: Nur Syarfeena Binte Mohd Fauzi

Typeset by Stallion Press
Email: enquiries@stallionpress.com

To Academician O.M. Belotserkovskii

Contents

Introduction

Entia non multiplicanda praeter necessitatem.

William Occam

The main aim of this book is the consideration of the physical bases of hydrodynamics and low-temperature plasma dynamics. The conceptual idea of the book is based on a course of lectures of one of the authors (V.M. Chechetkin); this course has been lectured for students and postgraduates in the Moscow State University. The goals of the course were as follows: (1) to present the basic ideas of fluid mechanics in a mathematically attractive manner (i.e., this manner ought to be mathematically correct but not extremely sophisticated); (2) to present the physical background and motivation for some constructions that have been used in recent theoretical and computational investigations connected with the Euler and Navier–Stokes equations and kinetic equations (in particular, Boltzmann, Vlasov-type and Kac equations); (3) to interest some of the students for applications of hydrodynamics in astrophysics and nonlinear problems of turbulence theory.

During the writing of the book, the authors in some cases compiled and essentially used the standard materials considered in [1–15] for the description of the well-known facts and methods applied in hydrodynamics. There were summarized methods and approaches of these works in addition to some new material connected with kinetics, theory of hydrodynamic instabilities and theory of coherent systems.

The book is divided into three parts and appendices. The first part contains an elementary derivation of the hydrodynamic equations, their connections with kinetic ones and an introduction to thermodynamics of gas/liquid. The second part contains a discussion of influence of nonlinearity in the given equations on the physical reality and possibility of the description of turbulence, convection, etc. in terms of Euler–Navier–Stokes dynamics. The third part contains an analysis of dynamics of coherent systems and its geometrical properties. In the appendices, we consider the special problems of hydrodynamics which can seem interesting only for experts, but now attract attention in connection with nonlinear dynamics and astrophysics: the properties of quasi-Chaplygin media, Richtmyer–Meshkov instability (for shock waves interacting with contact discontinuity), instabilities and turbulence in astrophysics.

This book does not claim the role of a fundamental course in hydrodynamics, such as [16–19], but we hope that this book will be useful to graduate students, post-graduate students and everyone who is interested in the physical aspects of hydrodynamics. The authors are grateful for the comments concerning the contents of the book.

The authors dedicate this book to the memory of academician Oleg Mikhailovich Belotserkovsky, cooperation with which in the field of hydrodynamics and coherent structures led the authors to write this book.

Part I
Bases of Hydrodynamics

Chapter 1

Equations of Motion

1.1. Euler's Equations

Let D_N be a region in N-dimensional space ($N = 2, 3$) filled with a matter (liquid or gas). Our purpose is to describe the motion of such a matter. Let $\vec{r} \in D$ (for $N = 2$: $\vec{r} = \{x, y\}$, for $N = 3$: $\vec{r} = \{x, y, z\}$) be a point in D and let us consider the particle of fluid moving through $\vec{r}$ at time t. Let $\vec{v}(\vec{r}, t)$ denote the velocity of the particle of fluid that is moving through $\vec{r}$ at time t. Thus, for each fixed time, $\vec{v}$ is a vector field named *velocity field of the fluid* on D. We also assume that the fluid has *mass density* $\rho(\vec{r}, t)$ defined by $m(V, t) = \int_V \rho(\vec{r}, t) dV$, where $V \subseteq D_N$ (V is any subregion in D_N), dV is the volume element and $m(V, t)$ is the mass of fluid in V at time t.

The assumption that ρ exists (*continuum assumption*) does not hold if the atomic structure of matter is taken into account; but for many applications, we can believe that continuum assumption is justified.

Our derivation of the matter motion equations is based on the following *basic principles*: (1) mass in the closed system is constant; (2) the rate of change of momentum of a portion of the fluid equals the force applied to it ($dp/dt = F$); and (3) the energy in the closed system is constant.

1.1.1. *Continuity equation*

Let us consider conclusions from the first above-mentioned *basic principles* (we assume that $\vec{v}$, ρ and other functions are smooth enough). The rate of change of mass in the fixed subregion $V \subseteq D_N$ ($V \neq V(t)$, i.e., V doesn't change with time) is

$$\frac{d}{dt} m(V, t) = \frac{d}{dt} \int_V \rho(\vec{r}, t) dV = \int_V \frac{\partial \rho(\vec{r}, t)}{\partial t} dV. \qquad (1.1)$$

Let $\vec{n}$ be the unit outward normal vector to ∂V (smooth boundary of volume V). The volume flow rate across ∂V per unit area is $\vec{v} \cdot \vec{n}$ and mass flow rate per unit area is $\rho \vec{v} \cdot \vec{n}$. From the first basic assumption, it follows that the rate of increase of mass in V equals the rate at which the mass is crossing ∂V in inward direction:

$$\frac{d}{dt} \int_V \rho dV = - \int_{\partial V} \rho \vec{v} \cdot \vec{n} d\mathcal{S}, \qquad (1.2)$$

where $d\mathcal{S}$ is the area element on ∂V. This is the *integral form of the law of conservation of mass*. By the Gauss–Ostrogradskii theorem, $\int_{\partial V} \rho \vec{v} \cdot \vec{n} d\mathcal{S} = \int_V \operatorname{div}(\rho \vec{v}) dV$ and, since it holds for any V and for ρ, $\vec{v}$ assuming smooth enough, equation (1.2) is equivalent to

$$\frac{\partial \rho}{\partial t} + \operatorname{div}(\rho \vec{v}) = 0 \qquad (1.3)$$

(it is the *differential form of the law of mass conservation* or *continuity equation*).

1.1.2. *Balance of momentum*

Now we consider the question of *momentum balance*. The velocity field in the fluid is given by $\vec{v}(\vec{r}(t), t) \equiv d\vec{r}/dt$, where $\vec{r}(t)$ is the path of the fluid particle. The acceleration of a fluid particle is given by

$$\vec{w}(t) \equiv \frac{d^2 \vec{r}(t)}{dt^2} = \frac{d\vec{v}(\vec{r}(t), t)}{dt} = \frac{\partial \vec{v}}{\partial t} + \sum_{k=1}^{N} \frac{\partial \vec{v}}{\partial r_k} \frac{\partial r_k}{\partial t}, \quad \vec{r} = \{r_k\}_{k=1,\dots,N},$$

$$(1.4)$$

where $r_1 = x, r_2 = y$ if $N = 2$ and $r_1 = x, r_2 = y, r_3 = z$ if $N = 3$. We can rewrite the last equation as

$$\vec{w}(t) = \frac{\partial v(\vec{r}, t)}{\partial t} + \vec{v} \cdot \nabla \vec{v} \equiv \frac{D}{Dt} \vec{v}(t), \tag{1.5}$$

where $D/Dt \equiv \partial/\partial t + \vec{v} \cdot \nabla$ is the *material derivative* (it takes into account that the field is moving and that the positions of fluid change with time). The material derivative may be applied to any scalar (then $\nabla = $ div) or vector (then $\nabla = $ grad) function. With the help of the material derivative, we can rewrite the continuity equation as

$$\frac{D\rho}{Dt} + \rho \nabla \vec{v} = 0. \tag{1.6}$$

Define an *ideal fluid* as one with the following property: for any motion of the ideal fluid, there exists a function $p(\vec{r}, t)$ (*pressure*) such that if S is a surface in the fluid with a chosen unit normal vector $\vec{n}$, the force of stress exerted across the surface S per unit area at time is $p(\vec{r}, t)\vec{n}$. If $S = \partial V$, where V is any subregion of D_N (or *volume of fluid* simply), then the total force exerted on fluid inside V by means of stress on its boundary is given by the Gauss–Ostrogradskii theorem

$$\vec{F}_s = -\int_{\partial V} p \vec{n} dS = -\int_V \text{grad}\, p dV.$$

If $\vec{f}_{\text{ex}}(\vec{r}, t)$ denotes the given external force (gravitational, electromagnetic, etc.) per unit mass, then the total external force is

$$\vec{F}_{\text{ex}}(\vec{r}, t) = \int_V \rho \vec{f}_{\text{ex}}(\vec{r}, t) dV. \tag{1.7}$$

The total force acting on volume is $\vec{F}_{\text{tot}} = \vec{F}_s + \vec{F}_{\text{ex}}$. From the second *basic principle*, the density of total force (per unit volume) is

$$\rho \frac{D\vec{v}}{Dt} = \vec{f}_{\text{tot}} = \vec{f}_s + \rho \vec{f}_{\text{ex}} = -\text{grad}\, p + \rho \vec{f}_{\text{ex}}. \tag{1.8}$$

This is the *differential form* of the *balance of momentum*. This equation is a vector one and therefore is equivalent to the system of N scalar equations for components of $\vec{v}$ (for planar case $N = 2$, for space case $N = 3$). If we rewrite (1.8) in the form

$$\rho\frac{\partial\vec{v}}{\partial t} + \rho(\vec{v}\cdot\nabla)\vec{v} = -\nabla p + \rho\vec{f}_{\mathrm{ex}}, \qquad (1.9)$$

then it is easy to see that the *integral form* of momentum balance is

$$\frac{d}{dt}\int_V \rho\vec{v}dV + \int_{\partial V}(p\vec{n} + \rho\vec{v}(\vec{v}\cdot\vec{n}))d\mathcal{S} = \int_V \rho\vec{f}_{\mathrm{ex}}dV. \qquad (1.10)$$

We have used an equation of continuity and Gauss theorem for fixed volume V. Let us consider some properties of momentum balance. Let $F(\vec{r}, t)$ be the trajectory followed by the fluid particle that is at point $\vec{r} \in D_N$ at time $t = 0$; we can consider the *fluid flow operator* $\widehat{F}[t]: \vec{r}(t) \to F(\vec{r}, t)$. This operator advances each fluid particle from its position at time $t = t_0$ to its position at the (fixed) time t (it is so-called *Lagrange approach* to hydrodynamics — see Remark 1.1 on page 16). The action of this operator cannot define for isolated points only: if one acts on each point of small volume (material element of fluid) δV, then we say that $\widehat{F}[t]$ *acts on the volume* δV:

$$\widehat{F}[t](\delta V[t_0]) = \delta V[t], \quad \delta V[t_0] \equiv \delta V,$$

since the material element δV moves with the fluid; $\delta V[t]$ is the *t-image of initial volume* δV.

We call a flow *incompressible* if for any t-image of fluid volume δV

$$|\delta V[t]| \equiv \int_{\delta V[t]} dV = \mathrm{const} \ \ (\text{i.e., } |\delta V[t]| = \mathrm{inv}(t)), \qquad (1.11)$$

or, in other words,

$$0 = \frac{d}{dt}\int_{\delta V[t]} dV = (*) = \int_{\delta V[t]} \mathrm{div}\,\vec{v}\,dV \ \ (\text{i.e., } \mathrm{div}\,\vec{v} = 0). \qquad (1.12)$$

The equality (∗) may be obtained with the help of basic properties of Jacobian $\widetilde{\mathfrak{I}}(\vec{r}, t)$ of the operator $\widehat{F}[t]$:

$$\frac{d}{dt} \int_{\delta V[t]} \rho \vec{v} dV = \frac{d}{dt} \int_{\delta V} (\rho \vec{v})(F(\vec{r}, t), t) \widetilde{\mathfrak{I}}(\vec{r}, t) dV,$$

$$\frac{\partial}{\partial t} \widetilde{\mathfrak{I}}(\vec{r}, t) = \widetilde{\mathfrak{I}}(\vec{r}, t) \ \mathrm{div}\, \vec{v}(F(\vec{r}, t), t)$$

(see Remark 1.2 on page 16).

From equation (1.6) and the evident fact of positivity of density, we obtain that fluid is incompressible if and only if $D\rho/Dt = 0$; if the fluid is *homogeneous* (i.e., $\rho = const$ in space), it follows that the fluid is incompressible if and only if density is not varying function of time (i.e., $\rho = \mathrm{inv}(t)$).

Now, let us consider motion of compressible fluid. The rate of change of material element of fluid, whose volume is δV is

$$\frac{d}{dt}(\delta V) = \int_{\delta V} \mathrm{div}\, \vec{v}\, dV = \mathrm{div}\, \vec{v} \cdot \delta V + o(\delta V), \qquad (1.13)$$

in which the rate of expansion $\mathrm{div}\, \vec{v}$ is evaluated at the instantaneous position of the material volume element and the symbol $o(\delta V)$ denotes a quantity of smaller order than δV. A convenient way of obtaining an exact relation from (1.13) is to consider the ratio of δV to its value at some initial instant, t_0 say, then $\delta V[t_0]$ being made indefinitely small. Thus,

$$\frac{d\widehat{\varkappa}}{dt} = \nabla \vec{v} \widehat{\varkappa}, \quad \widehat{\varkappa} = \lim_{\delta V[t_0] \to 0} \frac{\delta V[t]}{\delta V[t_0]}. \qquad (1.14)$$

The rate of change of the vector $d\vec{\ell}$ representing a material line element which remains approximately straight is simply the difference between the velocities at the two ends of the element, i.e.,

$$\frac{d}{dt}(d\vec{\ell}) = d\vec{\ell} \cdot \mathrm{div}\, \vec{v} + o(|\vec{\ell}|). \qquad (1.15)$$

Again, we can make this an exact relation by dividing by $|d\vec{\ell}(t_0)|$ and taking the limit as $|d\vec{\ell}(t_0)| \to 0$. The rate of change of volume of a material element depends on the magnitude of that volume, but not on the shape of the surface bounding it. We may therefore choose a material volume element in the form of a cylinder whose two end faces are identical material surface elements with vector area represented by $\delta\vec{S}$ and of which a generator is the material line element $d\vec{\ell}$; such a material volume element remains cylindrical under the action of pure straining and rigid-body rotation, although $\delta\vec{S}$, $d\vec{\ell}$ and the angle between these vectors all change, and

$$\delta V = d\vec{\ell} \cdot \delta\vec{S} + o(|\delta V|), \tag{1.16}$$

at all times. On substituting (1.16) in (1.13), we find with the help of (1.15) that

$$(d\ell)_i \left(\frac{d(\delta S)_i}{dt} + (\delta S)_j \frac{\partial v_j}{\partial r_i} - (\delta S)_i \frac{\partial v_j}{\partial r_j} \right) = o(\delta V)$$

(vector notation being less convenient here), and since this relation must hold for all choices of $d\vec{\ell}$, we have

$$\frac{d(\delta S)_i}{dt} = (\delta S)_i \frac{\partial v_j}{\partial r_j} - (\delta S)_j \frac{\partial v_j}{\partial r_i} + o(|\delta\vec{S}|). \tag{1.17}$$

Again, an exact relation may be obtained by dividing by $|\delta\vec{S}(t_0)|$ and taking the limit as $|\delta\vec{S}(t_0)| \to 0$. An alternative way of writing this expression for the rate of a change of a material surface element which follows from the continuity equation is

$$\frac{d(\rho\delta S)_i}{dt} = -\rho(\delta S)_j \frac{\partial v_j}{\partial r_i} + o(|\delta\vec{S}|). \tag{1.18}$$

In the particular case of an incompressible fluid, ρ and δV are each invariant for a material element, and the factor ρ drops out of the (1.18).

1.1.3. *Euler's system of equations for incompressible fluid*

There are $N + 2$ unknown functions $(\vec{v}, \rho, p)$ in the system of $N + 1$ equations (1.3)+(1.8); hence, for complete derivation of fluid motion, we need one more equation. From the third *basic principle*, it follows that this equation may be law of *conservation of energy.* The total energy of fluid can be written as $E_{\text{tot}} = E_{\text{kin}} + \epsilon$, where

$$E_{\text{kin}} = \frac{1}{2} \int_V \rho(\vec{v})^2 dV \qquad (1.19)$$

is the kinetic energy contained in a volume $V \subset D_N$ and ϵ is the internal energy of fluid (molecular interactions, etc.). For ideal fluid, we are interested in kinetic energy only.

The rate of change of kinetic energy of a moving volume $V[t]$ follows from (1.10):

$$\dot{E}_{\text{kin}} = \frac{d}{dt} \int_{V[t]} \frac{\rho}{2} (\vec{v})^2 dV$$

$$= \int_{V[t]} \frac{\rho}{2} \frac{D}{Dt} (\vec{v})^2 dV = \int_{V[t]} \rho \vec{v} \cdot \left(\frac{\partial \vec{v}}{\partial t} + (\vec{v} \cdot \nabla) \vec{v} \right) dV. \quad (1.20)$$

If div $\vec{v} = 0$, then the last equation can be rewritten as

$$\dot{E}_{\text{kin}} = \int_{V[t]} (-\vec{v} \cdot \operatorname{grad} p + \rho \vec{v} \cdot \vec{f}_{\text{ex}}) dV.$$

Therefore, the law of conservation of energy for incompressible flow is a consequence of balance of momentum (1.8).

The equations of *motion of incompressible fluid* (for which div $\vec{v} = 0$) are as follows:

$$\frac{D\rho}{Dt} = \frac{\partial \rho}{\partial t} + \rho \operatorname{div} \vec{v} = 0,$$

$$\rho \frac{D\vec{v}}{Dt} = -\operatorname{grad} p + \rho \cdot \vec{f}_{\text{ex}}, \qquad (1.21)$$

$$\rho \vec{v} \left(\frac{\partial \vec{v}}{\partial t} + \vec{v} \cdot \nabla \vec{v} \right) = -\vec{v} \cdot \operatorname{grad} p + \rho \vec{v} \cdot \vec{f}_{\text{ex}}.$$

The second equation of this system (momentum balance) is often called *Euler's equation (of motion).* System of equations (1.21) is called *Euler's system of equations* for the incompressible ideal fluid.

1.1.4. *Euler's system for compressible isentropic fluid*

If we consider a more realistic model of compressible *isentropic flow,* then Euler's equations have similar form. The flow is called *isentropic* (or adiabatic), if its entropy (per unit mass of fluid) s is a constant. The *first law of thermodynamics* for fluid can be written as

$$d\epsilon_s = Tds + \frac{p}{\rho^2}d\rho, \qquad (1.22)$$

where T is the temperature of fluid, ρ is the density, p is the pressure in flow, ϵ_s is specific *internal energy* (per unit mass), and h is the specific *enthalpy* or *heat function* of fluid:

$$h = \epsilon_s - \frac{p}{\rho}, \quad \operatorname{grad} h = \frac{\operatorname{grad} p}{\rho}.$$

If the pressure is a function of ρ only, then in the flow $s = \operatorname{const}$ (*isentropic flow*) and pressure is $p = \rho^2 \partial \epsilon_s / \partial \rho$ (see (1.22)). Consequently, we obtain for internal energy, $\epsilon_s = \int^\rho (\rho^\dagger)^{-2} \cdot p(\rho^\dagger) d\rho^\dagger$. The derivative of total energy by time can be written as

$$\dot{E}_{\text{tot}} = \frac{d}{dt} \int_{V[t]} \left(\frac{\rho}{2} \vec{v}^2 + \rho\epsilon_s \right) dV = \int_{V[t]} \rho\vec{v} \cdot \vec{f}_{\text{ex}} dV - \int_{\partial V[t]} p\vec{v} \cdot \vec{n} dS.$$

Let us consider the terms $\partial(\rho v^2/2)/\partial t$ and $\partial(\rho\epsilon_s)/\partial t$:

$$\frac{\partial}{\partial t} \frac{\rho v^2}{2} = \frac{v^2}{2} \frac{\partial \rho}{\partial t} + \rho\vec{v}\frac{\partial \vec{v}}{\partial t} = -\frac{v^2}{2}\operatorname{div}(\rho\vec{v}) - \vec{v} \cdot \operatorname{grad} p - \rho\vec{v}(\vec{v}\,\nabla)\vec{v}$$

$$= -\frac{v^2}{2}\operatorname{div}(\rho\vec{v}) - \rho\vec{v} \cdot \nabla(h + v^2/2) + \rho T\vec{v}\,\nabla s,$$

$$\frac{\partial(\rho\epsilon_s)}{\partial t} = h\frac{\partial \rho}{\partial t} + \rho T\frac{\partial s}{\partial t} = -h\operatorname{div}(\rho\vec{v}) - \rho T\vec{v}\,\nabla s.$$

Therefore, we obtain

$$\frac{\partial}{\partial t}\left(\frac{\rho v^2}{2} + \rho\epsilon_s \right) = -\operatorname{div}\left(\rho\vec{v}\left(\frac{v^2}{2} + h \right) \right) + \rho\vec{f}_{\text{ex}}\vec{v}, \qquad (1.23)$$

or, after integration over volume V,

$$\frac{\partial}{\partial t} \int_V \left(\frac{\rho v^2}{2} + \rho \epsilon_s\right) dV = -\int_V \text{div}\left(\rho \vec{v}\left(\frac{v^2}{2} + h\right)\right) dV + \int_V \rho \vec{f}_{\text{ex}} \vec{v} dV$$

$$= -\oint_{\partial V} \rho \vec{v}\left(\frac{\rho v^2}{2} + h\right) \vec{n} dS + \int_V \rho \vec{f}_{\text{ex}} \vec{v} dV.$$

The left-hand side of equation (1.23) is the rate of change of the total (ideal) fluid energy in the fixed volume V; the right-hand side is the rate of work being done by external forces, minus a term that must be interpreted as the rate of energy flow through the boundary surface of the volume V. Therefore, the quantity $\rho \vec{v}(v^2/2 + h)$ is the *energy flux density vector*.

So, Euler's equations for isentropic flow in differential form are

$$\frac{\partial \rho}{\partial t} + \text{div}(\rho \vec{v}) = 0 \quad \text{(continuity equation)},$$

$$\frac{\partial \vec{v}}{\partial t} + (\vec{v} \cdot \nabla)\vec{v} = -\text{grad}\, h + \vec{f}_{\text{ex}} \quad \text{(Euler's momentum equation)},$$

$$(1.24)$$

and equation (1.23), which can be interpreted as *equation of conservation of energy density*.

The *boundary conditions* for equations of motion of fluid are $v_n = \vec{v} \cdot \vec{n} = 0$ on ∂D_N (or $\vec{v} \cdot \vec{n} = \vec{v}_0 \cdot \vec{n}$ if ∂D is moving with velocity $\vec{v}_0$). In the ideal flow, the fluid does not cross the boundary but may move tangentially to the boundary.

1.1.5. *The vorticity equations*

If the velocity field of a fluid is $\vec{v} = \{v_1, v_2, v_3\}$, then in the three-dimensional case, $\vec{\omega}_{(3)} \equiv \text{rot}_{(3)}\, \vec{v}$ is called the *vorticity* (or *vorticity field* of flow):

$$\text{rot}_{(3)}\, \vec{v} = \begin{vmatrix} \vec{e}_x & \vec{e}_y & \vec{e}_z \\ \partial/\partial x & \partial/\partial y & \partial/\partial z \\ v_1 & v_2 & v_3 \end{vmatrix}. \tag{1.25}$$

For the two-dimensional case ($\vec{v} = \{v_1, v_2, 0\}$), the vorticity is defined as $\omega_{(2)} = |\vec{\omega}_{(3)}|_{\partial/\partial z = 0, v_3 = 0} = \partial v_2/\partial x - \partial v_1/\partial y$.

There are many purposes for which it is more convenient to think about the fluid motion in terms of vorticity rather than in terms of velocity, despite the simpler physical character of the latter quantity. It also proves to be possible and useful, in many important cases of fluid flow, to divide the flow field into two regions with different properties, one of them being characterized by the vorticity being approximately zero everywhere. Considerations of the way in which changes in the distribution of vorticity take place will therefore be given often in later chapters. We are not yet in a position to describe the effect that various forces acting on the fluid have on the vorticity, but we can note the purely kinematical consequences of the definition of $\omega_{(3)}$ as $\nabla \times \vec{v}$ or, equivalently, as twice the effective local angular velocity of the fluid. One consequence is the identity $\nabla \omega_{(3)} = 0$. A line in the fluid whose tangent is everywhere parallel to the local vorticity vector is termed a *vortex-line*. The surface in the fluid formed by all the vortex-lines passing through a given reducible closed curve drawn in the fluid is said to be a *vortex-tube*.

We will use the symbol curl instead of $\mathrm{rot}_{(N)}$, $N = 2, 3$ (in the case without exact definition of dimension).

The first equation of Euler's system (1.24) may be represented in *Lamb's form* for vorticities:

$$\frac{\partial(\mathrm{curl}\,\vec{v})}{\partial t} = \mathrm{curl}(\vec{v} \times \mathrm{curl}\,\vec{v}), \qquad (1.26)$$

where $\vec{v} \times \mathrm{curl}\,\vec{v}$ is an *outer product* of vectors $\vec{v}$ and $\mathrm{curl}\,\vec{v}$.

If $\partial\vec{v}/\partial t = 0$, then the flow is *stationary* and we can rewrite Euler's equation as

$$\frac{1}{2}\,\mathrm{grad}\,|\vec{v}|^2 - (\vec{v} \times \mathrm{curl}\,\vec{v}) = -\mathrm{grad}\,h.$$

The *streamlines of flow* are lines with the property: tangent vector to streamline has the same direction as that of the velocity vector (in the point $(\vec{r}, t)$, for fixed t). They are defined by system of differential equations

$$\frac{dx}{v_1} = \frac{dy}{v_2} = \frac{dz}{v_3} \quad \text{or} \quad \frac{d\vec{r}}{d\tau} = \vec{v}(\vec{r}(\tau), t),$$

where $\vec{r}(\tau)$ is a streamline (for fixed t) and τ is geodesic parameter on integral curve $\vec{v}$.

D. Bernoulli's theorem: In stationary adiabatic ($h = h_0$) and homogeneous incompressible ($h_0 = p/\rho$, $\rho = $ const) flows in the presence of conservative external forces only, the quantity $\frac{1}{2}|\vec{v}|^2 + h_0$ is constant along streamlines.

It is easy to prove this theorem: Since $b(\vec{v}) \equiv \frac{1}{2}\nabla|\vec{v}|^2 = (\vec{v} \cdot \nabla)\vec{v} + \vec{v} \times \operatorname{curl} \vec{v}$ and $(\vec{v} \cdot \nabla)\vec{v} = -\nabla h_0$, then $b(\vec{v}) + \nabla h_0 = \vec{v} \times \operatorname{curl} \vec{v}$ and

$$\frac{1}{2}(|\vec{v}|^2 + h_0)\Big|_{\vec{r}(\tau_1)}^{\vec{r}(\tau_2)} = \int_{\vec{r}(\tau_1)}^{\vec{r}(\tau_2)} \nabla\left(\frac{1}{2}|\vec{v}|^2 + h_0\right) \cdot d\vec{r}(\tau)$$

$$= \int_{\vec{r}(\tau_1)}^{\vec{r}(\tau_2)} (\vec{v} \times \operatorname{curl} \vec{v}) \cdot \frac{d\vec{r}(\tau)}{d\tau} d\tau = (*) = 0.$$

The equality $(*)$ is evident: vector $d\vec{r}/d\tau = \vec{v}(\vec{r}(\tau))$ is orthogonal to $\vec{v} \times \operatorname{curl} \vec{v}$.

For example, in the gravitational field (near surface of the Earth) for incompressible fluid, we should have $p + \rho|\vec{v}|^2/2 + \rho g z = $ const (here z is a height).

For irrotational flows, the velocity vector can be written as a gradient of a scalar potential ϕ (called *velocity potential*): $\vec{v} \equiv \nabla\phi$. The validity of this equation can be checked by noting that it automatically satisfies the conditions of irrotationality $\operatorname{curl} \vec{v} = \varepsilon_{ijk}\partial v_k/\partial r_j = 0$, where ε_{ijk} is the alternating tensor (permutation symbol). Euler's equation may be rewritten as

$$\frac{\partial \vec{v}}{\partial t} + \nabla\tilde{b} = \vec{v} \times \operatorname{curl} \vec{v}, \quad \tilde{b} \equiv \frac{|\vec{v}|^2}{2} + \int \frac{dp}{\rho} + gz, \quad |\vec{v}|^2 = \sum_{j=1}^{3} v_j^2.$$

$$(1.27)$$

Consequently, $\nabla(\partial\phi/\partial t + \tilde{b}) = 0$ and $\partial\phi/\partial t + \tilde{b} = b_0(t)$, where the integrating function $b_0(t)$ is independent of location (this is the so-called (generalized) *Bernoulli equation*).

Let γ be a simple closed contour in the fluid at $t = 0$ and $\gamma[t] = \widehat{F}[t]\gamma$, where $\widehat{F}[t]$ is the fluid flow operator. The *circulation of velocity* around $\gamma[t]$ is the integral by closed contour $\gamma[t]$:

$$\Gamma(\gamma[t]) = \oint \vec{v}(t) \cdot d\vec{\ell}.$$

W. Thomson's circulation theorem: For adiabatic (or incompressible) flow in the absence of external force, the circulation of velocity $\Gamma(\gamma[t]) = \mathrm{inv}(t)$ (i.e., is not a function of time).

For the proof of this theorem, we consider the derivative by the time of circulation:

$$\frac{d}{dt}\Gamma(\gamma[t]) = \oint \frac{d\vec{v}}{dt}d\vec{r} + \oint \vec{v}d\left(\frac{d\vec{r}}{dt}\right),$$

where $\vec{v}d(d\vec{r}/dt) = \vec{v}d\vec{v} = d(v^2)/2$; but $\oint d(v^2) = 0$ and $d\vec{v}/dt = -\mathrm{grad}\,h_0$ (h_0 is defined earlier). If we apply Stokes formula, then we obtain

$$\oint \frac{d\vec{v}}{dt}d\vec{r} = \oint \mathrm{curl}\left(\frac{d\vec{v}}{dt}\right)d\vec{S} = 0.$$

Here, $\vec{S}$ is an oriented element of square, $\partial\vec{S} = \gamma[t]$. Thus, $\frac{d}{dt}\oint \vec{v}d\vec{\ell} = 0$ or $\oint \vec{v}d\vec{\ell} = \int \mathrm{curl}\,\vec{v}d\vec{S} = \mathrm{const}$. The last equality may be rewritten for small contour as $\Gamma(\gamma[t]) \approx \mathrm{curl}\,\vec{v} \cdot d\vec{S} = \mathrm{const}$, i.e., vorticity moving with the flow of an adiabatic fluid.

The *vorticity equation* (for incompressible flow) in N-dimensional domain D_N is

$$\frac{D(\mathrm{rot}_{(N)}\,\vec{v})}{Dt} = \frac{\partial(\mathrm{rot}_{(N)}\,\vec{v})}{\partial t} + (\vec{v} \cdot \nabla)(\mathrm{rot}_{(N)}\,\vec{v}) = 0,$$

where for $N = 3$ (space case) $\vec{\omega}_{(3)} \equiv \mathrm{rot}_{(3)}\,\vec{v}$ is defined in (1.25), and for $N = 2$ (the planar case) $\omega_{(2)} \equiv \mathrm{rot}_{(2)}\,\vec{v} = \partial v_2/\partial x - \partial v_1/\partial y$; the boundary conditions are $\vec{v} \cdot \vec{n} = 0$, where $\vec{n}$ is the unit outward normal vector to ∂D_N.

Let us consider the case $N = 2$ for simplicity. We will assume that D_2 is simply connected; from $\mathrm{div}\,\vec{v} = 0$, we should have

$\partial v_1/\partial x = -\partial v_2/\partial y$. From vector calculus, there exists a scalar function $\psi(x, y, t)$ on D_2 unique up to additive constant such that

$$v_1 = \frac{\partial \psi(x, y, t)}{\partial y}, \quad v_2 = -\frac{\partial \psi(x, y, t)}{\partial x}. \tag{1.28}$$

The function $\psi(x, y)$ is the *stream function* for fixed t. The basic property of stream function is as follows: if $(x(\tau), y(\tau))$ is a streamline $(dx/d\tau = v_1, \ dy/d\tau = v_2)$, then

$$\frac{d}{d\tau}\psi(x(\tau), y(\tau), t) = -v_2 v_1 + v_1 v_2 = 0$$

(see formula (1.28)). The vorticity is now given by $\mathrm{rot}_{(2)}\, \vec{v} = -\partial^2\psi/\partial x^2 - \partial^2\psi/\partial y^2 = -\Delta\psi$. Hence, *vorticity* (or *Euler–Helmholtz*) *equation* has the following forms:

$$\frac{D(\mathrm{rot}_{(2)}\, \vec{v})}{Dt} \equiv \frac{\partial(\mathrm{rot}_{(2)}\, \vec{v})}{\partial t} + (\vec{v}\cdot\nabla)\,\mathrm{rot}_{(2)}\, \vec{v} = 0,$$

$$\Delta\psi = -\omega_{(2)}, \quad \psi = 0 \ \text{ on } \partial D_2$$

(for two-dimensional incompressible flow);

$$\frac{D(\mathrm{rot}_{(3)}\vec{v})}{Dt} - (\mathrm{rot}_{(3)}\, \vec{v}\cdot\nabla)\vec{v} = 0, \quad \Delta\vec{A} = -\omega_{(3)},$$

$$\mathrm{div}\,\vec{A} = 0, \quad \vec{v} = \mathrm{rot}_{(3)}\vec{A}$$

(for three-dimensional (3D) incompressible flow). Here, $\vec{A}$ is a *vector potential*. Use of the Green's function of the three-dimensional Laplace operator, in the absence of boundaries, yields

$$\vec{A}(\vec{r}) = \frac{1}{4\pi}\int \frac{\omega_{(3)}(\vec{r}^{\,\dagger})}{|\vec{r} - \vec{r}^{\,\dagger}|}d\vec{r}^{\,\dagger}, \quad \omega_{(3)}(\vec{r}^{\,\dagger}) \equiv \mathrm{rot}_{(3)}\vec{v}(\vec{r}^{\,\dagger}),$$

and thus

$$\vec{v} = \mathrm{rot}_{(3)}\vec{A} = -\frac{1}{4\pi}\int \frac{(\vec{r} - \vec{r}^{\,\dagger}) \times \omega_{(3)}(\vec{r}^{\,\dagger})}{|\vec{r} - \vec{r}^{\,\dagger}|^3}d\vec{r}^{\,\dagger}.$$

This is the *Biot–Savart law*.

Remark 1.1. There are two viewpoints on the consideration of the fluid motion: Euler and Lagrange ones. The Euler approach is based on the use of "fixed observation place" in the given point of space $\vec{r}$ (with Cartesian coordinates x, y, z). In this point, the observer measures hydrodynamic functions of flow $\rho = \rho(x, y, z; t)$, $\vec{v} = \vec{v}(x, y, z; t)$, $T = T(x, y, z; t)$ in any (fixed) time moment t. In the Lagrange approach, the observer moves with the particle of fluid in the flow and his placement is defined by coordinates of spacepoint, where the observer describes hydrodynamic quantities $\rho = \rho(x_0, y_0, z_0; t)$, $\vec{v} = \vec{v}(x_0, y_0, z_0; t)$, $T = T(x_0, y_0, z_0; t)$ at fixed moment t, where $\vec{r} = (x_0, y_0, z_0)$ is the initial $(t = t_0)$ placement of fluid particle (and observer) in the "laboratory" Cartesian coordinate system.

The analog of the first Euler's equation of fluid motion (1.15) for Lagrange approach to hydrodynamics is

$$\sum_{i=1}^{N} (\partial^2 r_i/\partial t^2)(\partial r_i/\partial (r_F))_i = \sum_{i=1}^{N} (f_{\text{ex}})_i \cdot (\partial r_i/\partial (x_F)_i) - (1/\rho)(\partial p/\partial (r_F)_i),$$

where r_i, $(r_F)_i$, $(f_{\text{ex}})_i$ are ith components of vectors $\vec{r}_F$, $\vec{f}_{\text{ex}}$, respectively $(i = 1, \ldots, N)$.

Lagrange analog of continuity equation has the following form (for $3D$ case):

$$\frac{\partial}{\partial t}\left(\rho \frac{\partial(x, y, z)}{\partial(x_F, y_F, z_F)}\right) = 0.$$

Remark 1.2. The Jacobian $\widetilde{\Im}(\vec{r}, t) = \frac{\partial(x_F, y_F, z_F)}{\partial(x, y, z)}$ characterizes the measure of change of volume $dx\,dy\,dz \to dx_F\,dy_F\,dz_F$; x_F, y_F, z_F are coordinates of observation point at time t (this point moves with the fluid).

$$\partial\widetilde{\Im}/\partial t = \begin{vmatrix} \partial^2 x_F/\partial t\partial x & \partial y_F/\partial x & \partial z_F/\partial x \\ \partial^2 x_F/\partial t\partial y & \partial y_F/\partial y & \partial z_F/\partial y \\ \partial^2 x_F/\partial t\partial z & \partial y_F/\partial z & \partial z_F/\partial z \end{vmatrix}$$

$$+ \begin{vmatrix} \partial x_F/\partial x & \partial^2 y_F/\partial x\partial t & \partial z_F/\partial x \\ \partial x_F/\partial y & \partial^2 y_F/\partial y\partial t & \partial z_F/\partial y \\ \partial x_F/\partial z & \partial^2 y_F/\partial z\partial t & \partial z_F/\partial z \end{vmatrix}$$

$$+ \begin{vmatrix} \partial x_F/\partial x & \partial y_F/\partial x & \partial^2 z_F/\partial x\partial t \\ \partial x_F/\partial y & \partial y_F/\partial y & \partial^2 z_F/\partial y\partial t \\ \partial x_F/\partial z & \partial y_F/\partial z & \partial^2 z_F/\partial z\partial t \end{vmatrix} = (\text{div } \vec{v})\widetilde{\Im}(\vec{r}, t).$$

Thus, the fluid is incompressible if $\widetilde{\Im}(\vec{r}, t) \equiv 1$.

1.2. Navier–Stokes Equations

Now we consider the second well-known model of fluid. This model (which is derived by the *Navier–Stokes equations* (NSE)) takes into account the molecular motion and its consequences. In the ideal fluid, Euler's equation may be written in tensor form as

$$\frac{\partial}{\partial t}\rho v_i = -\frac{\partial \Pi_{ik}}{\partial r_k}, \quad \Pi_{ik} = -\sigma_{ik} + \rho v_i v_k, \quad i, k = 1, \ldots, N,$$

or

$$\rho\left(\frac{\partial v_i}{\partial t} + v_k\frac{\partial v_i}{\partial r_k}\right) = -\frac{\partial p}{\partial r_i} + \sigma_{ik}^{(v)}, \tag{1.29}$$

where Π_{ik} is the symmetric tensor of momentum flow density, σ_{ik} is the stress tensor, and $\sigma_{ik}^{(v)}$ is the viscous stress tensor. The defined earlier momentum flow is invertible momentum transfer connected with mechanical motion of fluid parts (particles) and pressure forces acted in fluid. The *viscosity* (internal friction) in fluid comes to the origin of non-invertible momentum transfer in directions of decrease of velocity module.

We may obtain the equations of motion of viscous fluid if we add to the *ideal momentum flow* Π_{ik} the tensor term $\sigma_{ik}^{(v)}$ defined as the non-invertible transfer of momentum in fluid. Continuity equation does not change the form, but the transfer of momentum and energy are described by new equations (which are generalized Euler's ones).

1.2.1. *Obtaining NSE from physical assumptions*

The tensor of density of momentum flow in viscous fluid can be written as

$$\Pi_{ik} = -\sigma_{ik} + \rho v_i v_k, \quad \sigma_{ik} = -p\delta_{ik} + \sigma_{ik}^{(v)}, \tag{1.30}$$

where σ_{ik} is *stress tensor* and $\sigma_{ik}^{(v)}$ is *viscous stress tensor*. The general form of σ_{ik} can be obtained from the analysis of physical consequences of viscosity. Let's consider the following two consequences:

(a) processes of internal friction in the fluid arise when fluid volumes move with other velocities, since (if velocity gradients are not very large) $\sigma_{ik}^{(v)}$ must be a linear function of flow velocity derivatives by coordinates: $\sigma_{ik}^{(v)} = c_1 \cdot \partial v_i / \partial r_i + c_2$. But $\sigma_{ik}^{(v)}[\vec{v} = 0] = 0$ if $\vec{v} = 0$, then $c_2 = 0$;

(b) there is not internal friction in the fluid, if the fluid rotates with angular velocity $|\vec{\Omega}| = \text{const}$: $\sigma_{ik}^{(v)}[\Omega = \text{const}] = 0$. In this case, fluid velocity $\vec{v} = \vec{\Omega} \times \vec{r}$; the only linear combinations of the above-mentioned derivatives $\partial v_i / \partial r_k$, which are annulated if $\vec{v} = \vec{\Omega} \times \vec{r}$, are

$$\partial v_i / \partial r_k + \partial v_k / \partial r_i.$$

Consequently, $\sigma_{ik}^{(v)}$ must include these combinations.

The most general second rank tensor, which we can construct in a framework of assumptions (a) and (b), has the form of

$$\sigma_{ik}^{(v)} = \eta \left(\frac{\partial v_i}{\partial r_k} + \frac{\partial v_k}{\partial r_i} - \frac{2}{3} \delta_{ik} \frac{\partial v_j}{\partial r_j} \right) + \zeta \delta_{ik} \frac{\partial v_j}{\partial r_j}, \qquad (1.31)$$

where η is the *first viscosity coefficient*, or the *coefficient of dynamical (shear) viscosity*, ζ is the *second viscosity coefficient* (or the *coefficient of bulk viscosity*); $i, j, k = 1, \ldots, N$.

The equations of motion of viscous fluid can be obtained by adding the term $\partial \sigma_{ik}^{(v)} / \partial r_k$ to the right-hand side of Euler's equation (1.29):

$$\rho \left(\frac{\partial v_i}{\partial t} + v_k \frac{\partial v_i}{\partial r_k} \right) = -\frac{\partial p}{\partial r_i} + \frac{\partial}{\partial r_k} \left[\eta \left(\frac{\partial v_i}{\partial r_k} + \frac{\partial v_k}{\partial r_i} - \frac{2}{3} \delta_{ik} \frac{\partial v_j}{\partial r_j} \right) \right]$$

$$+ \frac{\partial}{\partial r_i} \left(\zeta \frac{\partial v_j}{\partial r_j} \right), \qquad (1.32)$$

or, equivalently,

$$\frac{\partial v_i}{\partial t} = -v_k \frac{\partial v_i}{\partial r_k} - \frac{1}{\rho} \frac{\partial p}{\partial r_i} + \frac{1}{\rho} \frac{\partial}{\partial r_k} \sigma_{ik}^{(v)}. \qquad (1.33)$$

This is the most general form of equations of motion of viscous fluid. The coefficients η, ζ are functions of temperature T and pressure p in fluid in the general case.

But if η, ζ do not depend on p, T in flow, then the tensor equation (1.32) can be rewritten in the vector form:

$$\rho \frac{D\vec{v}}{Dt} \equiv \rho \left(\frac{\partial \vec{v}}{\partial t} + (\vec{v}\,\nabla)\vec{v} \right) = -\operatorname{grad} p + \eta \Delta \vec{v} + \left(\zeta + \frac{\eta}{3} \right) \operatorname{grad} \operatorname{div} \vec{v}.$$

(1.34)

This equation is called the *Navier–Stokes equation (NSE)*.

One can simplify this equation if the fluid is incompressible $(\operatorname{div} \vec{v} = 0)$:

$$\frac{D\vec{v}}{Dt} \equiv \frac{\partial \vec{v}}{\partial t} + (\vec{v}\,\nabla)\vec{v} = -\frac{1}{\rho} \operatorname{grad} p + \nu \Delta \vec{v}, \quad \nu = \frac{\eta}{\rho}. \qquad (1.35)$$

It is possible to eliminate the pressure p from the last equation by applying to it the operation curl:

$$\frac{\partial}{\partial t} \operatorname{curl} \vec{v} + (\vec{v} \cdot \nabla) \operatorname{curl} \vec{v} - (\operatorname{curl} \vec{v} \cdot \nabla)\vec{v} = \nu \cdot \Delta \operatorname{curl} \vec{v}, \qquad (1.36)$$

where $\nu = \eta/\rho$ is the *kinematic viscosity coefficient*.

The boundary condition to Navier–Stokes equation is $\vec{v} = \vec{v}_0$ (but not $\vec{v} \cdot \vec{n} = \vec{v}_0 \cdot \vec{n}$ as for the ideal fluid). That is, the velocity of fluid on the boundary of region ∂D is equal to the velocity of this boundary ∂D (the new term $\nu \Delta \vec{v}$ raises the number of derivatives of $\vec{v}$. For example, on a solid wall at rest, one needs to add the condition that the tangential velocity be zero).

1.2.2. *The Reynolds parameter and similarity of flows*

Let us consider some scaling properties of the Navier–Stokes equations with the aim of introducing the *Reynolds parameter* (or Reynolds number) that measures the effect of viscosity on the flow.

Let us introduce in the region D_N ($N = 2, 3$) the *characteristic length* $\mathcal{L}$ and the *characteristic velocity* $\mathcal{V}_c$ for a given hydrodynamic problem ($\mathcal{T} = \mathcal{L}/\mathcal{V}_c$ determines the characteristic timescale in the system). These numbers are chosen in a somewhat arbitrary manner: e.g., if we consider flow past an ellipsoid, $\mathcal{L}$ can be connected with the length of one of the three semiaxes of this ellipsoid (we have to choose concrete semiaxis on the basis of physical sense of the investigated problem), and $\mathcal{V}_c$ can be the magnitude of the fluid velocity at infinity. $\mathcal{L}$ and $\mathcal{V}_c$ are merely reasonable length and velocity scales typical of the flow under consideration.

Now, we can introduce new dimensionless quantities:

$$v_j{}^\dagger = v_j/\mathcal{V}_c, \quad r_j{}^\dagger = r_j/\mathcal{L}, \quad t^\dagger = t/\mathcal{T}, \quad \vec{r} = \{r_j\}, \quad \vec{v} = \{v_j\},$$

$$j = 1, \ldots, N.$$

Let us rewrite equation (1.35) for incompressible fluid using these dimensionless variables:

$$\frac{\partial(v_1^\dagger \mathcal{V}_c)}{\partial t^\dagger} \frac{\partial t^\dagger}{\partial t} + \mathcal{V}_c v_1^\dagger \frac{\partial(v_1^\dagger \mathcal{V}_c)}{\partial r_1^\dagger} \frac{\partial r_1^\dagger}{\partial r_1} + \mathcal{V}_c v_2^\dagger \frac{\partial(v_1^\dagger \mathcal{V}_c)}{\partial r_2^\dagger} \frac{\partial r_2^\dagger}{\partial r_2}$$

$$+ \mathcal{V}_c v_3^\dagger \frac{\partial(v_1^\dagger \mathcal{V}_c)}{\partial r_3^\dagger} \frac{\partial r_3^\dagger}{\partial r_3}$$

$$= -\frac{1}{\rho} \frac{\partial p}{\partial r_1^\dagger} \frac{\partial r_1^\dagger}{\partial r_1} + \nu \left(\frac{\partial^2(v_1^\dagger \mathcal{V}_c)}{\partial(\mathcal{L} r_1^\dagger)^2} + \frac{\partial^2(v_1^\dagger \mathcal{V}_c)}{\partial(\mathcal{L} r_2^\dagger)^2} + \frac{\partial^2(v_1^\dagger \mathcal{V}_c)}{\partial(\mathcal{L} r_3^\dagger)^2} \right),$$

and the equations for the components of velocity v_2, v_3 are constructed in the similar manner. If we combine all three components, we obtain *dimensionless NSE* for incompressible fluid ($\mathrm{div}\,(\vec{v})^\dagger = 0$):

$$\frac{\partial \vec{v}^\dagger}{\partial t^\dagger} + (\vec{v}^\dagger \cdot \nabla^\dagger)\vec{v}^\dagger = -\nabla \left(\frac{p}{\rho \mathcal{V}_c^2} \right) + \frac{\nu}{\mathcal{L} \mathcal{V}_c} \Delta^\dagger \vec{v}^\dagger. \tag{1.37}$$

Let us define the *Reynolds parameter* Re to be the dimensionless number $Re = \mathcal{L} \mathcal{V}_c/\nu$ (consequently, on the right-hand side of formula (1.37), the coefficient of $\Delta \vec{v}$ is $1/Re$). The physical reason for introduction Re is that two flows *I* and *II* with the same geometry and the same Re are homotopic (similar) and velocity field $\vec{v}_{II}$ (for flow *II*) may be obtained from velocity field $\vec{v}_I$ (for flow *I*) by rescaling. The idea of the similarity of flows is used in the design of experimental models (for example, for investigation of behavior of a fluid flow around scaled-down version of the aircraft, spaceship, etc.).

Let us consider incompressible fluid and suppose Re is very small. Then $(\vec{v} \cdot \nabla)\vec{v} \propto \mathcal{V}_c^2/\mathcal{L} = f_1$, $(\eta/\rho)\Delta\vec{v} \propto \eta\mathcal{V}_c/\rho\mathcal{L}^2 = f_2$ and $f_1/f_2 = Re \ll 1$, i.e., *NSE* is converted to *Stokes' equation*:

$$\frac{\partial\vec{v}}{\partial t} = -\operatorname{grad} p + \frac{1}{Re}\Delta\vec{v} \quad (\text{and } \operatorname{div}\vec{v} = 0).$$

There are linear equations of parabolic type. For $Re \ll 1$ (slow velocity, large viscosity, small bodies), the solution of the Stokes equation provides good approximation to the solution of the NSE.

If Re is very large ($1/Re \ll 1$), then term $(\vec{v} \cdot \nabla)\vec{v}$ is dominant and one can say that viscous effects in the fluid are unimportant. The Reynolds number also determines whether a flow is *laminar* (i.e., smooth and orderly) or *turbulent* (i.e., disorderly and randomly fluctuating). It should be noted that there are other dimensionless parameters used in hydrodynamics (see Remark 1.3 below).

The simplest example of laminar shear flow occurs when a pressure head drives liquid through a channel of infinite depth in z-coordinate (*plane Poiseuille flow*). Let $\partial p/\partial x = \lambda_0 = \operatorname{const}$ be the pressure head per unit length. In steady state, the x-equation of motion reads $\rho v \cdot \partial v/\partial x = -\partial p/\partial x + \partial(\eta\partial v/\partial y)/\partial y$. When $\partial p/\partial x$ and η equal to constants, the velocity v does not depend on the variable x (flow remains invariant to translations down the length of the channel). Let boundary conditions (on the walls of the channel) for above-mentioned equation are $v(\pm H/2) = 0$. Then we obtain as the solution of this equation the following parabolic velocity profile:

$$v(y) = -\frac{|\lambda_0|}{2\eta}(y^2 - H^2/4).$$

Let the average velocity of the flow be $v_{av} = H^{-1}\int_{-H/2}^{H/2} v(y)dy = |\lambda_0|H^2/(12\eta)$, which equals 2/3 of the velocity at the center of the channel (see Fig. 1.1). The Reynolds number for the flow is $Re_{Pois} = v_{av}\mathcal{L}/\nu$, where $\mathcal{L} = H$ is the width of the channel and $\nu = \eta/\rho$ is the kinematic viscosity.

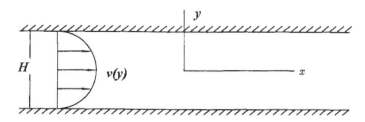

Fig. 1.1. The flow of liquid through a channel of width H, driven by a pressure head per unit length $|\lambda_0|$, has a parabolic velocity profile $v(y)$, which increases quadratically with distance from zero at the channel walls, $y = \pm H/2$, to a maximum value $|\lambda_0|H^2/(8\eta)$ in the middle of the channel, $y = 0$.

1.2.3. *Vorticity dynamics*

There exists the possibility to describe flow at large Reynolds number primarily in terms of the distribution of vorticity. The essential reason for this is that vorticity cannot be created or destroyed in the interior of a homogeneous fluid under normal conditions, and is produced only at boundaries. Rules about the way in which the vorticity associated with a material element of fluid changes as the element moves are available; and it is often possible to form a qualitative view of the distribution of vorticity throughout the fluid from inspection of the boundary conditions. Moreover, the net viscous force on an element of (incompressible) fluid is determined by the local gradients of vorticity. When the fluid viscosity is small, the net viscous force is significant only at places where the vorticity gradients are large; and if for some reason the vorticity is zero over a region of the flow, viscous stresses make no contribution to the net force on elements of fluid and may be ignored for most purposes.

The vorticity equation (1.36) for incompressible (div $\vec{v} = 0$) fluid can be rewritten as

$$\frac{\partial(\operatorname{curl}\vec{v})}{\partial t} = -(\vec{v}\cdot\nabla)\operatorname{curl}\vec{v} + (\operatorname{curl}\vec{v}\cdot\nabla)\vec{v} + \nu\cdot\nabla^2\operatorname{curl}\vec{v}, \quad (1.38)$$

since $\nabla(\operatorname{curl}\vec{v}) = \operatorname{div}(\operatorname{curl}\vec{v}) \equiv 0$.

We proceed now to interpret the contributions to the rate of change of curl $\vec{v}$ at a given point in space represented by the three terms on the right-hand side of (1.38). The first of these terms is the familiar rate of change due to convection of fluid in which the vorticity is non-uniform past the given point. The third term represents the rate of change of curl $\vec{v}$ due to molecular diffusion of vorticity in exactly the way that $\nu \nabla^2 \vec{v}$ represents the contribution to the acceleration from the diffusion of velocity. Vorticity, or angular velocity of the fluid, does not seem at first sight to be a transportable quantity, capable of being conveyed from one part of the fluid to another by molecular migration, but since all components of $\vec{v}$, at all points in the fluid, are transportable quantities, so too are spatial derivatives of $\vec{v}$ and so too, in effect, is vorticity. The second term on the right-hand side of (1.38) is the one that has no counterpart in the equation of momentum and that gives vorticity changes a distinctive character. Its meaning becomes evident if we write it in the form $|\text{curl}\,\vec{v}| \cdot \lim_{|AB| \to 0}(d\vec{v}/|AB|)$, where A and B are two neighboring points on the local vortex-line (see Fig. 1.2) and $d\vec{v}$ is the velocity of the fluid at the point B relative to that at A.

The corresponding contribution to the fractional rate of change of the vorticity $(|\text{curl}\,\vec{v}|^{-1} \cdot \partial(\text{curl}\,\vec{v})/\partial t)$ is thus identical with the fractional rate of change of the material line element vector extending from A to B.

In the case of two-dimensional motion, $\text{curl}\,\vec{v} \equiv \omega_{(2)}$ is everywhere normal to the plane of flow and $\text{curl}\,\vec{v} \cdot \nabla \vec{v} = 0$; then equation (1.38) reduces to the scalar equation $\partial \omega_{(2)}/\partial t + \vec{v} \cdot \nabla \omega_{(2)} = \nu \cdot \nabla^2 \omega_{(2)}$.

The fact that $\text{curl}\,\vec{v}$ changes, so far as the second term on the right-hand side of (1.38) is concerned, like the vector representing a material line element which coincides instantaneously with a portion of the local vortex-line may also be interpreted in terms of the behavior of the integral of the vorticity over a material surface. Let

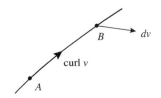

Fig. 1.2. Portion of a vortex-line.

us suppose that $\delta\Gamma(\gamma[t])$ is the circulation round the closed curve $\gamma[t]$ bounding a material plane surface element $\delta\vec{S}[t]$ located instantaneously at position $\vec{r}$; both $\delta\Gamma$ and $\mathrm{curl}\,\vec{v}\cdot\delta\vec{S}$ are equal to what we may term the vortex strength of the surface element $\delta\vec{S}$. The rate of change of this vortex strength is

$$\frac{d}{dt}(\delta\Gamma) = \frac{D\,\mathrm{curl}\,\vec{v}}{Dt}\cdot\delta\vec{S} + \frac{D\,\delta\vec{S}}{Dt}\cdot\mathrm{curl}\,\vec{v} = \nu\delta\vec{S}\cdot\nabla^2\,\mathrm{curl}\,\vec{v} + o(|\delta\vec{S}|),$$

which we see from (1.18) (with ρ constant) and (1.38) to reduce to

$$\frac{d}{dt}(\delta\Gamma) = \nu\delta\vec{S}\cdot\nabla^2(\mathrm{curl}\,\vec{v}) + o(\delta S). \tag{1.39}$$

It seems therefore that the vortex strength of a material surface element changes only as a consequence of molecular diffusion; changes in the magnitude and direction of $\delta\vec{S}$ have an effect on the strength which is exactly balanced by the changes due to the term $\mathrm{curl}\,\vec{v}\cdot\nabla\vec{v}$ on the right-hand side of (1.38). We can integrate (1.39) over an arbitrary open material surface and thereby find the rate of change of circulation round the bounding curve. However, it proves to be more useful to begin afresh. The circulation round a closed material curve drawn in the fluid is $\Gamma(\gamma[t]) = \oint_\gamma \vec{v}\cdot d\vec{\ell}$, and the typical element of integration can be regarded as a material line element $d\vec{\ell}$ which changes at a rate $d\vec{\ell}\cdot\nabla\vec{v}$. Then

$$\frac{d}{dt}\Gamma = \oint_\gamma\left(\frac{D\vec{v}}{Dt}\right)\cdot d\vec{\ell} + \oint_\gamma \vec{v}\cdot(d\vec{\ell}\cdot\nabla\vec{v})$$

$$= \oint_\gamma d\vec{\ell}\cdot\nabla\left(-\frac{p}{\rho} + \frac{1}{2}|\vec{v}|^2\right) + \nu\oint_\gamma(\nabla^2\vec{v})\cdot d\vec{\ell}$$

since ρ is uniform. For single-valued functions of position like p, ρ, $|\vec{v}|$, we have

$$\frac{d\Gamma}{dt} = \nu\oint_\gamma(\nabla^2\vec{v})\cdot d\vec{\ell} = -\nu\oint_\gamma(\nabla\times\mathrm{curl}\,\vec{v})\cdot d\vec{\ell}. \tag{1.40}$$

This relation is clearly consistent with (1.39). In the case of a material closed curve for which it is possible to find an open surface bounded by the curve and lying entirely in the fluid (such a closed curve is called *reducible* [18, 20]), integration of (1.39) over such a material surface and application of Stokes's theorem recovers (1.40) exactly. The relation (1.40) is actually a little stronger than (1.39), because in the case of material irreducible closed curves (say those which link a rigid cylindrical body of infinite length), integration of (1.39) over the open surface bounded by any two irreducible curves which are reconcilable with a one-to-one correspondence leads to the statement that $d\Gamma/dt + \nu \oint (\nabla \times \operatorname{curl} \vec{v}) \cdot d\vec{\ell}$ has the same value for all such closed curves, whereas (1.40) provides the result that this common value is zero (this is the generalization of the above-mentioned Thomson's circulation theorem).

1.2.4. *Dissipation energy rate in incompressible fluid*

There is a major difference between the ideal and viscous flow with regard to the energy of the fluid. The viscous terms provide a mechanism by which macroscopic energy can be converted into internal energy. General principles of thermodynamic state that this energy transfer is one-way. In particular, for incompressible flow, we should have $dE_{\text{kin}}/dt \leq 0$. Let's calculate the derivative kinetic energy by time:

$$\frac{\partial}{\partial t} \frac{\rho v^2}{2} = \rho v_i \frac{\partial v_i}{\partial t} \quad (\text{use } (1.24))$$

$$= -\rho \vec{v}(\vec{v} \cdot \nabla)\vec{v} - \vec{v} \cdot \nabla p + v_i \frac{\partial}{\partial t} \sigma_{ik}^{(v)}$$

$$= -\operatorname{div}\left[\rho \vec{v}\left(\frac{v^2}{2} + \frac{p}{\rho}\right) - v_i \sigma_{ik}^{(v)}\right] - \sigma_{ik}^{(v)} \frac{\partial v_i}{\partial r_k}.$$

Let's integrate last equation over the volume V:

$$\frac{\partial}{\partial t} \int_V \frac{\rho v^2}{2} dV = -\oint_{\partial V}\left[\rho \vec{v}\left(\frac{v^2}{2} + \frac{p}{\rho}\right) - v_i \sigma_{ik}^{(v)}\right] \vec{n} dS$$

$$- \int_V \sigma_{ik}^{(v)} \frac{\partial v_i}{\partial r_k} dV.$$

The first term on the right-hand side of this equation defines the change of kinetic energy of fluid in the volume V via energy flux across boundary $\mathcal{S} = \partial V$; the second term on the right-hand side characterizes dissipative decrease of kinetic energy (per time unit). If we integrate over $V = D_N$, then $\oint_{\partial V}[\ldots]\vec{n}d\mathcal{S} = 0$, and we obtain for dissipated energy (per second) in incompressible fluid:

$$\dot{E}_{\mathrm{kin}} = -\frac{1}{2} \int_V \sigma_{ik}^{(v)} \left(\frac{\partial v_i}{\partial r_k} + \frac{\partial v_k}{\partial r_i} \right) dV$$

$$= -\frac{\eta}{2} \int_V \left(\frac{\partial v_i}{\partial r_k} + \frac{\partial v_k}{\partial r_i} \right)^2 dV \leq 0.$$

Consequently, coefficient of viscosity $\eta > 0$.

1.2.5. *General equation of heat transfer*

Let us consider *conservation of total energy* in a viscous fluid for a material volume $V \subset D_N$:

$$\frac{D}{Dt} \int_V \rho \left(\frac{v^2}{2} + \epsilon_s \right) dV$$

$$= \int_V \rho \vec{f}_{\mathrm{ex}} \vec{v} dV + \int_{\partial V} v_i \sigma^{ij} n_i \cdot d\mathcal{S} - \int_{\partial V} \vec{q} \cdot \vec{n} d\mathcal{S}. \quad (1.41)$$

Here, the left-hand side gives the rate of change of the internal plus kinetic energy contained within volume V; the first term on the right-hand side is the rate at which work is done on the fluid by external volume forces, the second term is the rate at which work is done by surface forces arising from fluid stresses, the third term is the rate of energy loss out of the fluid element by means of a direct transport mechanism having an *energy flux* (or *thermal flow*, more precisely) $\vec{q} = -\kappa \nabla T$ (κ is the *coefficient of thermal conductivity*, T is the temperature of flow). The vector $\vec{q}$ is directed along the outward normal vector $\vec{n}$ of $\mathcal{S} = \partial V$, heat is lost from V. If we use Gauss–Ostrogradskii theorem to convert the surface integrals to volume ones, then (1.41) can be written in differential form as:

$$\frac{\partial}{\partial t} \rho \left(\frac{\rho v^2}{2} + \epsilon_s \right) + \frac{\partial}{\partial r_j} (\rho(h + v^2/2)v^i - v_i(\sigma^{(v)})^{ij} + q^i) = \rho v_i f_{\mathrm{ex}}^i.$$

$$(1.42)$$

This equation is the generalization of (1.23) to the case of viscous fluid. One may rewrite the last equation, if to use continuity equation and NSE (for change of the terms $\partial\rho/\partial t$ and $\partial\vec{v}/\partial t$, respectively):

$$\frac{\partial}{\partial t}\rho\left(\frac{\rho v^2}{2}+\epsilon_s\right) = -\frac{v^2}{2}\operatorname{div}(\rho\vec{v}) - \rho(\vec{v}\cdot\nabla)\frac{v^2}{2} - \vec{v}\nabla p + v_i\frac{\partial}{\partial r_k}(\sigma^{(v)})_{ik}$$

$$+ \rho\frac{\partial\epsilon_s}{\partial t} - \epsilon_s\operatorname{div}(\rho\vec{v}). \tag{1.43}$$

From thermodynamic equation $d\epsilon_s = Tds + p\rho^{-2}d\rho$, we obtain

$$\frac{\partial\epsilon_s}{\partial t} = T\frac{\partial s}{\partial t} + \frac{p}{\rho^2}\frac{\partial\rho}{\partial t} = T\frac{\partial s}{\partial t} - \frac{p}{\rho^2}\operatorname{div}(\rho\vec{v}).$$

Then (1.43) can be rewritten as

$$\frac{\partial}{\partial t}\left(\rho\epsilon_s + \frac{\rho v^2}{2}\right) = -\left(h + \frac{v^2}{2}\right)\operatorname{div}(\rho\vec{v}) - \rho(\vec{v}\cdot\nabla)\frac{v^2}{2} - \vec{v}\cdot\nabla p$$

$$+ \rho T\frac{\partial s}{\partial t} + \frac{\partial}{\partial x_k}(v_i\sigma_{ik}^{(v)}) - \sigma_{ik}^{(v)}\frac{\partial v_i}{\partial x_k}.$$

Finally, we have the equation for (specific) entropy balance:

$$\rho T\left(\frac{\partial s}{\partial t} + \vec{v}\cdot\nabla s\right) = \sigma_{ik}^{(v)}\frac{\partial v_i}{\partial r_k} + \operatorname{div}(\kappa T), \tag{1.44}$$

or, equivalently

$$\rho T\left(\frac{\partial s}{\partial t} + \vec{v}\cdot\nabla s\right) = \operatorname{div}(\kappa T) + \frac{\eta}{2}\left(\frac{\partial v_i}{\partial r_k} + \frac{\partial v_k}{\partial r_i} - \frac{2}{3}\delta_{ik}\frac{\partial v_j}{\partial r_j}\right)^2$$

$$+ \zeta(\operatorname{div}\vec{v})^2. \tag{1.45}$$

This is the *general equation of heat transfer*.

For incompressible fluid, this equation can be simplified:

$$\frac{\partial T}{\partial t} + \vec{v}\nabla T = \xi\Delta T + \frac{\nu}{2c_p}\left(\frac{\partial v_i}{\partial r_k} + \frac{\partial v_k}{\partial r_i}\right)^2, \tag{1.46}$$

where $\xi = \kappa/\rho c_p$ is the *thermal diffusivity* and c_p is the specific heat at constant pressure. If fluid does not move ($\vec{v} = 0$), then from (1.33) we obtain *Fourier's law*: $\partial T/\partial t = \xi\Delta T$.

Remark 1.3 Besides Reynolds parameter Re, there are other important similar parameters for fluid flows: Prandtl, Nusselt, Mach, Peclet, Grashof, Rayleigh numbers, etc.

The *Prandtl number* $Pr = c_p\eta/\kappa$, where c_p is the specific heat at $p = \text{const}$, η is the shear viscosity, and κ is the thermal conductivity. Since $c_p\eta/\kappa = (\eta/\rho)/(\kappa/\rho c_p)$, we see that Pr is the ratio of the kinematic viscosity (momentum diffusivity) to the thermal diffusivity ξ ($Pr = \nu/\xi$, since $\nu = \eta/\rho$ and $\xi = \kappa/(\rho c_p)$).

The *Peclet number* $Pe = \rho c_p v \mathcal{L}/\kappa = Re \cdot Pr$. Since the right-hand side in the formula of definition Pe may be written as $\rho c_p T v/(\kappa T/\mathcal{L})$, we see that the Peclet number is essentially the ratio of heat transported by the flow to heat transported by thermal conduction.

The *Nusselt number* is defined as $Nu = q\mathcal{L}/\kappa\Delta T$, where q is the heat flux per unit area and time through some surface area in the fluid, and $\Delta T/\mathcal{L}$ measures the temperature gradient near the surface. Thus, the Nusselt number gives a measure of the ratio of the total heat flux to the conductive heat flux through the surface.

The *Mach number* $M = \mathcal{V}_c/c$, where $\mathcal{V}_c$ is a characteristic flow speed and c is an adiabatic speed of sound in the fluid. Flows in which $M < 1$ are *subsonic* and flows in which $M > 1$ are *supersonic*.

The *Grashof number* is a dimensionless number in fluid dynamics which approximates the ratio of the buoyancy force to the viscous force acting on a fluid: $Gr = g\beta(T_S - T_\infty)\mathcal{L}^3/\nu^2$. Here g is an acceleration due to gravity, β is a volumetric thermal expansion coefficient, T_S is the source temperature, T_∞ is the quiescent temperature, $\mathcal{L}$ is the characteristic length and ν is a kinematic viscosity.

The *Rayleigh number* for a fluid is a dimensionless number associated with the heat transfer within the fluid. When the Rayleigh number is below the critical value for that fluid, heat transfer is primary in the form of conduction; when it exceeds the critical value, heat transfer is primarily in the form of convection. The Rayleigh number Ra is defined as the product of the Grashof number, which describes the relationship between buoyancy and viscosity within a fluid, and the Prandtl number, which describes the relationship between momentum diffusivity and thermal diffusivity. For free convection near a vertical wall, this number is: $Ra = Gr \cdot Pr = g\beta(T_S - T_\infty)\mathcal{L}^3/(\nu\xi)$, where (in the considered problem) T_S is a surface temperature, T_∞ is a fluid temperature far from the surface of the object and ξ is a thermal diffusivity.

1.3. Kinetic Equations and Gas Dynamics

A macroscopic sample of gas or fluid typically contains an enormous number of particles (molecules, atoms); the characteristic particle density is the *Avogadro number* $N_A = 6.022 \cdot 10^{23}$ mol^{-1}. Such large numbers immediately show that there is no hope in our attempt to develop an exact particle-by-particle description of the system. Instead, we require a *statistical* picture that gives the distribution of particles in space and over velocity. To a very high degree of approximation, we can consider the gas to be a dilute collection of classical point particles, so the interparticle collisions can be characterized by the classical *binary collision cross-sections*.

To describe the physical state of the gas statistically, we introduce the (*one-particle*) *distribution function* $f(\vec{r}, \vec{v}, t)$, defined as an average number of particles contained, at time t, in a volume element d^3r about $\vec{r}$ and a velocity-space element d^3v about $\vec{v}$ is $f d^3 r d^3 v$. We demand that $f(\vec{r}, \vec{v}, t) \geq 0$ everywhere, and that as $v_i \to \pm\infty$, $f \to 0$ sufficiently rapidly to guarantee that a finite number of particles has a finite energy.

The (local) particle density of gas is

$$N(\vec{r}, t) = \int_{-\infty}^{+\infty} \int_{-\infty}^{+\infty} \int_{-\infty}^{+\infty} f(\vec{r}, \vec{v}, t) dv_1 dv_2 dv_3, \qquad (1.47)$$

and the mass density is

$$\rho(\vec{r}, t) = m N(\vec{r}, t). \qquad (1.48)$$

The average (local) velocity of an element (macroscopic flow velocity) of gas is

$$\vec{u}(\vec{r}, t) = \langle \vec{v} \rangle = \frac{1}{N(\vec{r}, t)} \int_{-\infty}^{+\infty} \int_{-\infty}^{+\infty} \int_{-\infty}^{+\infty} \vec{v} f(\vec{r}, \vec{v}, t) dv_1 dv_2 dv_3. \qquad (1.49)$$

Let us consider the evolution of particle distribution function in detail.

1.3.1. *A Liouville's theorem*

A state of fluid can be specified by $3N$ canonical coordinates $r_1, \ldots, r_{3N} \in \mathcal{M}$ ($\mathcal{M}$ is the *configurational space*) and their momenta $p_1, \ldots, p_{3N}$ (N is the particle number in the system). The $6N$-dimensional space spanned by $\{\vec{r}_i, \vec{p}_i\}$ ($i = 1, \ldots, N$) is called the *phase space* $\mathcal{M}_2$ of the system. A point in phase space represents the state of the entire N-particle system, and is referred to as the *representative point*.

Through macroscopic measurements, we are not able to distinguish between two gases existing in different states (thus corresponding to two distinct representative points) but satisfying the same macroscopic conditions. Thus, when we speak of gas under certain macroscopic conditions, we are in fact referring not to a single state, but to *de facto* infinite number of states. In other words, we refer not to a single system, but to a collection of systems, identical in composition and macroscopic conditions but existing in different states.

This collection of systems is called an (*Gibbs'*) *ensemble*, which is geometrically represented by a distribution of representative points in $\mathcal{M}_2$. It may be described by a (generalized) density function

$$\mathcal{D}(r, p, t) \equiv \mathcal{D}(r_1, \ldots, r_{3N}, p_1, \ldots, p_{3N}, t)$$

$$= \prod_{i=1}^{N} \delta(\vec{r}_i - \vec{r}_i(t)) \delta\left(\vec{p}_i - m \frac{d\vec{r}_i(t)}{dt}\right),$$

so that $\mathcal{D}(r, p, t) d^{3N} r d^{3N} p$ is the number of representative points that at time t are contained in the infinitesimal volume element $d^{3N} r d^{3N} p$ of phase space centered about the point (r, p). An ensemble is completely specified by $\mathcal{D}(r, p, t)$. It is to be emphasized that members of an ensemble are mental copies of a system and do not interact with one another.

Let us consider the given $\mathcal{D}(r, p, t)$ at any time $t_\dagger$; the value of $\mathcal{D}(r, p, t)$ at $t > t_\dagger$ is determined by the dynamics of molecular motion. Let $\widehat{\mathsf{H}}(r_1, \ldots, r_{3N}, p_1, \ldots, p_{3N})$ be the Hamiltonian of a system in the ensemble. The equations of motion for a system are given by

$$\frac{dp_i}{dt} = -\frac{\partial \widehat{H}}{\partial r_i}, \quad \frac{dr_i}{dt} = \frac{\partial \widehat{H}}{\partial p_i}, \quad i = 1, \ldots, 3N. \quad (1.50)$$

These show how a representative point moves in phase space as time evolves. We assume that the Hamiltonian $\widehat{H}$ does not depend on any time derivative of r and q. Equations of motion are invariant under time reversal and determined by a unique manner the motion of a representative point for all times, when the position of one is given at any time. It follows immediately from these observations that the locus of a representative point is either a simple closed curve or a curve that never intersect itself. Furthermore, the loci of two distinct representative points never intersect.

Since the total number of systems in an ensemble is conserved, the number of representative points leaving any volume in phase space per unit time must be equal to the rate of decrease of the number of representative points in the same volume. Let Ω_{6N} be an arbitrary volume in $\mathcal{M}_2$ and let $\partial\Omega_{6N}$ be its surface. We denote by $\vec{\mathcal{V}}$ the $6N$-dimensional vector whose components are

$$\vec{\mathcal{V}} \equiv \left(\frac{dr_1}{dt}, \ldots, \frac{dr_{3N}}{dt}, \frac{dp_1}{dt}, \ldots, \frac{dp_{3N}}{dt} \right).$$

Let $\vec{n}$ be locally normal to the surface $\partial\Omega_{6N}$ vector. Then we have

$$\frac{d}{dt} \int_{\Omega_{6N}} \mathcal{D} d\Omega_{6N} + \int_{\partial\Omega_{6N}} \vec{n} \cdot \vec{\mathcal{V}} \mathcal{D} d\mathcal{S} = 0.$$

We can rewrite this equation as

$$\int_{\Omega_{6N}} \left(\frac{\partial \mathcal{D}}{\partial t} + \nabla_{6N} \left(\vec{\mathcal{V}} \mathcal{D} \right) \right) d\Omega_{6N}, \quad (1.51)$$

$$\nabla_{6N} \equiv \left(\frac{\partial}{\partial r_1}, \ldots, \frac{\partial}{\partial r_{3N}}, \frac{\partial}{\partial p_1}, \ldots, \frac{\partial}{\partial p_{3N}} \right).$$

Since Ω_{6N} is an arbitrary volume, the integrand of equation (1.51) must identically vanish. Hence

$$-\frac{\partial \mathcal{D}}{\partial t} = \nabla_{6N}\left(\vec{V}\mathcal{D}\right) = \sum_{i=1}^{3N}\left(\frac{\partial}{\partial r_i}\left(\mathcal{D}\frac{dr_i}{dt}\right) + \frac{\partial}{\partial p_i}\left(\mathcal{D}\frac{dp_i}{dt}\right)\right)$$

$$= \sum_{i=1}^{3N}\left(\frac{\partial \mathcal{D}}{\partial r_i}\frac{dr_i}{dt} + \frac{\partial \mathcal{D}}{\partial p_i}\frac{dp_i}{dt}\right)$$

$$+ \sum_{i=1}^{3N}\mathcal{D}\left(\frac{\partial}{\partial r_i}\left(\frac{dr_i}{dt}\right) + \frac{\partial}{\partial p_i}\left(\frac{dp_i}{dt}\right)\right).$$

By the equations of motion (1.50), we have

$$\frac{\partial}{\partial r_i}\left(\frac{dr_i}{dt}\right) + \frac{\partial}{\partial p_i}\left(\frac{dp_i}{dt}\right) = 0, \quad i = 1, \ldots, 3N.$$

Therefore, we have

$$\frac{\partial \mathcal{D}}{\partial t} + \sum_{i=1}^{3N}\left(\frac{\partial \mathcal{D}}{\partial r_i}\frac{dr_i}{dt} + \frac{\partial \mathcal{D}}{\partial p_i}\frac{dp_i}{dt}\right) \equiv \frac{d\mathcal{D}}{dt} = 0. \tag{1.52}$$

This relation is called the *Liouville's theorem*. Its geometrical interpretation is as follows. If we follow the motion of a representative point in phase space, we find that the density of representative points in its neighborhood is constant. Hence, the distribution of representative points moves in phase space like an incompressible fluid.

1.3.2. *The BBGKY hierarchy and the Boltzmann equation*

We can define *correlation functions* f_ℓ, which give the probabilities of finding ℓ particles having specified positions and momenta, in the systems forming an ensemble. The function f_1 is the familiar distribution function. The exact equations of motion for f_ℓ in classical mechanics show that to find f_1 we need to know f_2, which in turn depends on a knowledge of f_3, and so on until we come to the full N-body correlation function f_N. This system of equations is known as *Bogolyubov–Born–Green–Kirkwood–Yvon* (or *BBGKY*) *hierarchy*.

We shall derive it and show how the chain of equations can be truncated to yield the Boltzmann transport equation.

Let us consider an ensemble of systems, each being a gas of N molecules enclosed in volume V, with Hamiltonian $\widehat{H}$. Instead of general notation $\{r_i, p_i\}$ $(i = 1, \ldots, 3N)$ we shall denote the coordinates by the Cartesian vectors $\vec{r}_j$, $\vec{p}_j$:

$$\mu_j = (\vec{r}_j, \vec{p}_j), \quad d\mu_j = d^3 r_j d^3 p_j, \quad j = 1, \ldots, N.$$

The density function characterizing the ensemble is denoted $\mathcal{D}(1, \ldots, N; t)$, and assumed to be symmetric in $\mu_1, \ldots, \mu_N$; we can normalized $\mathcal{D}$ to unity: $\int \mathcal{D}(1, \ldots, N; t) \prod_{j=1}^{N} d\mu_j = 1$. Using the equations of motion in the form

$$\frac{\partial \mathcal{D}}{\partial t} = \sum_{j=1}^{N} \left(\frac{\partial}{\partial \vec{p}_j} \mathcal{D} \cdot \frac{\partial}{\partial \vec{r}_j} \widehat{H} - \frac{\partial}{\partial \vec{r}_j} \mathcal{D} \cdot \frac{\partial}{\partial \vec{p}_j} \widehat{H} \right), \qquad (1.53)$$

let us assume that the Hamiltonian has the form

$$\widehat{H} = \sum_{j=1}^{N} \frac{\vec{p}^2}{2m} + \sum_{j=1}^{N} U_i^{(\text{ex})} + \sum_{i<j} U_{ij}^{(\text{i-p})},$$

where $U_j^{(\text{ex})} = U^{(\text{ex})}(\vec{r}_j)$ is a potential of the external field in the point $\vec{r}_j$ and $U_{ij}^{(\text{i-p})} = U^{(\text{i-p})}(|\vec{r}_i - \vec{r}_j|)$ is an interparticle potential of interaction between particles in points $\vec{r}_i$ and $\vec{r}_j$. Then $\partial \widehat{H} / \partial \vec{r}_j = -\vec{F}_j^{(\text{ex})} - \sum_{i \neq j=1}^{N} F_{ij}^{(\text{i-p})}$, where $\vec{F}_j^{(\text{ex})} = -\partial U^{(\text{ex})} / \partial \vec{r}_j$, $\vec{F}_{ij}^{(\text{i-p})} = -\partial U^{(\text{i-p})}(|\vec{r}_i - \vec{r}_j|) / \partial \vec{r}_i$. Liouville's equation now has the form

$$\left(\frac{\partial}{\partial t} + \widetilde{\mathcal{D}}_N(1, \ldots, N) \right) \mathcal{D}(1, \ldots, N) = 0, \qquad (1.54)$$

$$\widetilde{\mathcal{D}}_N(1, \ldots, N) = \sum_{j=1}^{N} \widetilde{\mathcal{D}}_j^{(1)} + \frac{1}{2} \sum_{i,j=1, i \neq j}^{N} \widetilde{\mathcal{D}}_{ij}^{(2)}, \qquad (1.55)$$

$$\widetilde{\mathcal{D}}_j^{(1)} \equiv \frac{\vec{p}_j}{m} \frac{\partial}{\partial \vec{r}_j} + \vec{F}_j^{(\text{ex})} \frac{\partial}{\partial \vec{p}_j}, \qquad (1.56)$$

$$\widetilde{\mathcal{D}}_{ij}^{(2)} \equiv \vec{F}_{ij}^{(\text{i-p})} \frac{\partial}{\partial \vec{p}_i} + \vec{F}_{ji}^{(\text{i-p})} \frac{\partial}{\partial \vec{p}_j} = \vec{F}_{ij}^{(\text{i-p})} \left(\frac{\partial}{\partial \vec{p}_i} - \frac{\partial}{\partial \vec{p}_j} \right). \qquad (1.57)$$

The single-particle distribution function is defined by the *ensemble average* of $\sum_i \delta^3(\vec{r} - \vec{r}_j)\delta^3(\vec{p} - \vec{p}_j)$, that is

$$f_1(\vec{r}, \vec{p}, t) = \left\langle \sum_{i=1}^{N} \delta^3(\vec{r} - \vec{r}_j)\, \delta^3(\vec{p} - \vec{p}_j) \right\rangle$$

$$\equiv \frac{\int d^{3N}r\, d^{3N}p \sum_{i=1}^{N} \delta^3(\vec{r} - \vec{r}_j)\, \delta^3(\vec{p} - \vec{p}_j)\, \mathcal{D}(r, p, t)}{\int d^{3N}r\, d^{3N}p\, \mathcal{D}(r, p, t)}$$

$$= N \int d\mu_2 \ldots d\mu_N\, \mathcal{D}(1, \ldots, N; t). \qquad (1.58)$$

The factor N in the last form comes from the fact that all terms in the averaged sum have the same value, owing to the fact that $\mathcal{D}$ is symmetric in $\mu_1, \ldots, \mu_N$.

The general ℓ-particle distribution function (or *correlation function*) is defined by

$$f_\ell(1, \ldots, N, \mu, t) \equiv \frac{N!}{(N - \ell)!} \int d\mu_{\ell+1} \ldots d\mu_N \mathcal{D}(1, \ldots, N; t),$$

$$\ell = 1, \ldots, N. \qquad (1.59)$$

The equation of motion is

$$\frac{\partial f_\ell}{\partial t} = \frac{N!}{(N - \ell)!} \int d\mu_{\ell+1} \ldots d\mu_N \frac{\mathcal{D}(1, \ldots, N; t)}{\partial t}$$

$$= -\frac{N!}{(N - \ell)!} \int d\mu_{\ell+1} \ldots d\mu_N \tilde{\mathcal{D}}_N \mathcal{D} \qquad (1.60)$$

(see equation (1.54)). We isolate those terms in $\tilde{\mathcal{D}}_N$ involving only the coordinates $\mu_1, \ldots, \mu_\ell$:

$$\tilde{\mathcal{D}}_N(1, \ldots, N)$$

$$= \sum_{i=1}^{\ell} \tilde{\mathcal{D}}^{(1)} + \sum_{i=\ell+1}^{N} \tilde{\mathcal{D}}^{(1)} + \frac{1}{2} \sum_{i,j=1, i \neq j}^{\ell} \tilde{\mathcal{D}}^{(2)}$$

$$+ \frac{1}{2} \sum_{i,j=\ell+1,i\neq j}^{N} \widetilde{\mathcal{D}}^{(2)} + \sum_{i=1}^{\ell} \sum_{j=\ell+1}^{N} \widetilde{\mathcal{D}}^{(2)}$$

$$= \widetilde{\mathcal{D}}_\ell(1,\ldots,\ell) + \widetilde{\mathcal{D}}_{N-\ell}(\ell+1,\ldots,N) + \sum_{j=\ell+1}^{N} \widetilde{\mathcal{D}}^{(2)}. \qquad (1.61)$$

Note that

$$\int d\mu_{\ell+1} \ldots d\mu_N \widetilde{\mathcal{D}}_{N-\ell}(\ell+1,\ldots,N)\mathcal{D}(1,\ldots,N) = 0 \qquad (1.62)$$

because $\widetilde{\mathcal{D}}_{N-\ell}$ consists of gradient terms in $\vec{p}$ with $\vec{p}$-independent coefficients, and a gradient term in $\vec{r}$ with an $\vec{r}$-independent coefficient. Thus, the integral evaluates $\mathcal{D}$ on the boundary of phase space, where we assume $\mathcal{D}$ to vanish. Substituting (1.61) into (1.60), we obtain

$$\left(\frac{\partial}{\partial t} + \widetilde{\mathcal{D}}_\ell \right) f_\ell$$

$$= -\frac{N!}{(N-\ell)!} \int d\mu_{\ell+1} \ldots d\mu_N \sum_{i=1}^{\ell} \sum_{j=\ell+1}^{N} \widetilde{\mathcal{D}}_{ij}^{(2)} \mathcal{D}(1,\ldots,N)$$

$$= -\sum_{i=1}^{\ell} \int d\mu_{\ell+1} \widetilde{\mathcal{D}}_{i,\ell+1}^{(2)} \frac{N!}{(N-\ell+1)!} \int d\mu_{\ell+1} \ldots d\mu_N \mathcal{D}(1,\ldots,N)$$

$$= -\sum_{i=1}^{\ell} \int d\mu_{\ell+1} \widetilde{\mathcal{D}}_{i,\ell+1}^{(2)} f_{\ell+1}(1,\ldots,\ell+1).$$

In passing from the first to second equation, we have used the fact that the sum over j gives $N-\ell$ identical terms. Now, substitute $\widetilde{\mathcal{D}}^{(2)}$ from equation (1.57), and note that the second term there does not contribute, because it leads to a vanishing surface term. Hence, we have

$$\left(\frac{\partial}{\partial t} + \tilde{\mathcal{D}}_\ell\right) f_\ell(1, \ldots, \ell)$$

$$= -\sum_{i=1}^{\ell} \int d\mu_{\ell+1} \vec{F}_{i,\ell+1}^{(i-p)} \cdot \frac{\partial}{\partial \vec{p}_i} f_{\ell+1}(1, \ldots, \ell+1), \quad (1.63)$$

which is the BBGKY hierarchy ($\ell = 1, \ldots, N$). The left-hand side of each of the equations is a "streaming term", involving only the ℓ particles under consideration. For $\ell > 1$, it includes the effect of intermolecular scattering among the ℓ particles. The right-hand side is the "collision integral", which describes the effect of scattering between the particles under consideration with an "outsider", thus coupling f_ℓ to $f_{\ell+1}$.

The first two equations in the hierarchy read

$$\left(\frac{\partial}{\partial t} + \frac{\vec{p}_1}{m}\frac{\partial}{\partial \vec{r}_1} + \vec{F}_1\frac{\partial}{\partial \vec{p}_1}\right) f_1(\mu_1, t)$$

$$= -\int d\mu_2 \vec{F}_{1,2}^{(i-p)} \cdot \frac{\partial f_2(\mu_1, \mu_2, t)}{\partial \vec{p}_1}, \quad (1.64)$$

$$\left(\frac{\partial}{\partial t} + \frac{\vec{p}_1}{m}\frac{\partial}{\partial \vec{r}_1} + \frac{\vec{p}_2}{m}\frac{\partial}{\partial \vec{r}_2} + \vec{F}_1\frac{\partial}{\partial \vec{p}_1} + \vec{F}_2\frac{\partial}{\partial \vec{p}_2} + \frac{1}{2}\vec{F}_{1,2}^{(i-p)}\right.$$

$$\left.\cdot\left(\frac{\partial}{\partial \vec{p}_1} - \frac{\partial}{\partial \vec{p}_2}\right)\right) \times f_2(\mu_1, \mu_2, t)$$

$$= -\int d\mu_3 \left(\vec{F}_{1,3}^{(i-p)}\frac{\partial}{\partial \vec{p}_1} + \vec{F}_{2,3}^{(i-p)}\frac{\partial}{\partial \vec{p}_2}\right) f_3(\mu_1, \mu_2, \mu_3, t). \quad (1.65)$$

The terms in the equations have dimensions $f_\ell/$time, and different time scales are involved:

$$\vec{F}^{(i-p)} \cdot \frac{\partial}{\partial \vec{p}} \sim \frac{1}{\tau_1}, \quad \vec{F}^{(ex)} \cdot \frac{\partial}{\partial \vec{p}} \sim \frac{1}{\tau_2}, \quad \frac{\vec{p}}{m} \cdot \frac{\partial}{\partial \vec{r}} \sim \frac{1}{\tau_3}, \quad (1.66)$$

where τ_1 is the duration of a collision, τ_2 is the time for a molecule to traverse a characteristic distance over which the external potential varies significantly, and τ_3 is the time for a molecule to traverse a characteristic distance over which the correlation function varies significantly. The time τ_1 is the shortest and τ_2 is the longest.

The equation for f_1 is unique in the hierarchy, in that "streaming" sets a rather slow timescale, for it does not involve intermolecular scattering. The collision integral, which has more rapid variations, sets the timescale of f_1. This is why the equilibrium condition is determined by the vanishing of the collision integral.

The equation for f_2 (and higher ones as well) contains a collision term of the order τ_1^{-1} on the left side. The collision integral on the right side is smaller by a factor of the order nr_0^3 (where n is the density and r_0 is the range of the intermolecular potential) because the integration of $\vec{r}_3$ extends only over a volume of radius r_0. Now, $r_0 \approx 10^{-8}$ cm and $n \approx 10^{19}$ cm^{-3} under standard conditions; hence $nr_0^3 \approx 10^{-5}$. Thus, for f_2 (and higher equations too), the timescale is set by the streaming terms instead of collision integral, which will neglect.

With neglect of the right-hand side of (1.65), the hierarchy is truncated at f_2, and we have only two coupled equations for f_1 and f_2:

$$\left(\frac{\partial}{\partial t} + \frac{\vec{p}_1}{m} \frac{\partial}{\partial \vec{r}_1} \right) f_1(\mu_1, t)$$

$$= - \int_{4\pi r_0^3/3} d\mu_2 \, \vec{F}_{12}^{(i-p)} \frac{\partial f_2(\mu_1, \mu_2, t)}{\partial \vec{p}_1} \equiv St(f_1), \qquad (1.67)$$

$$\left(\frac{\partial}{\partial t} + \frac{\vec{p}_1}{m} \frac{\partial}{\partial \vec{r}_1} + \frac{\vec{p}_2}{m} \frac{\partial}{\partial \vec{r}_2} + \frac{1}{2} \vec{F}_{12}^{(i-p)} \left(\frac{\partial}{\partial \vec{p}_1} - \frac{\partial}{\partial \vec{p}_2} \right) \right) f_2(\mu_1, \mu_2, t)$$

$$= 0, \qquad (1.68)$$

where we have set all external forces to zero, for simplicity. We also assume for simplicity that the force $\vec{F}^{(i-p)}$ vanishes outside the range r_0.

The salient qualitative features of (1.67) and (1.68) are that f_2 varies in time with characteristic period τ_1, and in space with characteristic distance r_0, while f_1 varies much less rapidly, by a factor nr_0^3. Thus, f_1 measures space and time with much coarser scales than f_2. The correlations in f_2 are produced by collisions between particles 1 and 2. When their positions are so far separated as to be out of molecular interaction range, we expect that there will be no correlation between 1 and 2, and f_2 will assume a product form (neglecting possible correlations produced by collisions with a third particle):

$$\lim_{|\vec{r}_1 - \vec{r}_2| \gg r_0} f_2(\mu_1, \mu_2, t) \longrightarrow f_1(\mu_1, t) f_1(\mu_2, t). \qquad (1.69)$$

To evaluate $St(f_1)$, however, we need f_2 not in the uncorrelated region, but in the region where two particles are colliding. To look at this region, it is convenient to use total and relative coordinates:

$$\vec{p}_{(+)} = \vec{p}_1 + \vec{p}_2, \quad \vec{p}_{(-)} = \frac{\vec{p}_2 - \vec{p}_1}{2}, \quad \vec{r}_{(+)} = \frac{\vec{r}_1 + \vec{r}_2}{2}, \quad \vec{r}_{(-)} = \vec{r}_2 - \vec{r}_1.$$

Then (1.68) becomes

$$\left(\frac{\partial}{\partial t} + \frac{\vec{p}_{(-)}}{m} \frac{\partial}{\partial \vec{r}_{(+)}} + \frac{\vec{p}_{(-)}}{m} \frac{\partial}{\partial \vec{r}_{(+)}} + \vec{F}^{(i-p)}(\vec{r}_{(-)}) \frac{\partial}{\partial \vec{p}_{(-)}} \right)$$

$$\times f_2(\vec{r}_{(+)}, \vec{p}_{(+)}, \vec{r}_{(-)}, \vec{p}_{(-)}, t) = 0, \quad \vec{F}^{(i-p)}(\vec{r}_{(-)})$$

$$= -\frac{\partial U^{(i-p)}(|\vec{r}_{(-)}|)}{\partial \vec{r}_{(-)}}. \qquad (1.70)$$

Transform to the center-of-mass system by putting $\vec{p}_{(+)} = 0$. The equation can then be rewritten, to first order in dt, as the streaming condition:

$$f_2\left(\vec{r}_{(-)} + \frac{\vec{p}_{(-)}}{m} dt, \vec{p}_{(-)} + \vec{F}^{(i-p)}(\vec{r}_{(-)}) dt, t + dt \right) = f_2(\vec{r}_{(-)}, \vec{p}_{(-)}, t). \qquad (1.71)$$

The equilibrium situation, for which $\partial f_2 / \partial t = 0$, is a steady-state scattering, by the force field $\vec{F}^{(i-p)}$, of a beam of particles consisting of all momenta, at all impact parameters. Outside the sphere of interaction (with radius r_0), the uncorrelated factorized form (1.69) holds. However, boundary values of the momenta are correlated through the fact that momenta entering the sphere at a specific impact parameter must leave the sphere at the correct scattering angle, and vice versa.

To "derive" the Boltzmann transport equation (for one-particle function f_1), we assume that, since f_2 has a shorter timescale than f_1, it reaches equilibrium earlier than f_1. Thus, we set $\partial f_2/\partial t = 0$, and assume f_2 has attained the equilibrium form described earlier. Similarly, we assume that the range of force r_0 is essentially zero from the point of view of f_1. Thus, in the factorized form of f_2 just before and after collision, we can put $\vec{r}_1$ and $\vec{r}_2$ both equal to the same value.

With this in mind, we substitute (1.68) into (1.67) to obtain

$$
\begin{aligned}
St(f_1) &= -\int_{4\pi r_0^3/3} d\mu_2 \vec{F}_{12}^{(i-p)} \frac{\partial f_2(\mu_1, \mu_2, t)}{\partial \vec{p}_1} \\
&= -\int_{4\pi r_0^3/3} d\mu_2 \vec{F}_{12}^{(i-p)} \left(\frac{\partial}{\partial \vec{p}_1} - \frac{\partial}{\partial \vec{p}_2} \right) f_2(\mu_1, \mu_2, t) \\
&= \frac{1}{m} \int_{4\pi r_0^3/3} d\mu_2 \left(\vec{p}_1 \frac{\partial}{\partial \vec{r}_1} + \vec{p}_2 \frac{\partial}{\partial \vec{r}_2} \right) f_2(\mu_1, \mu_2, t). \quad (1.72)
\end{aligned}
$$

Using coordinates with subscripts $(\pm)$, and neglecting the gradient with respect to $\vec{r}_{(+)}$, we have

$$
\begin{aligned}
St(f_1) &= -\frac{1}{m} \int d^3 p_2 \int_{r<r_0} d^3 r_{(-)} (\vec{p}_1 - \vec{p}_2) \frac{\partial f_2}{\partial \vec{r}_{(-)}} \\
&= \frac{1}{m} \int d^3 p_2 |\vec{p}_1 - \vec{p}_2| \int d\varphi b db \int_{x_1}^{x_2} \frac{\partial f_2}{\partial x} dx,
\end{aligned}
$$

$$
f_2(x_1) = f_1(\vec{p}_1) f_1(\vec{p}_2), \qquad f_2(x_2) = f_1(\vec{p}_1^\dagger) f_1(\vec{p}_2^\dagger),
$$

where $\vec{p}_1^\dagger$, $\vec{p}_2^\dagger$ are the final momenta in the scattering process when the initial momenta are $\vec{p}_1, \vec{p}_2$, φ is azimuthal angle; the geometry of collision is defined by the impact parameter b (the perpendicular distance of particle of the beam from the straight-line extension of the incident relative velocity vector $\vec{g} = 2\vec{p}_{(-)}/m$) and parameters x_1, x_2, which are x-coordinates of points X_1, X_2 on the circle $O(O_1, r_0)$, x-axis is parallel $\vec{p}_{(-)}$ and $O_1 \in x$-axis. This circle is the projection of sphere with radius r_0 (range of correlation of two particles) on the plane of two particles moving; the points X_1, X_2 are "incoming" and "outcoming" points of intersection of "falling" particle trajectory with $\partial O(O_1, r_0)$.

Finally, we have

$$St(f_1) = \int d^3 p_2 d\Omega |\vec{g}| \frac{d\sigma(\vartheta, |\vec{g}|)}{d\Omega}$$

$$\times \left(f(\vec{r}, \vec{v}_1^{\,\dagger}, t) f(\vec{r}, \vec{v}_2^{\,\dagger}, t) - f(\vec{r}, \vec{v}_1, t) f(\vec{r}, \vec{v}_2, t) \right)$$

$$\equiv \int \frac{d\sigma(\chi, g)}{d\Omega} g \left(f_1^{\dagger} f_2^{\dagger} - f_1 f_2 \right) d\vec{v}_2 \, d\Omega, \quad d\sigma = b \, db \, d\varphi,$$

$$\vec{g} = \vec{v}_1 - \vec{v}_2, \tag{1.73}$$

where $d\sigma(\chi, g)/d\Omega$ is the *differential cross-section*, χ is an *angle of deflection*.

Hence, we can write the equation of evolution of $f \equiv f_1$ (*Boltzmann equation*) in the standard form (see (1.67)):

$$\left(\frac{\partial}{\partial t} + \vec{v} \frac{\partial}{\partial \vec{r}} \right) f(\vec{r}, \vec{v}, t) = \int_{\mathbb{R}^3} \sigma(\chi, g) g \left(f^{\dagger} f_*^{\,\dagger} - f f_* \right) d\vec{v}_* d\Omega,$$

$$\sigma(\chi, g) = \frac{b(\chi)}{\sin \chi} \left| \frac{db(\chi)}{d\chi} \right|, \quad g = |\vec{v} - \vec{v}_*|, \tag{1.74}$$

where $\sigma(\chi, g)$ is an *effective cross-section*, $\vec{v}^{\,\dagger}$, $\vec{v}_*^{\,\dagger}$ are the final velocities of particles in the scattering process, when the initial velocities are $\vec{v}$, $\vec{v}_*$, $f_*^{\dagger} = f(\vec{r}, \vec{v}_*^{\,\dagger}, t)$, etc. If external field $\vec{F}$ is present, we must add the term $(\vec{F}/m)(\partial f/\partial \vec{v})$ in the left-hand side of this equation. Of course, we assume that the external forces are absent.

One may rewrite the Boltzmann equation in other forms:

$$\left(\frac{\partial}{\partial t} + \vec{v} \frac{\partial}{\partial \vec{r}} \right) f(\vec{r}, \vec{v}, t)$$

$$= \int d\vec{v}_* \int_0^\infty b \, db \int_0^{2\pi} g \left(f^{\dagger} f_*^{\,\dagger} - f f_* \right) d\varphi$$

$$= \int_{\mathbb{R}^3} d\vec{v}_* \int_0^{2\pi} d\varphi \int_0^{\pi/2} d\vartheta B(\vartheta, g) \left(f^{\dagger} f_*^{\,\dagger} - f f_* \right), \tag{1.75}$$

$$B(\vartheta, g) = g \, b \left| \frac{db}{d\vartheta} \right|, \quad \vartheta = \frac{\pi}{2} - \frac{\chi}{2}, \tag{1.76}$$

where $b(\chi)$ is defined from the effective cross-section $\sigma(\chi, g)$. For potential $U^{(\text{i-p})}(r) = c_0/r^n$ ($c_0 = \text{const}$, $n \in \mathbb{N}^+$), we have

$$b\left|\frac{db}{d\chi}\right| = \left(\frac{2nc_0}{m/2}\right)^{2/n} \cdot g^{-4/n} q(\chi) \left|\frac{dq}{d\chi}\right|, \quad q = \left(\frac{m/2}{2nc_0}\right)^{1/n} \cdot b\, g^{2/n},$$

and for $n = 4$, we obtain the case of the so-called *Maxwell molecules* (in this case B does not depend on g). For potential of "hard spheres" (with radius R_0):

$$U^{(\text{i-p})} = \begin{cases} \infty & \text{if } r < R_0, \\ 0, & \text{if } r > R_0, \end{cases}$$

we obtain $b = R_0 \cos(\chi/2)$ and $\sigma = R_0^2/4$. The collision kernel $B(\vartheta, g) = (g/2)\sin(2\vartheta)$ for this potential.

In conclusion, we note that the Boltzmann equation can be obtained by a more simple method with the help of special physical assumptions without using the BBGKY hierarchy directly. This method is based on the "molecular chaos hypothesis": the number of interacted particle pairs situated in the element $\Delta\vec{r}$ with the velocities on the ranges $(\vec{v}, \vec{v} + \Delta\vec{v})$ and $(\vec{v}_*, \vec{v}_* v + \Delta\vec{v}_*)$ is

$$f(\vec{r}, \vec{v}, t)\Delta\vec{r}\Delta\vec{v} f(\vec{r}, \vec{v}_*, t)\Delta\vec{r}\Delta\vec{v}_*. \tag{1.77}$$

One can prove that this hypothesis is *de facto* the consequence of considered earlier approach.

1.3.3. *The properties of the Boltzmann equation*

Let's consider the properties of the Boltzmann equation.

- The one-particle distribution function $f(\vec{r}, \vec{v}, t)$ (defined as the number of particles in the unit volume V_μ of phase space) is non-negative: $f \geq 0$. The Boltzmann equation keeps this property and its solutions are non-negative too.

- Let us rewrite collisional term of the Boltzmann equation as bilinear operator: $St(f) \equiv J(f, f)$. There is

$$J(y, h) = \frac{1}{2} \int d\vec{v}_1 \int \sigma(\chi; g) g(y^\dagger h_1^\dagger + y_1^\dagger h^\dagger - yh_1 - y_1 h) d\Omega$$
$$= J(h, y),$$

where y, h are any functions of $(\vec{r}, \vec{v}, t)$ and we noted $y_1^\dagger = y(\vec{r}, (\vec{v}_1)^\dagger, t)$, etc. We consider the functional

$$\mathcal{I} = \int J(y, h) \psi(\vec{v}) d\vec{v}$$
$$\equiv \frac{1}{2} \int d\vec{v} d\vec{v}_1 \int \sigma(\chi; g) g(y^\dagger h_1^\dagger + y_1^\dagger h^\dagger - yh_1 - y_1 h) \psi(\vec{v}) d\Omega,$$
$$\forall \psi(\vec{v}) \in L^2(\mathbb{R}^3). \tag{1.78}$$

One can prove that $\mathcal{I}$ is equal to

$$\mathcal{I}_\Sigma = \frac{1}{8} \int d\vec{v} d\vec{v}_1 \int \sigma(\chi; g) g(y^\dagger h_1^\dagger + y_1^\dagger h^\dagger - yh_1 - y_1 h)$$
$$\times \left(\psi(\vec{v}) + \psi(\vec{v}_1) - \psi(\vec{v}^\dagger) - \psi(\vec{v}_1^\dagger) \right) d\Omega = \mathcal{I}. \tag{1.79}$$

Consequently, $\mathcal{I} = 0$ for any function $\psi(\vec{v})$ satisfying the condition $\psi(\vec{v}^\dagger) + \psi(\vec{v}_1^\dagger) = \psi(\vec{v}) + \psi(\vec{v}_1)$. The solutions of this equation are called *summational* (or *collision*) *invariants*. The most general form of collision invariant is the linear combination

$$\psi(\vec{v}) = \sum_{i=0}^{4} c_i \psi_i(\vec{v}), \quad c_i \in \mathbb{R}^1,$$

where basic invariants are as follows: $\psi_0 = 1$, $\psi_i = v_i$ $(i = 1, 2, 3)$, $\psi_4 = v^2/2$.

From the existence of collision invariants, one can obtain the solution of the nonlinear equation $J(f, f) = 0$. We assume $\psi(\vec{v}) = \ln f(\vec{v})$

and $y = h = f$ in the formulae (1.78)–(1.79), then

$$\int \mathcal{J}(f,f)\ln f d\vec{v}$$

$$= \frac{1}{4}\int d\vec{v}d\vec{v}_1 \int \sigma(\chi;g)g \left(f(\vec{v}^\dagger)f(\vec{v}_1{}^\dagger) - f(\vec{v})f(\vec{v}_1)\right)$$

$$\times \ln\left(\frac{f(\vec{v})f(\vec{v}_1)}{f(\vec{v}^\dagger)f(\vec{v}_1{}^\dagger)}\right)d\Omega. \tag{1.80}$$

Functions $\sigma(\chi;g)g$ and f are non-negative; for any $k_1, k_2 > 0$, we have the inequality $d(k_1, k_2) \equiv (k_1 - k_2)\ln(k_2/k_1) \leq 0$, and $d(k_1, k_2) = 0$ if and only if $k_1 = k_2$. Then $\int \mathcal{J}(f,f)\ln f d\vec{v} \leq 0$ and the inequality becomes equality if and only if $k_2 = f(\vec{v})f(\vec{v}_1) = f(\vec{v}^\dagger)f(\vec{v}_1{}^\dagger) = k_1$, i.e.,

$$\ln f(\vec{v}) + \ln f(\vec{v}_1) = \ln f(\vec{v}^\dagger) + \ln f(\vec{v}_1{}^\dagger).$$

Consequently, the quantity $\ln f$ is the linear combination of collision invariants:

$$\ln f = \ln A - \alpha v_0^2 + 2\alpha\vec{v}_0 \cdot \vec{v} - \alpha v^2, \quad \text{i.e.,} \quad f = A\exp(-\alpha(\vec{v} - \vec{v}_0)^2),$$

where $A, \alpha, \vec{v}_0$ are parameters (the some functions of variables $\vec{r}, t$).

From the physical viewpoint, these parameters can be expressed in terms of macroscopic observable quantities. Let us introduce the *local kinetic energy of heat moving* $\bar{\epsilon}$ as the average kinetic energy in the system of observation moving with the local velocity $\vec{u}$:

$$\bar{\epsilon} = \left\langle \frac{m}{2}(\vec{v} - \vec{u})^2 \right\rangle = \frac{1}{N(\vec{r},t)}\int \frac{m(\vec{v} - \vec{u})^2}{2}f(\vec{r},\vec{v},t)d\vec{v}. \tag{1.81}$$

In a dilute gas, this quantity is defined as *local temperature*:

$$\bar{\epsilon} = \frac{3}{2}k_B T(\vec{r},t),$$

$$k_B = 1.38 \cdot 10^{-23} \text{ J/K is the Boltzmann constant.} \tag{1.82}$$

Finally, using formulas (1.47), (1.49) and (1.82) with $f = A\exp(-\alpha(\vec{v} - \vec{v}_0)^2)$, we obtain

$$N(\vec{r},t) = A\left(\frac{\pi}{\alpha}\right)^{3/2}, \quad \vec{u} = \vec{v}_0, \quad k_B T = \frac{m}{2\alpha}.$$

These results permit one to rewrite $f = A\exp(-\alpha(\vec{v} - \vec{v}_0)^2)$ in the form of *local Maxwell distribution function* or *Maxwellian*:

$$f = f^{(0)} = N(\vec{r}, t)\left(\frac{m}{2\pi k_B T(\vec{r}, t)}\right)\exp\left(-\frac{m(\vec{v} - \vec{u}(\vec{r}, t))^2}{2k_B T(\vec{r}, t)}\right),$$

$$(1.83)$$

- From the inequality $\int \mathcal{J}(f, f)\ln f\,d\vec{v} \leq 0$ (see (1.80) and the following text) follows *Boltzmann's H-theorem*: an evolution of one-particle distribution function to Maxwellian is non-reversible.

Suppose that the gas is thermally isolated, is homogeneous so that f is independent of $\vec{r}$, and external forces are absent. At $t = 0$, there exists $f(\vec{v}, t = 0)$, a described system with finite density $\mathcal{I}_1 = \int f(\vec{v}, 0)d\vec{v} < \infty$ and finite kinetic energy density $\mathcal{I}_2 = \int v^2 f(\vec{v}, 0)d\vec{v} < \infty$. These inequalities are true for $t > 0$ (this fact follows from the properties of the Boltzmann equation $\partial f/\partial t = \mathcal{J}(f, f)$). Consider the functional $H(t) \equiv \int f(\vec{v}, t)\ln f(\vec{v}, t)d\vec{v}$. We have

$$\frac{dH(t)}{dt} = \int \frac{\partial f}{\partial t}(1 + \ln f)d\vec{v} = \int \mathcal{J}(f, f)\ln f\,d\vec{v} \leq 0,$$

since "1" is collision invariant. From the property of finite density, $f(\vec{v}) \in L^1(\mathbb{R}^3)$ and $|H(t)| \to \infty$ if and only if the integration region is infinite: $f \to 0$, $\ln f \to -\infty$ if $v \to \infty$. The quantity $H(t) \to \infty$ if and only if $|\ln f| > v^2$, $v \to \infty$ (see inequality for $\mathcal{I}_2$ above); but then $f < \exp(-v^2)$, and consequently, $|H(t)| < \infty$. Hence, $H(t)$ cannot decrease infinitely and for $t \to \infty$, $dH/dt = 0$ and the solution of the Boltzmann equation $f \to f^{(0)}$.

For the generalization of the H-theorem for inhomogeneous systems, we consider evolution equation for $H(\vec{v}, t)$ (obtained from the Boltzmann equation):

$$\frac{d\mathcal{H}(t)}{dt} + \oint_S \vec{J}_S \cdot \vec{n}dS = \int_V \sigma_H(\vec{r}, t)d\vec{r},$$

$$\mathcal{H}(t) = \int_\omega H(\vec{r}, t) d\vec{r}, \quad \vec{J}_H(\vec{r}, t) = \int \vec{v} f \ln f d\vec{v},$$

$$\sigma_H(\vec{r}, t) = \int J(f, f) \ln f d\vec{v} \leq 0, \tag{1.84}$$

where $d\mathcal{S}$ is the infinitesimal element of the surface $\mathcal{S} = \partial V$, V is the volume element. The right-hand side of (1.84) is non-positive and if the system is closed ($\vec{J}_H = 0$), then $d\mathcal{H}/dt \leq 0$, i.e., solution $f \to f^{(0)}$ and $\sigma_H \equiv 0$.

1.3.4. *The linear Boltzmann equation*

If the distribution function is nearly a global Maxwellian $f^{(0)}$ (i.e., $f^{(0)}$ is constant on $\vec{r}$ and t), then the collision operator $J(F, F)$ can be linearized around that Maxwellian. Let us write F as

$$F(\vec{r}, \vec{v}, t) = f^{(0)}(\vec{v}) + \sqrt{f^{(0)}(\vec{v})} f(\vec{r}, \vec{v}, t). \tag{1.85}$$

The Boltzmann equation (with solution f) can be rewritten in the form of

$$\left(\frac{\partial}{\partial t} + \vec{v} \frac{\partial}{\partial \vec{r}}\right) f = \widehat{L}(f) + \nu_B \widehat{M}(f, f),$$

$$\widehat{L}(f) = \frac{2}{\sqrt{f^{(0)}}} J(f^{(0)}, \sqrt{f^{(0)}} f), \tag{1.86}$$

$$\nu_B \widehat{M}(f, f) = \frac{1}{\sqrt{f^{(0)}}} J(\sqrt{f^{(0)}}, \sqrt{f^{(0)}}).$$

If f is small, then we expect that $\nu_B \widehat{M}(f, f)$ is even smaller, and we can approximate (1.86) by the *linear Boltzmann equation*

$$\left(\frac{\partial}{\partial t} + \vec{v} \frac{\partial}{\partial \vec{r}}\right) f = \widehat{L}(f). \tag{1.87}$$

The linearized collision operator $\widehat{L}$ is self-adjoint in $L^2(\vec{v} \in \mathbb{R}^3)$ and non-positive, i.e., $\langle f, \widehat{L}(f) \rangle \leq 0$, where $\langle a(\vec{v}), b(\vec{v}) \rangle = \int_{\mathbb{R}^3} ab \, d\vec{v}$ is the scalar product in the Hilbert space $L^2(\mathbb{R}^3)$. Its null space is five-dimensional and is spanned by $\{\widetilde{\chi}_k\}_{k=0}^4$ with $\widetilde{\chi}_k(\vec{v}) = \sqrt{f^{(0)}(\vec{v})} \psi_k(\vec{v})$, where $\psi_k(\vec{v})$ $(k = 0, \ldots, 4)$ are basic collision invariants. Furthermore, $\widehat{L}$ can be written as the sum of a multiplication operator and integral operator $\widehat{K}_B$:

$$\widehat{L}(f) = -\nu_B(\vec{v}) \cdot f(\vec{v}) + \widehat{K}_B(f(\vec{v})),$$
$$\nu_1(1 + |\vec{v}|)^\gamma \leq \nu_B(\vec{v}) \leq \nu_2(1 + |\vec{v}|)^\gamma, \tag{1.88}$$
$$\widehat{K}_B f(\vec{v}) = 2\pi \int B(\vartheta, g)(f(\vec{v}^\dagger) + f(\vec{v}_1^\dagger) - f(\vec{v}_1))d\vec{v}_1 d\vartheta$$

(see the following details). Here, ν_1, ν_2 are positive constants and $\gamma = 1 - 4/n$ for the interparticle potential $U^{\text{(i-p)}}(r) = c_0/r^n$ $(n > 0)$.

1.3.5. *Fluid dynamic and boundary layer limits of the Boltzmann equation*

In a gas which is not too rarefied, the mean free path between molecular collisions is very small (for liquid, this fact is obvious). As it approaches zero, a description of the gas as a continuous fluid is reached. This limit is therefore important as a foundation for fluid dynamics and as a theory for correcting the fluid dynamic description of a gas.

The dimensionless mean free path δ enters the Boltzmann equation as a collision rate; i.e., the equation becomes

$$\frac{\partial f}{\partial t} + \vec{v}\frac{\partial f}{\partial \vec{r}} = \frac{1}{\delta}\mathcal{J}(f, f). \tag{1.89}$$

Since δ is small, f must approximately satisfy $\mathcal{J}(f, f) = 0$, which implies that approximately $f = f^{(0)}$. The fluid dynamic variables $\rho, \vec{v}, T$ will depend on $\vec{r}$ and t. By applying the conservation laws $\mathcal{I} = \langle \psi_i, \mathcal{J}(f_1, f_2) \rangle = 0$ (see (1.78)) to (1.89) and assuming that f is Maxwellian, one can derive the compressible Euler equations of fluid dynamics for $\rho, \vec{v}, T$. This fluid dynamic limit is non-uniform and is not valid near boundaries, around shocks, or for a short time after the imposition of arbitrary initial conditions. In these regions, we must consider the problems of initial layers, boundary layers and shock layers.

Let us consider the initial as the boundary layers. If the initial prescription of f is not Maxwellian, then f will rapidly change to become Maxwellian. In this initial layer of rapid evolution, the time derivative of f must be large to balance the collision process. It follows that f must vary on the fast timescale $\tau = t/\delta$ and that the dominant terms in the Boltzmann equation (1.89) are

$$\frac{\partial f}{\partial \tau} = \mathcal{J}(f, f). \tag{1.90}$$

One can prove that for a fairly general class of initial data, the solution $f(\vec{v}, t)$ of (1.90) converges to a Maxwellian distribution in the weak sense in $L^1(\vec{v} \in \mathbb{R}^3)$ as $\tau \to \infty$. This indicates that during the initial layer, the distribution changes to the form required by $\mathcal{J}(f, f) = 0$ and $f = f^{(0)}$.

If the Boltzmann equation is solved in the spatial domain V (which does not change in time), there is a condition at each point $\vec{r}_b$ of the boundary ∂V that

$$f(\vec{r}, \vec{v}, t) = \int \vec{v}_* \cdot \vec{n} b(\vec{v}, \vec{v}_*, \vec{r}) f(\vec{r}, \vec{v}_*, t) d\vec{v}_*, \quad b(\vec{v}, \vec{v}_*, t) \text{ is fixed kernel} \tag{1.91}$$

for $\vec{v} \cdot \vec{n} < 0$, in which $\vec{n}$ is outward normal to $\vec{r}_b$. If there is a local Maxwellian $f^{(0)}$, satisfying (1.91), then near the wall $f \approx f^{(0)}$. If the boundary condition (1.91) does not admit a Maxwellian solution, then there will be large spatial gradients in f in the direction (call it r_1) normal to the boundary. The resulting leading order boundary layer equation is a half-space problem $\vec{v}_1 \cdot \partial f / \partial r_1 = \mathcal{J}(f, f)$, $r_1 > 0$ with boundary condition (1.91) at $r_1 = 0$ and matching condition $f \to f^{(0)}$, as $r_1 \to \infty$.

1.3.6. *The Hilbert and Chapman–Enskog methods*

D. Hilbert proved the following result: if the Boltzmann equation can be written as

$$\frac{\partial f}{\partial t} + \vec{v}\frac{\partial f}{\partial \vec{r}} = \frac{1}{\delta}\mathcal{J}(f, f), \quad \delta \ll 1, \tag{1.92}$$

and if the solution of (1.92) can be written as

$$f = \sum_{n=0}^{\infty} \delta^n f^{(n)},$$ (1.93)

then this solution f is defined by a unique manner as follows:

$$\rho(\vec{r}, 0) = mN(\vec{r}, t = 0), \quad \vec{u}(\vec{r}, 0) = \frac{1}{N(\vec{r}, 0)} \int \vec{v} f(\vec{r}, \vec{v}, 0) d\vec{v},$$

$$T(\vec{r}, 0) = \frac{2}{3k_B N(\vec{r}, 0)} \int \frac{m(\vec{v} - \vec{u})^2}{2} f(\vec{r}, \vec{v}, 0) d\vec{v}.$$

Substituting an expansion (1.93) into (1.92), we obtain the chain of equations for functions $f^{(n)}$:

$$J(f^{(0)}, f^{(0)}) = 0,$$ (1.94)

$$\left(\frac{\partial}{\partial t} + \vec{v} \cdot \frac{\partial}{\partial \vec{r}}\right) f^{(0)} = 2J(f^{(1)}, f^{(0)}),$$ (1.95)

$$\left(\frac{\partial}{\partial t} + \vec{v} \cdot \frac{\partial}{\partial \vec{r}}\right) f^{(n-1)} = \sum_{k=1}^{n-1} J(f^{(n-k)}, f^{(k)}) + 2J(f^{(n)}, f^{(0)}),$$

$$n > 1.$$ (1.96)

From equation (1.93), similar expansions for functions follow:

$$\rho_i(\vec{r}, t) = \begin{pmatrix} \rho(\vec{r}, t) \\ \rho(\vec{r}, t)\vec{u}(\vec{r}, t) \\ \rho(\vec{r}, t)(u^2(\vec{r}, t)/2 + \epsilon_s(\vec{r}, t)) \end{pmatrix}, \quad i = 0, \ldots, 4,$$

which satisfy the conditions $\rho_i = m \int f(\vec{r}, \vec{v}, t)\psi_i(\vec{v})d\vec{v}$, $i = 0, \ldots, 4$. These expansions are

$$\rho_i = \sum_{n=0}^{\infty} \delta^n \rho_i^{(n)}, \quad \rho_i^{(n)} = m \int f^{(n)}(\vec{r}, \vec{v}, t)\psi_i(\vec{v})d\vec{v}.$$ (1.97)

The function $f^{(0)}$ is Maxwellian (1.83) with parameters $N = N^{(0)}$, $\vec{u} = \vec{u}^{(0)}$, $T = T^{(0)}$ and corresponding $\rho_i^{(0)}$ (it follows from (1.94)). If function $f^{(0)}$ is known, then (1.95) is linear non-homogeneous integral equation for function $f^{(1)}$. The non-trivial solution of this equation exists if and only if

$$\int m\psi_i(\vec{v})\left(\frac{\partial f^{(0)}}{\partial t} + \vec{v}\cdot\frac{\partial f^{(0)}}{\partial \vec{r}}\right)d\vec{v} = 0, \quad i = 0,\ldots,4 \qquad (1.98)$$

(it is the consequence of a *Fredholm alternative*).

Equation (1.98) may be rewritten as $\partial\rho_i^{(0)}/\partial t = \widehat{D}_i^{(0)}(\{\rho_j^{(0)}\})$, where $D_i^{(0)}$ are nonlinear hydrodynamic operators:

$$\widehat{D}_0^{(0)}(\{\rho_j^{(0)}\}) = \frac{\partial}{\partial\vec{r}}\rho^{(0)}\vec{u}^{(0)},$$

$$\widehat{D}_i^{(0)}(\{\rho_j^{(0)}\}) = -\left(\frac{\partial}{\partial\vec{r}}\left(\rho^{(0)}\vec{u}^{(0)}\vec{u}^{(0)} + p^{(0)}\vec{U}\right)\right)_i, \quad i = 1,2,3,$$

$$\widehat{D}_4^{(0)}(\{\rho_j^{(0)}\}) = -\frac{\partial}{\partial\vec{r}}\left(\rho^{(0)}\vec{u}^{(0)}\left(\frac{(u^{(0)})^2}{2} + \epsilon_s^{(0)} + \frac{p^{(0)}}{\rho^{(0)}}\right)\right),$$

$$p^{(0)} = \frac{2\rho^{(0)}\epsilon_s^{(0)}}{3}, \quad j = 0,\ldots,4.$$

Hence, from equation (1.98) (for $f^{(1)}$) the parameters of $f^{(0)}$ are defined and it follows that the corresponding momenta $\rho_j^{(0)}$ are obeyed for ideal fluid equations. Besides, we can write

$$f^{(1)} = f_s^{(1)}(\vec{r},\vec{v},t) + \sum_{j=0}^{4}(\alpha_j^{(1)}(\vec{r},t)m\psi_j(\vec{v})f^{(0)}(\vec{r},\vec{v},t)), \quad j = 0,\ldots,4,$$

where $f_s^{(1)}$ is the special solution of non-homogeneous integral equation (1.95), and $\alpha_j^{(1)}$ are new unknown functions (the term $\sum_{j=0}^{4}(\ldots)$ is the general solution of the corresponding homogeneous equation).

The Hilbert method is based on the definition of $\alpha_j^{(1)}$ from equation (1.98) and parameters $\rho_j^{(n)}$ from (1.97). Equation (1.98) can be written in the form of the differential equation $\partial\rho_j^{(n)}/\partial t = \widehat{\pi}(\rho_j^{(k)})$, $k \le n$, where $\widehat{\pi}$ is the operator depending on $\rho_j^{(k)}$ only. The solution of these equations with initial data $\rho_j^{(n)}(\vec{r},t=0) = \rho_j(\vec{r},t=0)\delta_{n,0}$ is defined as the function $f_j^{(n)}$ by unique manner.

Let us consider the physical sense of the small parameter δ. The quantity $\mathcal{J}(f, f)$ can be estimated as

$$\mathcal{J}(f, f) \approx N(\vec{r}, t) r_0^2 \langle v \rangle (f - f^{(0)}).$$

Let us introduce dimensionless variables $t^{\dagger} = t/\tau$, $(\vec{r})^{\dagger} = \vec{r}/\ell$, $(\vec{v})^{\dagger} = \vec{v}/\langle v \rangle$, where ℓ is the macroscopic length characterizing the change of f in the space, $\langle v \rangle \approx \sqrt{k_B T/m}$ is typical heat velocity, $\tau \approx \ell/\langle v \rangle$. At the time intervals $\Delta t < \tau_{\text{rel}} \approx (N r_0^2 \langle v \rangle)^{-1}$ (relaxation time), the function of distribution f changes fastly; if $\partial f/\partial \vec{r} = 0$, then the equilibrium in the system arises for $t \gtrsim \tau_{\text{rel}}$. Then dimensionless Boltzmann equation has the form of

$$\frac{\partial f}{\partial t^{\dagger}} + \vec{v}^{\dagger} \cdot \frac{\partial f}{\partial (\vec{r})^{\dagger}} = \frac{1}{\delta}(f - f^{(0)}),$$

where $\delta = \tau_{\text{rel}}/\tau = Kn \ll 1$ is the Knudsen number (it may be defined as $Kn = \lambda/\ell$ too, λ is the mean free path of particle).

If we consider the function f as functional of $\rho_i(\vec{r}, t)$ for $t > \tau_{\text{rel}}$: such that $f = f(\vec{r}, \vec{v} | \{\rho_i(\vec{r}, t)\})$ (i.e., f does not depend on time directly), then such function is called the *normal solution* of the Boltzmann equation.

These solutions are used in the *Chapman–Enskog method* too. We rewrite the Boltzmann equation in the form of

$$\sum_{i=0}^{4} \frac{\partial f}{\partial \rho_i} \frac{\partial \rho_i}{\partial t} + \vec{v} \cdot \frac{\partial f}{\partial \vec{r}} = \frac{1}{\delta} \mathcal{J}(f, f). \tag{1.99}$$

From the identity $\int \psi_i(\vec{v}) \mathcal{J}(f, f) d\vec{v} = 0$ $(i = 0, \ldots, 4)$, we obtain

$$\frac{\partial}{\partial t} \int \psi_i f d\vec{v} + \frac{\partial}{\partial \vec{r}} \int \vec{v} \psi_i f d\vec{v} = 0, \tag{1.100}$$

or, in terms of ρ_i:

$$\frac{\partial \rho_i}{\partial t} = \widehat{D}_i(\{\rho_k\}), \quad i, k = 0, \ldots, 4, \tag{1.101}$$

where operator $\widehat{D}_i$ is defined as

$$\widehat{D}_i(\{\rho_k\}) = -\frac{\partial}{\partial \vec{r}} \int m\vec{v}\psi_i(\vec{v}) f(\vec{r}, \vec{v}|\{\rho_k(\vec{r}, t)\}) d\vec{v}.$$

Therefore, we can rewrite equation (1.99) in the form without derivative $\partial/\partial t$:

$$\sum_{i=0}^{4} \frac{\partial f}{\partial \rho_i} \widehat{D}_i(\{\rho_k\}) + \vec{v} \cdot \frac{\partial f}{\partial \vec{r}} = \frac{1}{\delta} \mathcal{J}(f, f). \tag{1.102}$$

Let us consider the series $f = \sum_{n=0}^{\infty} \delta^n f^{(n)}$, where $f^{(n)} = f^{(n)}(\vec{r}, \vec{v}|\{\rho_k(\delta)\})$; we assume that operator $\widehat{D}_i$ is analytical:

$$\widehat{D}_i(\{\rho_k\}) = \sum_{n=0}^{\infty} \delta^n \widehat{D}_i^{(n)}(\{\rho_k\}), \tag{1.103}$$

$$\widehat{D}_i^{(n)}(\{\rho_k\}) = -\frac{\partial}{\partial \vec{r}} \int m\vec{v}\psi_i(\vec{v}) f^{(n)}(\vec{r}, \vec{v}|\{\rho_k(\vec{r}, t)\}) d\vec{v}, \tag{1.104}$$

$$\rho_k = m \int \psi_k f^{(0)} d\vec{v}, \quad m \int \psi_k f^{(n)} d\vec{v} = 0, \quad n \geq 1, \ k = 0, \ldots, 4.$$

Now, $\rho_k(\vec{r}, t \,|\, \delta)$ is not analytical by δ (as in the Hilbert method)! We consider these coefficients as variables of $f^{(n)}$ only.

If we use the series for f and formula (1.103) in equation (1.102), then we take the hierarchy of the Chapman–Enskog approximation in powers δ:

$$\mathcal{J}(f^{(0)}, f^{(0)}) = 0 \quad \text{(the power } \delta^{-1}), \tag{1.105}$$

$$\sum_{i=0}^{4} \frac{\partial f^{(0)}}{\partial \rho_i} D_i^{(0)}(\{\rho_k\}) + \vec{v} \cdot \frac{\partial f^{(0)}}{\partial \vec{r}} = 2\mathcal{J}(f^{(1)}, f^{(0)}) \quad \text{(the power } \delta^0), \tag{1.106}$$

$$\sum_{j=0}^{n-1}\sum_{i=0}^{4} \frac{\partial f^{(j)}}{\partial \rho_i} D_i^{(n-j-1)}(\{\rho_k\}) - \sum_{j=1}^{n-1} \mathcal{J}(f^{(j)}, f^{(n-j)})$$

$$+ \vec{v} \cdot \frac{\partial f^{(n-1)}}{\partial \vec{r}} = 2\mathcal{J}(f^{(n)}, f^{(0)}) \quad \text{(the power } \delta^{n-1}, \ n > 1). \tag{1.107}$$

The solution of equation (1.105) (*the first-order approximation of the Chapman–Enskog method*) is the Maxwellian $f^{(0)}$ (see formula (1.83)), and conditions (1.104) are defined as parameters $N, \vec{u}, T$ in terms of $\rho_i(\vec{r}, t)$ $(i = 0, \ldots, 4)$:

$$N(\vec{r}, t) = \frac{\rho_0(\vec{r}, t)}{m}, \quad u_\alpha(\vec{r}, t) = \frac{\rho_\alpha}{\rho_0(\vec{r}, t)} \quad (\alpha = 1, 2, 3 \equiv x, y, z),$$

$$k_B T(\vec{r}, t) = \frac{2m}{3} \left(\frac{\rho_4}{\rho_0} - \frac{\rho_1^2 + \rho_2^2 + \rho_3^2}{2\rho_0^2} \right).$$

By equation (1.103), operators $\widehat{D}_i^{(0)}$ $(i = 0, \ldots, 4)$ have the following forms:

$$\widehat{D}_1^{(0)}(\{\rho_k\}) = -\frac{\partial}{\partial \vec{r}}(\rho \vec{v}), \tag{1.108}$$

$$\widehat{D}_i^{(0)}(\{\rho_k\}) = \left(-\vec{u}\frac{\partial}{\partial \vec{r}}(\rho \vec{u}) - \frac{\partial}{\partial \vec{r}} P_{ij} \right)_i, \quad \frac{\partial}{\partial \vec{r}} P_{ij} = \nabla p,$$

$$P_{ij} = m \int \xi_i \xi_j f d\vec{v}, \quad \vec{v} = \vec{u} + \vec{\xi}, \quad \langle \vec{v} \rangle = \vec{u}, \tag{1.109}$$

$$\langle \vec{\xi} \rangle = 0 \quad (i, j = 1, 2, 3 = x, y, z)$$

$$\widehat{D}_5^{(0)}(\{\rho_k\}) = -\frac{\partial}{\partial \vec{r}} \left\langle \vec{v}\frac{v^2}{2} \right\rangle$$

$$= -\frac{\partial}{\partial \vec{r}}(\rho \vec{u}(u^2/2 + \epsilon + p/\rho)), \quad \epsilon = \langle \xi^2/2 \rangle. \tag{1.110}$$

Therefore, we obtain the right-hand side of equation (1.101), which is equivalent of the Euler system (1.24). Consequently, the first approximation of Chapman–Enskog method leads to the equations of Euler ideal hydrodynamics.

Let us consider the linear integral equation (1.106) (with the right-hand side unequal to zero) — the *second-order approximation of Chapman–Enskog method*. Let us rewrite this equation in the form of

$$\sum_{i=0}^{4} \frac{\partial(\ln f^{(0)})}{\partial \rho_i} \left(\widehat{D}_i^{(0)}(\{\rho_k\}) + \vec{v} \cdot \frac{\partial \rho_i}{\partial \vec{r}} \right) = \frac{2}{f^{(0)}} \mathcal{J}(f^{(0)}, f^{(0)}). \tag{1.111}$$

By the definition $f^{(0)}$, we have

$$\frac{\partial(\ln f^{(0)})}{\partial n} = \frac{1}{N(\vec{r},t)}, \qquad \frac{\partial(\ln f^{(0)})}{\partial u_i} = \frac{m\xi_i}{k_B T(\vec{r},t)},$$

$$\frac{\partial(\ln f^{(0)})}{\partial T} = \frac{1}{T(\vec{r},t)}\left(\frac{m\xi^2}{2k_B T(\vec{r},t)} - \frac{3}{2}\right)$$

and, consequently,

$$\frac{\partial(\ln f^{(0)})}{\partial \rho_0} = \frac{1}{\rho}\left(1 - \frac{m\vec{\xi}\cdot\vec{u}}{k_B T} + \left(\frac{m\xi^2}{2k_B T(\vec{r},t)} - \frac{3}{2}\right)\left(\frac{mu^2}{3k_B T} - 1\right)\right),$$

$$(1.112)$$

$$\frac{\partial(\ln f^{(0)})}{\partial \rho_i} = \frac{m}{\rho k_B T}\left(\xi_i - \frac{2u_i}{3}\left(\frac{m\xi^2}{2k_B T(\vec{r},t)} - \frac{3}{2}\right)\right), \qquad i=1,2,3,$$

$$(1.113)$$

$$\frac{\partial(\ln f^{(0)})}{\partial \rho_4} = \frac{2m}{3\rho k_B T}\left(\frac{m\xi^2}{2k_B T(\vec{r},t)} - \frac{3}{2}\right). \qquad (1.114)$$

Now, we use the last formulas (1.112)–(1.114) on the left-hand side of (1.111):

$$\frac{2}{f^{(0)}}\mathcal{J}(f^{(1)},f^{(0)}) = \sum_{i,j=x,y,z}\frac{m}{2k_B T}\left(\xi_i\xi_j - \frac{\xi^2\delta_{ij}}{3}\right)\left(\frac{\partial u_i}{\partial r_j} + \frac{\partial u_j}{\partial r_i}\right)$$

$$+ \sum_{j=x,y,z}\frac{\xi_j}{T}\left(\frac{m\xi^2}{2k_B T} - \frac{5}{2}\frac{\partial T}{\partial r_j}\right). \qquad (1.115)$$

The operator $\mathcal{J}(f^{(1)},f^{(0)})$ on the left-hand side of this equation is linear (by $f^{(1)}$) and isotropic (i.e., it commutates with rotation operators). Consequently, we can construct the solution of (1.115) in the form of

$$f^{(1)} = f^{(0)}\left(\sum_{i,j=x,y,z}\frac{m}{2k_B T}a_\natural(\xi)\left(\xi_i\xi_j - \frac{\xi^2}{3}\delta_{ij}\right)\left(\frac{\partial u_i}{\partial r_j} + \frac{\partial u_j}{\partial r_i}\right)\right.$$

$$\left. + \sum_{j=x,y,z}\frac{1}{T}\xi_j b_\natural(\xi)\left(\frac{m\xi^2}{2k_B T} - \frac{5}{2}\right)\frac{\partial T}{\partial r_j}\right), \qquad (1.116)$$

where scalar coefficients $a_\natural$, $b_\natural$ depend on $\xi = |\vec{\xi}|$ only. If we use formula (1.116) for calculating $\widehat{D}_i^{(1)}$ (see (1.103)), we obtain

$$\widehat{D}_0^{(1)} = 0,$$

$$\widehat{D}_i^{(1)} = \sum_{j=x,y,z} \frac{\partial}{\partial r_j} \eta \left(\left(\frac{\partial u_i}{\partial r_j} + \frac{\partial u_j}{\partial r_i} \right) \right.$$

$$\left. - \frac{2}{3} \delta_{ij} \left(\sum_{k=x,y,z} \frac{\partial u_k}{\partial r_k} \right) \right), \quad i = 1, 2, 3 = x, y, z,$$

$$\widehat{D}_4^{(1)} = \sum_{j=x,y,z} \frac{\partial}{\partial r_j} \kappa \frac{\partial T}{\partial r_j}$$

$$+ \sum_{i,j=x,y,z} \frac{\partial}{\partial r_j} \eta \left(\left(\frac{\partial u_i}{\partial r_j} + \frac{\partial u_j}{\partial r_i} \right) - \frac{2}{3} \delta_{ij} \left(\sum_{k=x,y,z} \frac{\partial u_k}{\partial r_k} \right) \right) u_i,$$

where the quantities η and κ are

$$\eta = -\frac{m^2}{k_B T} \int \xi_i^2 \xi_j^2 a_\natural(\xi) f^{(0)} d\xi, \quad i, j = x, y, z,$$

$$\kappa = -\frac{m^2}{k_B T^2} \int \xi_i^2 \frac{\xi^2}{2} \left(\frac{\xi^2}{2} - \frac{5 k_B T}{2m} \right) b_\natural(\xi) f^{(0)} d\xi.$$

Here, coefficient η is a microscopical dynamical viscosity, and coefficient κ is thermal conductivity. Therefore, equation (1.106) with the obtained $\widehat{D}_k^{(1)}$ ($k = 0, \ldots, 4$) is equivalent to Navier–Stokes hydrodynamical equations.

The *third-order approximation of the Chapman–Enskog method* leads to the *Burnett (post-Stokesian) equations*. These equations have complicated structures and properties, and are used for the analysis of fast hydrodynamical processes.

We ought to note that the derivation of Euler equations from the Boltzmann equation is exact, but for obtaining the NSE, we need phenomenologic coefficients (or exact interparticle potential). Thus, Navier–Stokes equations are not rigorous consequences of kinetic theory.

Besides, we want the readers to pay attention to the fact that *collision integral $\mathcal{J}$* is a *local quantity*. Thus, $\mathcal{J}$ does not have stimulating influence on the origin and development of large structures in hydrodynamic flows. In particular, one may assume that occurrence of turbulence is not connected with collisional processes described by this integral.

1.3.7. *The validity conditions of the Hilbert method for the Boltzmann equation*

Now, we show the validity of the Hilbert expansion for nonlinear Boltzmann equation. Let us start with the initial data $F_0 = f^{(0)} + \sqrt{f^{(0)}} f$, where $f^{(0)}$ is a global Maxwellian. Let us define spaces B_ℓ and B_{m,ℓ,z_0} with norms

$$\|b_1(\vec{k})\|_\ell = \left(\int_{\mathbb{R}^3} (1 + |k|^2)^\ell |b(\vec{k})|^2 |d\vec{k}| \right)^{1/2}, \tag{1.117}$$

$$b_1 \in B_\ell = \{b_1 : \|b_1\|_\ell < \infty\},$$

$$\|b_2(\vec{r}, \vec{v})\|_{m,\ell, z_0} = \sup_{\vec{v}} (1 + |\vec{v}|^2)^{m/2} \| \exp(z_0|k|) \widehat{f}(\vec{k}, \vec{v}) \|_\ell, \tag{1.118}$$

$$b_2 \in B_{m,\ell,z_0} = \{b_2 : \|b_2\|_{m,\ell,z_0} < \infty\}.$$

If $f \in B_{m,\ell,z_0}$ then, in each of its spatial variables, it is analytic in the region $|\Im(z)| < z_0$ and belongs to $L^2(x + iy)$ for any fixed $|y| < z_0$. The index m measures the decay as $|\vec{v}| \to \infty$.

Next, we define

$$Q_\Delta(f) = \sup_{0 \le z_0 \le z_1} \sup_{0 \le t < (z_1 - z_0)/\Delta} \|f(t)\|_{m,\ell,z_0} \left(1 - \frac{t}{\Delta(z_0 - z)} \right). \tag{1.119}$$

If $Q_\Delta(f) < \infty$ then at time t, f is analytic (in each spatial variable) in the region $|\Im(z)| < z_0$ with $z_0 = z_1 - t/\Delta$. This includes functions with singularities on $|\Im(z)| = z_0$ at $t = 0$, which then move into the considered region at speed Δ^{-1}.

Let ϕ be the solution of equation

$$\left(\frac{\partial}{\partial t} + \vec{v} \cdot \frac{\partial}{\partial \vec{r}} \right) \phi = \frac{1}{\delta} (\widehat{L}(\phi) + \nu_B \widehat{M}(\phi, \phi_1)) \tag{1.120}$$

with $Q_\Delta(\phi)$, $Q_\Delta(\phi_1) < \infty$. Then

$$Q_\Delta(\phi) \le c \cdot \widetilde{S} Q_\Delta(\phi_1) \le c \cdot \widetilde{S} Q_{\Delta_1}(\phi_1), \quad \Delta < \Delta_1,$$

$$\widetilde{S} = \sup_{0 \le z_0 \le z_1} \sup_{0 \le t < (z_1 - z_0)/\Delta} \|f\|_{m,\ell,z_0}. \tag{1.121}$$

The abstract Cauchy–Kovalevskaya theorem is then applied to obtain the following existence theorem: if the initial data $f_0 \in B_{m,\ell,z_1}$ for some $z_1 > 0$, $\ell \ge 2$, $m \ge 3$ and $\|f_0\|_{m,\ell,z_1}$ is sufficiently small, then there is a unique solution $f^{(\delta)}$ of the nonlinear Boltzmann equation

$$\left(\frac{\partial}{\partial t} + \vec{v} \cdot \frac{\partial}{\partial \vec{r}}\right) f^{(\delta)} = \frac{1}{\delta} \widehat{L}(f^{(\delta)}) + \nu_B \widehat{M}(f^{(\delta)}, f^{(\delta)}) \tag{1.122}$$

for any $0 < \delta < 1$. Also, there is $\Delta > 0$ so that for any $t \in [0, \Delta(z_1 - z_0)]$ we have $\|f^{(\delta)}(t)\|_{m,\ell,z_0} \le \|f_0^{(\delta)}\|_{m,\ell,z_1}$. The constants are independent of δ. To get the convergence of $f^{(\delta)}$ as $\delta \to 0$, we need uniform smoothness. Let us define the Hölder norm:

$$\|f\|_\Delta = \sup_{0 < t_1 < t < \Delta(z_1 - z_0)} \sup_{0 \le z_0 < z_1} \|f(t) - f(t_1)\|_{m-\sigma, \ell-\sigma, z_0}$$

$$\times \left(\frac{t_1}{t - t_1} \cdot \left(1 - \frac{t}{\Delta(z_1 - z_0)}\right)\right)^\sigma.$$

Then under the above-mentioned conditions, we have $\|f^{(\delta)}\|_\Delta \le c\|f_0\|_{m,\ell,z_1}$, $c \ne c(\delta)$.

Finally, using the Arzela–Ascoli theorem, we can show that $f^{(\delta)}$ converges to f in B_{m,ℓ,z_1} for $t \in [0, \Delta(z_1 - z_0)]$ and obtain the following results:

(1) Let the initial data for the nonlinear Boltzmann equation

$$\partial F/\partial t + \vec{v} \cdot \partial F/\partial \vec{r} = \delta^{-1} \mathcal{J}(F, F) \tag{1.123}$$

be $F_0 = f^{(0)} + \sqrt{f^{(0)}} f_0 > 0$ and $f_0 \in B_{m,\ell,z_1}$ for some $z_1 > 0$, $\ell \ge 2$, $m \ge 3$. Then the unique solution of the above-mentioned equation converges to F as $\delta \to 0$ in B_{m,ℓ,z_0} for $t \in [0, (z_1 - z_0)/\Delta]$, where F is a local Maxwellian distribution with fluid dynamic variables ρ, $\vec{u}$, T satisfying Euler's equations for $t \in [0; z_1/\Delta]$ (see [21]).

(2) For sufficiently small δ, there is a solution $F^{(\delta)}$ of the non-linear Boltzmann equation (1.123) with $\|f^{(0)} - F^{(\delta)}\| \leq \widetilde{\zeta} \cdot \delta$ ($\widetilde{\zeta} = \text{const} > 0$) for $t \in [0, t_0]$ (here, $f^{(0)} = f^{(0)}(\vec{r}, \vec{v}, t)$ is the Maxwellian distribution corresponding to the smooth solution $(\rho, \vec{u}, t)$ of Euler's system for $t \in [0, t_0]$) (see [22]).

This result shows the validity of the first term in the Hilbert expansion. By demanding more smoothness for $(\rho, \vec{u}, t)$ and including more Hilbert expansion terms in $f^{(0)}(\vec{r}, \vec{v}, t)$, the difference in $\|f^{(0)} - F^{(\delta)}\| \leq \widetilde{\zeta} \cdot \delta$ could be made arbitrarily small. If there are smooth solutions (uniformly in δ) of the NSE, the same can be done for the Chapman–Enskog expansion.

1.3.8. *The properties of the linear Boltzmann equation*

The Cauchy problem for linearized Boltzmann equation with a small mean free path is

$$\left(\frac{\partial}{\partial t} + \vec{v} \cdot \frac{\partial}{\partial \vec{r}}\right) f = \frac{1}{\delta}\widehat{L}(f), \quad f(\vec{r}, \vec{v}, t = 0) = f_0(\vec{r}, \vec{v}). \quad (1.124)$$

This can be solved with the Fourier–Laplace transform method by first looking for solutions $\omega = \omega(\vec{k})$, $\widehat{f} = \widehat{f}(\vec{k}, \vec{v})$ of the equation

$$(\omega - i\vec{k} \cdot \vec{v})\widehat{f} = \widehat{L}(\widehat{f}), \quad (1.125)$$

$$\widehat{f}(\vec{k}, \vec{v}, \omega) = \int_0^\infty dt \int_{\mathbb{R}^3} f(\vec{r}, \vec{v}, t) \cdot \exp(-\omega t - i\vec{k}\vec{v}) d\vec{v},$$

in which ω and $\vec{k}$ have been scaled by δ. Since $\widehat{L}$ operates on $\vec{v}$, we expect that for each $\vec{k}$, there is an infinite number or a continuum of solutions of (1.125), which span the space $H_1 = \{h(\vec{v}) : h, |\vec{v}|h \in L^2(\mathbb{R}^3)\}$.

One can see from [23] that there are five solutions $\omega(\vec{k})$ and corresponding eigenfunctions $\Omega_\omega(\vec{k}, \vec{v})$ such that $\omega(\vec{k}) \to 0$ as $\vec{k} \to 0$. The real part of each of these solutions $\omega(\vec{k})$ is non-positive and for large $|\vec{k}|$, they stay bounded away from the imaginary axis. The remaining values of ω, both discrete and continuous, are all bounded away from the imaginary axis with negative real part.

The solution f of the linearized equation (1.124) is composed of the solutions $\hat{f}$ of (1.125) each multiplied by $\exp(\omega t/\delta)$. Thus, if ω stays bounded away from the imaginary axis with the negative real part, the corresponding term in f will decay exponentially in time at the rate $O(\omega/\delta)$. It then does not belong to the Hilbert or Chapman–Enskog expansions, but rather to the initial layer.

The expansions come only from the five eigenvalues $\omega(\vec{k})$ which go to 0 as $\vec{k} \to 0$. We'll study the eigenvalues and their relation to the Hilbert and Chapman–Enskog expansion, throughout assuming that the interparticle force law comes from a hard cut-off potential.

The spectral values ω of $\widehat{L} + i\vec{k} \cdot \vec{v}$ have non-positive real part since $\widehat{L}$ is non-positive. The operator $\widehat{L}$ can be written as $\widehat{L} = -\nu_B + \widehat{K}_B$, where $\widehat{K}_B$ is compact. Hence, the continuous spectrum of $\widehat{L} + i\vec{v} \cdot \vec{k}$ is the same as that of $\widehat{L}_0 \equiv -\nu_B + i\vec{v} \cdot \vec{k}$, which is

$$Co(\widehat{L}_0) = \{\omega = -\nu(\vec{v}) + i\vec{v} \cdot \vec{k} \text{ for some } \vec{v}\}$$

by Weyl's theorem (see, for example, [24]). All ω in $Co(\widehat{L}_0)$ have $\Re(\omega) < -\nu_1 < 0$ (see equation (1.88)). The spectra of interest, with vanishing small real part, must then be discrete eigenvalues of finite multiplicity. One can show that the only null eigenfunctions of $\widehat{L} + i\vec{v} \cdot \vec{k}$ occur at $\vec{k} = 0$ and are functions $\widetilde{\chi}_i(\vec{v})$, $i = 0, \ldots, 4$. Thus, we look for eigenvalues $\omega^{(i)}$ and eigenfunctions $\Omega_\omega^{(i)}$ as power series expansions:

$$\omega^{(i)}(\vec{k}) = \sum_{n=1}^{\infty} \omega_n^{(i)} |k|^n,$$

$$\Omega_\omega^{(i)}(\vec{k}, \vec{v}) = \sum_{n=0}^{\infty} (\Omega_\omega^{(i)})_n \left(\vec{v}, \frac{\vec{k}}{|\vec{k}|}\right) |k|^n, \quad i = 0, \ldots, 4, \quad (1.126)$$

$$(\Omega_\omega^{(i)})_0 \left(\vec{v}, \frac{\vec{k}}{|\vec{k}|}\right) \in \text{Ker}\,(\widehat{L}).$$

The terms $\omega_1^{(i)}$ are found as solutions of the generalized eigenvalue problem

$$\det \langle \widetilde{\chi}_m, (\omega + i\vec{v} \cdot \vec{k})\widetilde{\chi}_n \rangle = 0. \quad (1.127)$$

The resulting expansion for $\omega^{(i)}$ is (with altered notation)

$$\omega^{(i)}(\vec{k}) = i\omega_1^{(i)}|k| - \omega_2^{(i)}|k|^2 + \cdots, \quad \omega_2^{(i)} > 0, \quad \omega_1^{(i)} \text{ is real.}$$
(1.128)

This expansion is convergent if the interparticle force law is that of hard spheres; for soft potentials, the continuous spectrum touches the imaginary axis for each $\vec{k}$ and the existence of such eigenvalues is unknown. For these potentials, the Hilbert and Chapman–Enskog expansions are not known to be asymptotically correct.

The solution of the linearized Boltzmann equation (1.124) is

$$f(\vec{r}, \vec{v}, t) = \sum_{j=0}^{5} \int a_{(j)}(\vec{k}) \exp(\omega^{(j)}(\delta\vec{k})t/\delta + i\vec{k}\cdot\vec{r})\Omega_\omega^{(j)}(\delta\vec{k}, \vec{v})d\vec{k}$$

$$+ O(\exp(\nu_1 t/\delta)),$$
(1.129)

$$a_{(j)}(\vec{k}) = \int \Omega_\omega^{(j)}(\delta\vec{k}, \vec{v})(\hat{f}_0)(\vec{k}, \vec{v})d\vec{v} \equiv \langle \Omega_\omega^{(j)}, (\hat{f}_0) \rangle.$$

Next, we show the connection with the Hilbert and Chapman–Enskog expansions. The leading terms, $i\omega_1^{(j)}|k|$, are exactly the eigenvalues of the linearized Euler equations [25]. The first two terms, given by

$$\lambda_{(j)}(\vec{k}) = i\omega_1^{(j)}|k| - \delta\omega_2^{(j)}|k|^2,$$
(1.130)

are exactly the eigenvalues of the linearized Navier–Stokes equations. Furthermore, the eigenvalues of the Boltzmann and the fluid equations agree. The eigenfunctions for the Fourier transform of the Euler equation are

$$\hat{s}^{(j)}(k) = \alpha_0^{(j)}(\vec{k})\hat{\rho}(\vec{k}) + \sum_{i=1}^{3} \alpha_i^{(j)}(\vec{k})\widehat{\vec{u}}(\vec{k}) + \alpha_4^{(j)}(\vec{k})\widehat{T}(k).$$
(1.131)

The Boltzmann eigenfunctions have the leading term

$$(\Omega_\omega^{(j)})_0(\vec{k}, \vec{v}) = \alpha_0^{(j)}(\vec{k})\widetilde{\chi}_0(\vec{v}) + \sum_{i=1}^{3} \alpha_i^{(j)}(\vec{k})\widetilde{\chi}_i(\vec{k}) + \alpha_4^{(j)}(\vec{k})\widetilde{\chi}_4(\vec{v}),$$
(1.132)

so that

$$\langle \Omega_0^{(j)}(k), \widehat{f}(k) \rangle = s^{(j)}(k), \tag{1.133}$$

with $\widehat{\rho}, \widehat{\vec{u}}, \widehat{T}$ replaced by $\widehat{\rho}(f), \widehat{\vec{u}}(f), \widehat{T}(f)$. The eigenfunctions for the Navier–Stokes equations are

$$^{(NS)}\widehat{s}^{(j)}(k) = \widehat{s}^{(j)}(k) + O(|k|/\delta). \tag{1.134}$$

The Hilbert expansion is obtained from (1.129) by replacing $\exp(\omega^{(j)}(\delta\vec{k}) \cdot t/\delta)$ with

$$\exp(i\omega_1^{(j)}|k|t) \cdot (1 - \delta\omega_2^{(j)}|k|^2 t + \delta^2\omega_3^{(j)}|k|^3 t + \cdots),$$

and substituting the expansion (1.126) for $\Omega_\omega^{(j)}$ in equation (1.129). The expansion is thus seen to be asymptotic to the solution f of (1.124).

The first term of the Chapman–Enskog expansion is found to agree with f by replacing $\exp(\omega^{(j)}(\vec{k}/\delta) \cdot t/\delta)$ with

$$\exp(i\omega_1^{(j)}|k|t - \delta\omega_2^{(j)}|k|^2 t) \cdot (1 + \delta^2\omega_3^{(j)}|k|^2 t + \cdots),$$

and substituting for $\Omega_\omega^{(j)}$ in (1.129) the function corresponding to $^{(NS)}s^{(j)}$.

Now, we consider the agreement of the linear and nonlinear Boltzmann equation and the linear compressible Navier–Stokes equations. In the expression (1.129) for f, the term $O(\exp(-\nu_1 t))$ (with $\delta = 1$) is negligible and the size of f is

$$\|f(t)\| \le c \sum_{j=0}^{4} \left(\int_{\mathbb{R}^3} |\exp(\omega^{(j)}(\vec{k})t)|^2 d\vec{k} \right)^{1/2} \|f_0\|$$

$$\le c \sum_{j=0}^{4} \left(\int_{\mathbb{R}^3} |\exp(-2\omega_2^{(j)}|\vec{k}|^2 t) d\vec{k} \right)^{1/2} \|f_0\| \le ct^{-3/4}\|f_0\|,$$

$$\tag{1.135}$$

in which

$$\|f\|^2 = \int_{\mathbb{R}^3} d\vec{r} \int_{\mathbb{R}^3} |f(\vec{r}, \vec{v})|^2 d\vec{v}.$$

The final term in (1.135) can also be shown to be a lower bound on $\|f(t)\|$, so that this establishes the dominant asymptotic decay rate of the solution of the linearized Boltzmann equation.

Next, consider the difference $\widetilde{\Delta}f = f_N - f_L$ between the solution f_N of the nonlinear Boltzmann equation (1.86) and the solution f_L of the linear Boltzmann equation (1.87) with the same initial conditions. It satisfies

$$\left(\frac{\partial}{\partial t} + \vec{v} \cdot \frac{\partial}{\partial \vec{v}}\right)(\widetilde{\Delta}f) = \widehat{L}(\widetilde{\Delta}f) + \nu_B \widehat{M}(f_N, f_N). \tag{1.136}$$

We will show that $\widetilde{\Delta}f$ decays faster than f_L, because the nonlinear term contributes no fluid dynamic part to $\widetilde{\Delta}f$. In other words, since $\langle \widetilde{\chi_i}, \nu_B \widehat{M}(f_N, f_N) \rangle = 0$, then using $(\Omega_\omega^{(i)})_0(\vec{v}, \vec{k}) \in Ker(\widehat{L})$, we obtain

$$\langle (\Omega_\omega^{(0)})_0(\vec{k}), \nu_B \widehat{M}(f_N, f_N)(\vec{k}) \rangle = 0,$$

$$|\langle (\Omega_\omega^{(0)})(\vec{k}), \nu_B \widehat{M}(f_N, f_N)(\vec{k}) \rangle| \le c|k| = 0. \tag{1.137}$$

If we consider $\nu_B \widehat{M}(f_N, f_N)$ as an inhomogeneity in (1.136), then by Duhamel's principle, the solution $\widetilde{\Delta}f$ has an expression analogous to (1.129):

$$\widetilde{\Delta}f(\vec{r}, \vec{v}, t) = \sum_{j=0}^{5} \int_0^t \int_{\mathbb{R}^3} a_{(j)}(k, t-s)$$

$$\times \exp(\omega^{(j)})(k)(s + i\vec{k} \cdot \vec{r})\Omega_\omega^{(j)}(k)d\vec{k}ds + O(\exp(-\nu_1 t)),$$

$$a_{(j)}(t)(\vec{k}, t) = \langle \Omega_\omega^{(j)}(\vec{k}), \nu_B \widehat{M}(f_N, f_N)(\vec{k}, t) \rangle. \tag{1.138}$$

Assume that $\|\nu_B \widehat{M}(f_N, f_N)\| \le c(1 + t)^{-3/2}$; this is consistent with (1.135) and can be proved by iteration. Then we estimate $\widetilde{\Delta}f$ using (1.137) by

$$\|\widetilde{\Delta}f\| \le c \sum_{j=0}^{5} \int_0^t \left(\int_{\mathbb{R}^3} \exp(-2\omega_2^{(j)}|k|^2 s)|k|^2(1 + t - s)^{-3}d\vec{k}\right)^{1/2}$$

$$ds \le ct^{-5/4}. \tag{1.139}$$

A comparison with the estimate (1.135) for f_L shows that the dominant part of f_N is the same as that of f_L.

Now, we compare the properties of the linear Boltzmann equation and the linear compressible Navier–Stokes equations. The agreement between their eigenvalues and eigenfunctions is given by

$$|\omega^{(j)}(\vec{k}) - \lambda_{(j)}(\vec{k})| \le c|k|^3, \quad |\langle \Omega_\omega^{(j)}(k), \widehat{f}_0(k) \rangle - {}^{(\mathrm{NS})}s^{(j)}(\vec{k})\rangle| \le c|k|$$

$$(1.140)$$

(see equations (1.126), (1.130), (1.133) and (1.134)). Let $f_{\mathrm{NS}}^{(0)}$ denote the Maxwellian whose fluid dynamic variables solve the linear Navier–Stokes equations, and let $\tilde{f}_{\mathrm{NS}}^{(0)}$ denote the linearization of $f_{\mathrm{NS}}^{(0)}$ about the uniform state. Let f denote the linear Boltzmann solution. It follows that

$$\|f - \tilde{f}_{\mathrm{NS}}^{(0)}\| \le \sum_{j=0}^{4} \Big(\big| \big(\exp(\omega^{(j)}t) - \exp(\lambda_{(j)}(k)t) \big) {}^{(\mathrm{NS})}s^{(j)}(\vec{k})$$

$$+ \exp(\omega^{(j)}(k)t)(\langle \Omega_\omega^{(j)}(k), \widehat{f}_0(k) \rangle - {}^{(\mathrm{NS})}s^{(j)}(\vec{k}))\big|^2 d\vec{k} \Big)^{1/2}$$

$$\le \sum_{j=1}^{5} \Big(\int_{\mathbb{R}^3} \exp(-2\omega_0^{(j)}|k|^2 t)^2 (|k|^3 t + |k|)^2 \Big)^{1/2} \le t^{-5/4}.$$

$$(1.141)$$

This shows that the dominant terms in the two solutions are identical.

1.3.9. *The Grad method for the solution of the Boltzmann equation*

In this section, we will consider the *H. Grad method* for solution of the Boltzmann equation.

Let us multiply the Boltzmann equation (1.75) (with additional force term in the left-hand side for generality) by $m\psi_i$, $i = 0, \ldots, 4$. From the identity $\mathcal{I} = 0$ (see (1.79)) for collision invariants, we obtain hydrodynamic equations

$$\frac{\partial \rho}{\partial t} + \frac{\partial}{\partial r_i}(\rho u_i) = 0, \qquad (1.142)$$

$$\left(\frac{\partial}{\partial t} + u_j \frac{\partial}{\partial r_j}\right) u_i = -\frac{1}{\rho}\frac{\partial P_{ij}}{\partial r_j} + \frac{F_i}{m}, \qquad (1.143)$$

$$\frac{3\rho R}{2}\left(\frac{\partial}{\partial t} + u_j \frac{\partial}{\partial r_j}\right) T = -\frac{\partial q_j}{\partial r_j} - P_{ij}\frac{\partial u_i}{\partial r_j}, \qquad (1.144)$$

where

$$P_{ij} \equiv p_{ij} + p\delta_{ij}$$

$$= m \int \xi_i \xi_j \cdot f(\vec{r}, \vec{v}, t) d\vec{v}, \quad q_i = \frac{m}{2}\int \xi^2 \xi_i \cdot f(\vec{r}, \vec{v}, t) d\vec{v},$$

$\vec{\xi} = \vec{v} - \vec{u}$ is the heat velocity of molecules, $\langle \vec{v} \rangle = \vec{u}$, $\langle \vec{\xi} \rangle = 0$, $p = N k_B T$,

$$\frac{3N k_B T}{2} = \int \frac{m\xi^2 f}{2} d\vec{v},$$

$$\vec{u}(\vec{r}, t) = \frac{1}{N(\vec{r}, t)}\int \vec{v} \cdot f(\vec{r}, \vec{v}, t) d\vec{v}, \quad i, j = 1, 2, 3.$$

So, we have five equations for 13 unknown variables ρ, u_i, P_{ij}, q_i. To close this system, we use *method of moments*. The moment of Kth order of the distribution function f is tensor

$$M_{\alpha_1, \ldots, \alpha_K}^{(K)} = \int \prod_{\beta=1}^{K} v_{\alpha_\beta} f d\vec{v}, \quad \mathcal{M}_{\alpha_1, \ldots, \alpha_K}^{(K)} = \int \prod_{\beta=1}^{K} \xi_{\alpha_\beta} f d\vec{v},$$

$$\xi_{\alpha_\beta} = v_{\alpha_\beta} - u_{\alpha_\beta}, \quad \alpha_\beta = 1, 2, 3.$$

Our hydrodynamical variables are

$$N(\vec{r}, t) = \int f d\vec{v} = M^{(0)}, \quad N u_i(\vec{r}, t) = \int v_i f d\vec{v} = M_i^{(1)} \quad (i = 1, 2, 3),$$

$$P_{ij}(\vec{v}, t) = m \int \xi_i \xi_j f d\vec{v} = m \mathcal{M}_{ij}^{(2)},$$

$$q_i = \int \frac{m}{2}\xi_i \xi^2 f d\vec{v} = \frac{m}{2}\mathcal{M}_{ijj}^{(3)} = \frac{m}{2}(\mathcal{M}_{i11} + \mathcal{M}_{i22} + \mathcal{M}_{i33}).$$

Let us represent the distribution function as $f(\vec{r}, \vec{v}, t) = \check{f}(\vec{v}, M^{(0)},$ $M^{(1)}, \ldots)$, where moments $M^{(i)}$ are functions of $\vec{r}$ and t. If we substitute $\check{f}$ into (1.75), we obtain that the Boltzmann equation is equivalent to infinitive-order system of equations (in general case). If we want to find the approximate solution of the Boltzmann equation, we use in $\check{f}$ $K_0 < \infty$ moments $M^{(0)}, \ldots, M^{(K_0)}$.

Let us consider f as series

$$f = f^{(0)} \left(a^0 H^0 + a^{(1)}_{\alpha_1} H^{(1)}_{\alpha_1} + \cdots + \frac{1}{K!} a^{(K)}_{\alpha_1 \ldots \alpha_K} H^{(K)}_{\alpha_1 \ldots \alpha_K} + \cdots \right),$$

where $a^{(K)}_{\alpha_1 \ldots \alpha_K}$ are coefficients depending on $\vec{r}$ and t,

$$f^{(0)} = N \left(\frac{m}{2\pi k_B T} \right)^{3/2} \exp\left(-\frac{m\xi^2}{2k_B T} \right) = N \left(\frac{m}{k_B T} \right)^{3/2} \omega(\vec{v}_\flat)$$

is the local Maxwellian, $\omega(\vec{v}_\flat) = (2\pi)^{-3/2} \exp(-v_\flat^2/2)$, $\vec{v}_\flat = \sqrt{(m/k_B T)}\xi$,

$$H^{(K)}_{\alpha_1 \ldots \alpha_K}(v_1, v_2, v_3) = \frac{(-1)^K}{\omega} \frac{\partial^K \omega}{\partial v_{\alpha_1} \ldots \partial v_{\alpha_K}}, \quad \alpha_i = 1, 2, 3,$$

$$H^{(0)} = 1, \quad H^{(1)}_i = (v_\flat)_i, \quad H^{(2)}_{ij} = (v_\flat)_i (v_\flat)_j - \delta_{ij}, \ldots$$

are (tensorial) Hermite polynomials. Using orthogonality property of these polynomials, one may obtain coefficients $a^{(K)}_{\alpha_1 \ldots \alpha_K}$:

$$a^{(K)}_{\alpha_1 \ldots \alpha_K} = \frac{1}{N} \int f H^{(K)}_{\alpha_1 \ldots \alpha_K} d\vec{v}, \tag{1.145}$$

$$a^{(0)} = 1, \quad a^{(1)}_i = 0, \quad a^{(2)}_{ij} = \frac{p_{ij}}{p}, \quad a^{(3)}_{ijk} = \frac{m\mathcal{M}_{ijk}}{p} \sqrt{\frac{m}{k_B T}}, \ldots$$

If we approximate f as

$$f = f^{(0)} \left(1 + \frac{1}{2} a^{(2)}_{ij} H^{(2)}_{ij} + \frac{1}{6} a^{(3)}_{ijk} H^{(3)}_{ijk} \right), \tag{1.146}$$

or, equivalently,

$$f = f^{(0)} \left(1 + \frac{p_{ij}}{2p} \frac{m\xi_i \xi_j}{k_B T} + \frac{m\mathcal{M}_{ijk}}{6p} \left(\frac{m}{k_B T} \right)^2 \xi_i \xi_j \xi_k - \frac{q_i}{p} \left(\frac{m}{k_B T} \right)^2 \xi_i \right), \tag{1.147}$$

we obtain the so-called twentieth-moment approximation (i.e., there are 20 moments n, u_i, T, p_{ij}, $\mathcal{M}_{ijk}$ in the approximate function).

One can obtain the equations for moments if we multiply the Boltzmann equation by $H^{(K)}_{\alpha_1...\alpha_K}$ and integrate over all $\vec{v}$:

$$\int H^{(K)} \left(\frac{\partial f}{\partial t} + v_i \frac{\partial f}{\partial r_i} \right) d\vec{v}$$

$$= \int \left(\frac{df\, H^{(K)}}{dt} + \sqrt{\frac{k_B T}{m}} \frac{\partial(v_\flat)_i H^{(K)} f}{\partial r_i} \right.$$

$$\left. - f \left(\frac{dH^{(K)}}{dt} + \sqrt{\frac{k_B T}{m}} \frac{\partial(v_\flat)_i H^{(K)}}{\partial r_i} \right) \right) d\vec{v} = \mathcal{J}[H^{(K)}],$$

$$d/dt \equiv \partial/\partial t + u_i \partial/\partial r_i, \tag{1.148}$$

$$\mathcal{J}[H^{(K)}] = \int H^{(K)} (f^\dagger f^\dagger_* - f f_*) g b\, db\, d\vec{v}_* d\varphi d\vec{v}$$

$$= \frac{1}{2} \int f f_* \left(H^{(K)}(\vec{v}^\dagger) + H^{(K)}(\vec{v}^\dagger_*) - H^{(K)}(\vec{v}) \right.$$

$$\left. - H^{(K)}(\vec{v}_*) \right) g b\, db\, d\varphi d\vec{v}_* d\vec{v}.$$

For twentieth-moment approximation, we use (1.146)–(1.147) ($a^{(K)} = 0$ for $K > K_0 = 3$) and equations for moments (1.148) become

$$\frac{\partial \rho}{\partial t} + \frac{\partial}{\partial r_m}(\rho u_m) = 0, \tag{1.149}$$

$$\frac{\partial u_i}{\partial t} + u_m \frac{\partial u_i}{\partial r_m} + \frac{1}{\rho} \frac{\partial P_{ij}}{\partial r_m} = 0, \tag{1.150}$$

$$\frac{\partial p_{ij}}{\partial t} + \frac{\partial(u_m p)}{r_m} + \frac{2}{3} P_{im} \frac{\partial u_i}{\partial r_m} + \frac{2}{3} \frac{\partial q_m}{\partial r_m} = 0, \tag{1.151}$$

$$\frac{\partial p_{ij}}{\partial t} + \frac{\partial(u_m p_{ij})}{r_m} + m \frac{\partial \mathcal{M}_{ijm}}{\partial r_m} - \frac{2}{3} \delta_{ij} \frac{\partial q_m}{\partial r_m} + p_{im} \frac{\partial u_j}{\partial r_m} + p_{jm} \frac{\partial u_i}{\partial r_m}$$

$$- \frac{2}{3} \delta_{ij} p_{ms} \frac{\partial u_m}{\partial r_s} + p \left(\frac{\partial u_i}{\partial r_j} + \frac{\partial u_j}{\partial r_i} - \frac{2}{3} \delta_{ij} \frac{\partial u_m}{\partial r_m} \right)$$

$$= \frac{p}{N} \mathcal{J}[H^{(2)}_{ij}], \tag{1.152}$$

$$m\frac{\partial \mathcal{M}_{ijk}}{\partial t} + m\frac{\partial u_m \mathcal{M}_{ijk}}{\partial r_m}$$

$$+ m\left(\mathcal{M}_{mjk}\frac{\partial u_i}{\partial r_m} + \mathcal{M}_{imk}\frac{\partial u_j}{\partial r_m} + \mathcal{M}_{ijm}\frac{\partial u_k}{\partial r_m}\right)$$

$$+ \frac{\partial}{\partial r_i}\left(p_{jk}\frac{k_B T}{m}\right) + \frac{\partial}{\partial r_j}\left(p_{jk}\frac{k_B T}{m}\right) + \frac{\partial}{\partial r_k}\left(p_{ij}\frac{k_B T}{m}\right)$$

$$+ (p_{im}\delta_{jk} + p_{jm}\delta_{ik} + p_{km}\delta_{ij})\frac{\partial}{\partial r_m}\left(\frac{k_B T}{m}\right)$$

$$- \frac{1}{\rho}\left(p_{ij}\frac{\partial P_{km}}{\partial r_m} + p_{ik}\frac{\partial P_{jm}}{\partial r_m} + p_{jm}\frac{\partial P_{im}}{\partial r_m}\right)$$

$$+ p\left(\delta_{jk}\frac{\partial}{\partial x_i} + \delta_{ik}\frac{\partial}{\partial x_j} + \delta_{ij}\frac{\partial}{\partial x_k}\right)\frac{k_B T}{m} = \frac{p}{N}\sqrt{\frac{k_B T}{m}}\,\mathcal{J}[H_{ijk}^{(3)}].$$

$$(1.153)$$

The most useful approximation in the Grad's method is the *thirteen-moment* one. All moments in this approximation have clear physical sense and can be measured: N, u_j, T, p_{ij}, q_i. We represent the distribution function as

$$f = f^{(0)}\left(1 + \frac{1}{2}a_{ij}^{(2)}H_{ij}^{(2)} + b_i H_i^{(3)}\right), \qquad (1.154)$$

where $H_i^{(3)} = (v_\flat)_i(v_\flat^2 - 5)$ is the convolution of $H_{ijk}^{(3)}$ by two indices. If we multiply (1.154) by $H_i^{(3)}(\vec{v}_\flat)$ and integrate over $\vec{v}_\flat$, then we obtain $b_i = a^{(3)}/10$ and

$$f = f^{(0)}\left(1 + \frac{1}{2}a_{ij}^{(2)}H_{ij}^{(2)} + \frac{1}{10}a_i^{(3)}H_i^{(3)}\right)$$

$$= f^{(0)}\left(1 + \frac{p_{ij}}{2p}\left(\frac{m}{k_B T}\right)\xi_i\xi_j - \frac{q_i}{p}\left(\frac{m}{k_B T}\right)\left(1 - \frac{\xi^2}{5}\frac{m}{k_B T}\right)\xi_i\right).$$

$$(1.155)$$

Let us substitute (1.155) into (1.145), then we obtain $a_{ij}^{(3)} = \frac{1}{5}(a_i^{(3)}\delta_{jk} + a_j^{(3)}\delta_{ik} + a_k^{(3)}\delta_{ij})$ or, equivalently:

$$m\mathcal{M}_{ijk} = \frac{2}{4}(q_i\delta_{jk} + q_j\delta_{ik} + q_k\delta_{ij}). \qquad (1.156)$$

Hence, changing moments $\mathcal{M}_{ijk}$ in equations (1.152) and (1.153) in accordance with (1.156), we obtain the thirteen-moment system of Grad's equations, in which the first three equations are (1.149), (1.150), (1.151), and the fourth and fifth equations are (for Maxwell molecules)

$$\frac{\partial p_{ij}}{t} + \frac{\partial(u_m p_{ij})}{\partial r_m} + \frac{2}{5}\left(\frac{\partial q_i}{\partial r_j} + \frac{\partial q_j}{\partial r_i} - \frac{2}{3}\delta_{ij}\frac{\partial q_m}{\partial r_m}\right)$$

$$+ p_{im}\frac{\partial u_j}{\partial r_m} + p_{jm}\frac{\partial u_i}{\partial r_m} - \frac{2}{3}\delta_{ij}p_{ms}\frac{\partial u_m}{\partial r_s}$$

$$+ p\left(\frac{\partial u_i}{\partial r_j} + \frac{\partial u_j}{\partial r_i} - \frac{2}{3}\delta_{ij}\frac{\partial u_m}{\partial r_m}\right) + \frac{6A}{m}\left(\frac{8c_0}{m}\right)^{1/2}\rho p_{ij} = 0,$$

$$(1.157)$$

$$\frac{\partial q_i}{\partial t} + \frac{\partial(u_m q_i)}{\partial r_m} + \frac{7}{5}p_{im}\frac{\partial u_i}{\partial r_m} + \frac{2}{5}q_m\frac{\partial u_m}{\partial r_i} + \frac{2}{5}q_i\frac{\partial u_m}{\partial r_m}$$

$$+ \frac{k_B T}{m}\frac{\partial p_{im}}{\partial r_m} + \frac{7}{2}p_{im}\frac{\partial}{\partial r_m}\left(\frac{k_B T}{m}\right) - \frac{p_{im}}{\rho}\frac{\partial P_{ms}}{\partial r_m}$$

$$+ \frac{5}{2}p\frac{\partial}{\partial r_i}\left(\frac{k_B T}{m}\right) + \frac{4A}{m}\left(\frac{8c_0}{m}\right)\rho q_i = 0, \qquad (1.158)$$

where c_0 is a constant from interparticle potential $U^{(i-p)} = c_0/r^n$ and for Maxwell molecules $n = 4$, $A = 0.343$ and

$$\mathcal{J}[H_{ij}^{(2)}] = -6A\sqrt{8c_0/m}\,N^2 a_{ij}^{(2)},$$

$$\mathcal{J}[H_{ijk}^{(3)}] = -A\sqrt{8c_0/m}\,N^2(9a_{ijk}^{(3)} - a_i^{(3)}\delta_{jk} - a_j^{(3)}\delta_{ik} - a_k^{(3)}\delta_{ij}).$$

This system includes measurable quantities only.

Let us rewrite equations (1.157) and (1.158) in the form of

$$\frac{\partial p_{ij}}{\partial t} + A_{ij} + \frac{p_{ij}}{\tau_0} = 0, \qquad (1.159)$$

$$\frac{\partial q_i}{\partial t} + B_i + \frac{2}{3}\frac{q_i}{\tau_0} = 0, \qquad \tau_0^{-1} = \frac{6A}{m}\sqrt{\frac{8c_0}{m}}\rho, \qquad (1.160)$$

where τ_0 is the relaxation time (for $t \sim \tau_0$, the quantities p_{ij} and q_i do not depend on the initial data).

The solutions of this system of equations (we consider A_{ij}, B_i, τ_0 as given functions of time) are as follows:

$$p_{ij}(t) = p_{ij}(0) \exp\left(-\int_0^t \frac{d\tau}{\tau_0}\right)$$

$$-\int_0^t A_{ij}(s) \exp\left(-\int_s^t \frac{d\tau}{\tau_0}\right) ds \quad (\tau_0 \to 0)$$

$$= \left(p_{ij} + \tau_0 A_{ij} - \tau_0 \frac{\partial A_{ij}\tau_0}{\partial t} + \cdots\right)\Bigg|_{t=0}$$

$$\times \exp\left(-\int_0^t d\tau/\tau_0\right) - \left(\tau_0 A_{ij} - \tau_0 \frac{\partial A_{ij}\tau_0}{\partial t} + \cdots\right),$$

$$(1.161)$$

$$q_i(t) = q_i(0) \exp\left(-\frac{2}{3}\int_0^t \frac{d\tau}{\tau_0}\right)$$

$$-\int_0^t B_i(s) \exp\left(-\int_s^t \frac{d\tau}{\tau_0}\right) ds \quad (\tau_0 \to 0)$$

$$= \left(q_i + \frac{3\tau_0 B_i}{2} - \frac{9\tau_0}{4}\frac{\partial B_i\tau_0}{\partial t} + \cdots\right)\Bigg|_{t=0}$$

$$\times \exp\left(-\int 2d\tau/3\tau_0\right) - \left(\frac{3\tau_0 B_i}{2} - \frac{9\tau_0}{4}\frac{\partial B_i\tau_0}{\partial t} + \cdots\right).$$

$$(1.162)$$

For small Kn (when characteristic time of flow $\tau \gg \tau_0$), the influence of history of collision processes on the quantities p_{ij} and q_i is negligibly small; it is the consequence of factors $\exp(-\cdots)$ in equations (1.161) and (1.162). Therefore, we may write

$$p_{ij} \approx -\tau_0 A_{ij}, \quad q_i \approx -\frac{3}{2}\tau_0 B_i. \qquad (1.163)$$

Using definition A_{ij} and B_i, from these equations, we obtain approximately

$$p_{ij} = -\tau_0 p\left(\frac{\partial u_i}{\partial r_j} + \frac{\partial u_j}{\partial r_i} - \frac{2}{3}\delta_{ij}\frac{\partial u_m}{\partial r_m}\right), \quad q_i = -\frac{15}{4}\tau_0 p\frac{\partial}{\partial r_i}\left(\frac{k_B T}{m}\right).$$

$$(1.164)$$

If we substitute (1.164) into equations (1.149)–(1.151), we obtain the Navier–Stokes equations (for Maxwell gas) with the viscosity coefficient

$$\eta = \tau_0 p = \frac{1}{6A}\sqrt{\frac{m}{8c_0}}k_B T,$$

and the coefficient of thermal conductivity

$$\kappa = \frac{15 k_B}{4T}p\tau_0 = \frac{5}{8A}\frac{k_B^2 T}{\sqrt{8c_0 m}}.$$

If we conserve in the expansions (1.161) and (1.162) two terms, we obtain the above-mentioned Burnett's approximation:

$$p_{ij} = -\tau_0 A_{ij} + \tau_0 \frac{\partial \tau_0 A_{ij}}{\partial t}$$

$$= -2\eta \left[\frac{\partial u_i}{\partial r_j}\right] + k_1 \frac{\eta^2}{p}\frac{\partial u_k}{\partial r_k}\left[\frac{\partial u_i}{\partial r_j}\right]$$

$$+ \frac{k_2 \eta^2}{p}\left(\left[\frac{\partial}{\partial r_i}\left(\frac{F_i}{m} - \frac{1}{\rho}\frac{\partial p}{\partial r_j}\right)\right] - \left[\frac{\partial u_k}{\partial r_i}\frac{\partial u_j}{\partial r_k}\right]\right)$$

$$- 2\left[\left[\frac{\partial u_i}{\partial r_k}\right]\cdot\frac{\partial u_k}{\partial r_j}\right] + \frac{k_3\eta^2}{\rho T}\left[\frac{\partial^2 T}{\partial r_i \partial r_j}\right] + \frac{k_4\eta^2}{\rho p T}\left[\frac{\partial p}{\partial r_j}\frac{\partial T}{\partial r_j}\right]$$

$$+ \frac{k_5\eta^2}{\rho T^2}\left[\frac{\partial T}{\partial r_i}\frac{\partial T}{\partial r_j}\right] + \frac{k_6\eta^2}{p}\left[\left[\frac{\partial u_i}{\partial r_k}\right]\left[\frac{\partial u_k}{\partial r_j}\right]\right],$$

$$q_i = -\frac{3}{2}\tau_0 B_i + \frac{9\tau_0}{4}\frac{\partial \tau_0 B_i}{\partial t}$$

$$= -\kappa\frac{\partial T}{\partial r_i} + \frac{\tilde{k}_1\eta^2}{\rho T}\frac{\partial u_j}{\partial r_j}\frac{\partial T}{\partial r_i} + \frac{\tilde{k}_2\eta^2}{\rho T}\left(\frac{2}{3}\frac{\partial}{\partial r_i}\left(T\frac{\partial u_j}{\partial r_j}\right) + 2\frac{\partial u_j}{\partial r_i}\frac{\partial T}{\partial r_j}\right)$$

$$+ \left(\frac{\tilde{k}_3\eta^2}{\rho p}\frac{\partial p}{\partial r_j} + \frac{\tilde{k}_4\eta^2}{\rho T}\frac{\partial T}{\partial r_j}\right)\left[\frac{\partial u_j}{\partial r_i}\right],$$

$$[C_{ij}] \equiv \frac{1}{2}(C_{ij} + C_{ji}) - \frac{1}{3}\delta_{ij}C_{kk},$$

where (for Maxwell molecules):

$$k_1 = \frac{14}{3} - \frac{4T}{3\eta}\frac{d\eta}{dT}, \quad k_2 = 2, \quad k_3 = 3, \quad k_4 = 0,$$

$$k_5 = \frac{3T}{\eta}\frac{d\eta}{dT}, \quad k_6 = 8,$$

$$\tilde{k}_1 = \frac{105}{8} - \frac{15T}{4\eta}\frac{d\eta}{dT}, \quad \tilde{k}_2 = -\frac{45}{8}, \quad \tilde{k}_3 = -3, \quad \tilde{k}_4 = 3,$$

$$\tilde{k}_5 = \frac{105}{4} + \frac{3T}{\eta}\frac{d\eta}{dT}.$$

For particles with other interparticle potentials, the coefficients k_i $(i = 1,\ldots,6)$ and $\tilde{k}_j$ $(j = 1,\ldots,5)$ have the similar form (for example, for hard spheres $k_1^{(hs)} = 1.014k_1$, $k_2^{(hs)} = 1.014k_2$, $k_3^{(hs)} = 0.806k_3$, etc.)

1.4. Summary

In this chapter, the basic properties of Euler and Navier–Stokes hydrodynamical equations, and also their connection with the kinetic Boltzmann equation were considered.

In Section 1.1, on a basis of *a priori* postulated main principles (laws of conservation of mass, momentum and energy in a closed system), we obtain the equations of continuity, transfer of momentum and energy in differential and integral forms for ideal incompressible and isentropic liquid (in the general case at the presence of a field of external forces). The possibility of statement boundary conditions to the obtained (Euler) system of the hydrodynamical equations was shown. The validity of Bernoulli's and Thomson's theorems was shown. The concept of stream function was introduced, and forms of vorticity equations for 2D and 3D incompressible flows were obtained. In Section 1.2, the Navier–Stokes equations taking into account molecular movement in a liquid have been considered. On the basis of reasonable physical assumptions, the equation movement for the viscous liquid, generalizing Euler's equation of transport of momentum, was obtained. Concepts of the first and second viscosities were introduced in an appropriate way. Expression for dissipation energy rate in incompressible fluid was analyzed. On the basis of the law of conservation of the total energy in a viscous liquid, the general equation of heat transfer (entropy transfer) was deduced.

In Section 1.3, properties of kinetic Boltzmann equation were considered. Validity of the general Liouville theorem (conservation of phase volume) was shown. Concepts of BBGKY hierarchy and correlation function were introduced. The deduction of the Boltzmann equation for single-particle function of distribution from Bogolyubov's chain (with the help of an assumption about factorization of 2-particle function f_2) was demonstrated. The properties of collisional term of the Boltzmann equation considered as bilinear integral operator were analyzed. The properties of collisional invariants were considered, the H-theorem was formulated and proved. Hilbert's methods, Chapman–Enskog and Grad approaches to the solving of the Boltzmann equation were considered. On their basis, the technique of obtaining hydrodynamical Euler's, Navier–Stokes, Burnett equations was shown. Exact expressions for the earlier introduced phenomenological coefficients of viscosity and heat conductivity in the equations of hydrodynamics were obtained. The properties of linear Boltzmann equation were considered and the connection of this equation with Navier–Stokes equations was demonstrated.

During the work on the material of this chapter, authors used the following books and papers: [3, 7–9, 12, 15, 26, 28, 29, 36], etc. We especially recommend for readers Refs. [8, 9, 16] for a detailed study of the problems mentioned in this chapter.

Chapter 2

Gravity Waves and Solitons

2.1. Wave Equation and Free Surface Gravity Waves

Waves in fluid are generated due to the existence of restoring (inertial) forces that tend to bring the medium back to its undisturbed state. If restoring forces are caused by compressibility or elasticity of the material medium, then the resulting motion is caused by compression and elastic waves; in these waves, the vector of fluid particles motion is collinear of the vector of direction of wave propagation. Another kind of wave motion is the so-called *gravity waves* which occur at the free surface of a liquid or at the interface between two fluids of different density; the particle motion in gravity wave can have components both along and perpendicular to the direction of wave propagation. In this case, gravity forces play the role of restoring ones.

Now, we consider free surface gravity waves in detail.

2.1.1. *The wave equation and (co-)sinusoidal waves*

The equation describing wave motions of small amplitude is

$$\frac{\partial^2 \xi}{\partial t^2} = c^2 \triangle \xi, \tag{2.1}$$

where $\xi(\vec{r}, t)$ is the function connected with the disturbance of some hydrodynamical or thermodynamical quantity and c is the scalar parameter. Wave traveling only in the x-direction is described by

$$\frac{\partial^2 \xi}{\partial t^2} = c^2 \frac{\partial^2 \xi}{\partial x^2}. \tag{2.2}$$

This equation has a general solution of the form $\xi(x,t) = f_-(x - ct) + f_+(x + ct)$, where f_- and f_+ are arbitrary functions of arguments $x - ct$ and $x + ct$ correspondingly; the function $f_-(x - ct)$ represents a wave propagating in the positive x-direction at speed c and $f_+(x + ct)$ propagates in the negative x-direction at speed c.

The Cauchy problem for the equation (2.2) with initial conditions

$$\xi(x, t = 0) = \xi_1(x), \quad \frac{\partial \xi(x, t = 0)}{\partial t} = \xi_2(x)$$

has the solution $\xi(x,t) = f_-^{(1)}(x - ct) + f_+^{(2)}(x + ct)$, where

$$f_-^{(1)}(y) = \frac{\xi_1(y)}{2} - \frac{1}{2c} \int_{y_0}^{y} \xi_2(\widetilde{y}) d\widetilde{y},$$

$$f_+^{(2)}(y) = \frac{\xi_1(y)}{2} + \frac{1}{2c} \int_{y_0}^{y} \xi_2(\widetilde{y}) d\widetilde{y}. \tag{2.3}$$

We can rewrite this solution in the form of *d'Alembert formula*:

$$\xi(x, t) = \frac{\xi_1(x - ct) + \xi_1(x + ct)}{2} + \frac{1}{2c} \int_{x-ct}^{x+ct} \xi_2(y) dy. \tag{2.4}$$

According to Fourier's principle, any arbitrary disturbance can be decomposed into sinusoidal wave components of different wavelengths and amplitudes. Thus, it is important to study real (or imaginary) part of the wave of the form

$$\xi = \xi_k \exp\left(i s(k, x, t)\right), \quad s(k, x, t) \equiv kx - \omega t,$$

$$k = \frac{2\pi}{\lambda}, \quad \omega = 2\pi\nu, \tag{2.5}$$

where ξ_k is an *amplitude of the wave*, k is a *wavevector* or *wavenumber vector* (for one-dimensional case), ω is an *angular frequency*, and λ is a *wavelength*. This wave is called (co)sinusoidal (i.e., $\Re(\xi) \propto \cos(s)$ and $\Im(\xi) \propto \sin(s)$). The velocity of the wave is $c = \lambda\nu = \omega/k$. If two waves with the same magnitude ξ_0 but different wavevectors

$(k_0 + \Delta k$ and $k_0 - \Delta k)$ both travel in the x-direction, then we obtain *superposition* of these waves, and the resulting wave has the form of

$$\xi = A_0 \exp\left(i(k_0 x - \omega_0 t)\right), \quad A_0 = 2\xi_0 \cos\left(\Delta k \cdot x - \Delta\omega \cdot t\right),$$

$$\Delta\omega = c\Delta k, \qquad\qquad \omega_0 = ck_0.$$

We may introduce the speed c_g of change of "amplitude" $A_0(x,t)$:

$$\frac{\Delta\omega}{\Delta k} \approx \left(\frac{d\omega}{dk}\right)_{k_0} = c_g.$$

If the dispersion law for waves $\omega = \omega(k)$ is not linear (i.e., $\omega(k) \neq ck$), then c_g (*group velocity*) is not equal to c (*phase velocity*).

For three-dimensional waves, equation (2.5) is generalized to

$$\xi = \xi_k \exp(i(k_x x + k_y y + k_z z - \omega t)) = \xi_k \exp(i(\vec{k} \cdot \vec{x} - \omega t)), \quad (2.6)$$

where $\vec{k} = (k_x, k_y, k_z)$ is a *wavenumber vector* for the three-dimensional case $(|\vec{k}|^2 = k_x^2 + k_y^2 + k_z^2)$, and the wavelength of (2.6) is $\lambda = 2\pi/|\vec{k}|$. The phase velocity is the vector $\vec{v} = \omega\vec{k}/|\vec{k}|^2 = \{\omega/k_x; \omega/k_y; \omega/k_z\}$; the magnitude of phase velocity is $c = \omega/|\vec{k}|$.

2.1.2. *Free surface gravity waves*

Let us consider the free surface of liquid of uniform depth d, which may be large or small compared to the wavelength λ; we assume that the amplitude ξ_k of oscillations of the free surface is small, i.e., $|\xi_k|/\lambda \ll 1$, $|\xi_k|/d \ll 1$. The condition $|\xi_k|/\lambda \ll 1$ implies that the slope of the liquid surface is small, and the condition $|\xi_k|/d \ll 1$ implies that the instantaneous depth does not differ essentially from the undisturbed depth. These conditions allow us to linearize the problem, which will be formulated further.

We will consider a case where the waves propagate in the x-direction only, and the motion is two-dimensional in the xz-plane; the vertical coordinate z is perpendicular to the undisturbed free surface (xy-plane). The free surface displacement is $\xi(x,t)$. The motion is irrotational, and we can use in consideration a velocity potential. The condition of irrotationality of planar flow $rot_{(2)}\vec{v} = \partial v_2/\partial x - \partial v_1/\partial z = 0$ guarantees the existence of scalar function ϕ (a *velocity potential*) such as $v_1 \equiv \partial\phi/\partial x$, $v_2 \equiv \partial\phi/\partial z$. Substitution into the continuity equation in the form $(v_1)_x + (v_2)_z = 0$ gives the Laplace equation

$$\triangle_2\phi = \frac{\partial^2\phi}{\partial x^2} + \frac{\partial^2\phi}{\partial z^2} = 0. \tag{2.7}$$

Boundary conditions for this equation are as follows: (I)$\partial\phi/\partial z = 0$ at $z = -d$ (at the bottom), (II)$\partial\xi/\partial t + v_1\partial\xi/\partial x = (v_2)_\xi$ at $z = \xi$, where $(v_2)_\xi$ is the vertical component of fluid velocity (at the free surface). We can simplify the second (*kinematic*) condition for small amplitude waves: $\partial\xi/\partial t = (\partial\phi/\partial z)|_{z=0}$ (II†). In addition to the kinematic condition at the surface, there is a *dynamic condition* that the pressure just below the free surface is equal to ambient pressure (with surface tension neglected). Taking the ambient pressure to be zero, this condition is $p = 0$ at $z = \xi$ (III); for small-amplitude waves from the unsteady Bernoulli equation, $\partial\phi/\partial t + \frac{1}{2}(v_1^2 + v_2^2) + p/\rho + gz = B_0(t)$ (see (1.27) and further text in Chapter 1) is applicable. The function $B_0(t)$ can be absorbed in $\partial\phi/\partial t$ by redefining ϕ. Neglecting the non-linear term $v_1^2 + v_2^2$ for small-amplitude waves, the linearized form of the Bernoulli equation is $\partial\phi/\partial t + p/\rho + gz = 0$. The substitution into the surface boundary of the condition $p|_{z=\xi} = 0$ gives the simplified condition (III†): $\partial\phi/\partial t = -g\xi$ at $z = 0$ (instead of $z = \xi$).

To solve equation (2.7) with boundary conditions (I) + (II†) + (III†), we assume that waves have sinusoidal dependence upon phase $\mathsf{s}(k, x, t)$ (i.e., $\xi = \xi_0\cos(kx - \omega t)$) and potential ϕ may be represented in the following form:

$$\phi(z, x, t) = \phi_0(z) \cdot \sin(kx - \omega t),$$

where $\phi_0(z)$ and $\omega(k)$ are to be determined. The substitution of above $\phi(z, x, t)$ into the Laplace equation gives $d^2\phi_0/dz^2 = k^2\phi_0$, whose general solution is $\phi_0(z) = \phi_0^{(1)}\exp(kz) + \phi_0^{(2)}\exp(-kz)$. The constants $\phi_0^{(1,2)}$ are determined from the boundary conditions (I) and (II†):

$$\phi_0^{(1)} = \frac{\xi_0\omega}{k(1 - \exp(-2kd))}, \quad \phi_0^{(2)} = \frac{\xi_0\omega\exp(-2kd)}{k(1 - \exp(-2kd))}. \tag{2.8}$$

Then for velocity components, we have

$$v_1 = \xi_0 \omega \frac{\text{ch}\,k(z+d)}{\text{sh}\,kd} \cos(kx - \omega t),$$

$$v_2 = \xi_0 \omega \frac{\text{sh}\,k(z+d)}{\text{sh}\,kd} \sin(kx - \omega t). \tag{2.9}$$

We obtain the *dispersion relation* between k and ω by the substitution of $\phi = (\xi_0 \omega/k)\text{ch}(kz + kd)\sin(kx - \omega t)/\text{sh}(kd)$ and $\xi = \xi_0 \cos(kx - \omega t)$ into the condition (III†) (i.e., $(\partial\phi/\partial t + g\xi)|_{z=0} = 0$):

$$\omega = \sqrt{gk \cdot th(kd)}. \tag{2.10}$$

Consequently, the phase velocity has the form:

$$c = \omega/k = \sqrt{\lambda g/(2\pi) \cdot th(2\pi d/\lambda)} \quad (\text{so } k = 2\pi/\lambda). \tag{2.11}$$

Surface gravity waves possess kinetic energy due to the motion of the fluid and potential energy due to the deformation of the free surface. Kinetic energy per unit of horizontal area is found by integrating over the depth and averaging over the wavelength:

$$E_k = \frac{\rho}{2\lambda} \int_0^\lambda \int_{-d}^0 (v_1^2 + v_2^2)dzdx. \tag{2.12}$$

Here, the z-integral is taken up to $z = 0$ because the integral up to $z = \xi$ gives a higher order term. The substitution of the velocity components from (2.9) gives

$$E_k = \frac{1}{2}\rho g\langle \xi^2 \rangle = \frac{1}{4}\rho g\xi_0^2. \tag{2.13}$$

2.1.3. *Deep and shallow water*

The above analysis is applicable whatever the magnitude of λ is in relation to the water depth d. Interesting simplifications result for $d \gg \lambda$ (*deep water*) and $d \ll \lambda$ (*shallow water*).

In the "deep water" approximation:

- $th(2\pi d/\lambda) \approx 1$, and from the formula (2.11), it follows that $c = \sqrt{g/k}$; consequently, longer waves in deep water propagate faster;

- the velocity components (see (2.9)): $v_1 = \xi_0 \omega \exp(kz) \cos(kx - \omega t)$, $v_2 = \xi_0 \omega \exp(kz) \sin(kx - \omega t)$;
- the perturbation pressure $p^\dagger \equiv p + \rho g z = -\rho \partial \phi / \partial t = \rho g \xi_0 \exp(kz)$ $\cos(kx - \omega t)$.

In the "shallow water" approximation:

- $th(2\pi d/\lambda) \approx 2\pi d/\lambda$, and from the formula (2.11), it follows that $c = \sqrt{gd}$; consequently, the wave speed is independent of wavelength and increases water depth;
- the velocity components: $v_1 = \frac{\xi_0 \omega}{kd} \cos(kx - \omega t)$, $v_2 = \xi_0 \omega$ $(1 + \frac{z}{d}) \sin(kx - \omega t)$ (the vertical component is much smaller than the horizontal component);
- the perturbation pressure $p^\dagger \equiv p + \rho g z = \rho g \xi_0 \cos(kx - \omega t) = \rho g \xi$; consequently, the pressure change at any point is independent of depth, and equals the hydrostatic increase of pressure due to the surface elevation change ξ. The pressure field is therefore completely hydrostatic in shallow water waves. Vertical accelerations are negligible because of the small v_2-field. Therefore, shallow water waves are also called *hydrostatic waves*.

2.2. Waves in Dispersive Media

Above we assume that the wave amplitude is small (i.e., we can neglect the higher order terms in the Bernoulli equation for simplification boundary conditions at the free surface $z = \xi$). The consequence of such linear analysis was that the waves of arbitrary shape propagate unchanged in form if the system is non-dispersive, such as shallow water waves. The unchanging form is a result of the fact that all wavelengths, of which the initial wave form is composed, propagate at the same speed $c = \sqrt{gd}$, provided all the sinusoidal components satisfy the shallow water approximation $dk \ll 1$. We now see that the unchanging wave form result is no longer valid if *finite amplitude* (nonlinear) effects are considered.

Nonlinear effects in non-dispersive media continually accumulate and add up until they become large changes (blow-up effect, hydraulic jump, etc.). Such an accumulation is prevented in a dispersive medium because the different Fourier components propagate at different speeds and become separated from each other. In a dispersive system, then, nonlinear steepening could cancel out the dispersive spreading, resulting in finite amplitude waves of constant form.

2.2.1. *The stokes waves*

The surface elevation of irrotational waves in deep water (Stokes, 1847) is given by

$$\xi = \xi_0 \cos(kx - ckt) + \frac{1}{2}k\xi_0^2 \cos(2kx - 2ckt)$$

$$+ \frac{3}{8}k^2\xi_0^3 \cos(3kx - 3ckt) + \cdots , \qquad (2.14)$$

where the speed of propagation is $c = \sqrt{g/k + gk\xi_0^2}$. Equation (2.14) is the Fourier series for the wave form ξ. The addition of Fourier components of different wavelengths in this equation shows that the wave profile ξ is no longer exactly sinusoidal. The arguments in the cosine terms show that all the Fourier components propagate at the same speed c, so that the wave profile propagates unchanged in time. It has now been established that the existence of periodic wavetrains of unchanging form is a typical feature of nonlinear dispersive systems; another result, generally valid for nonlinear systems, is that the wave speed depends on the amplitude: $c = c(\xi_0)$.

Periodic finite-amplitude irrotational waves in deep water are called *Stokes waves*.

2.2.2. *Solitary waves*

Let us consider nonlinear waves in a slightly dispersive system. These waves approximately satisfy the nonlinear *Korteweg–de Vries equation*

$$\frac{\partial \xi}{\partial t} + c_0 \frac{\partial \xi}{\partial x} + \frac{3c_0\xi}{8d}\frac{\partial \xi}{\partial x} + \frac{c_0 d^2}{6}\frac{\partial^3 \xi}{\partial x^3} = 0, \qquad (2.15)$$

where $c_0 = \sqrt{gd}$. The first two terms appear in the linear non-dispersive limit. The third term is due to finite amplitude effects and the fourth term results from the weak dispersion due to the water depth being not shallow enough. This agrees with the first two terms in the Taylor series expansion of the dispersion relation $c = \sqrt{(g/k) \cdot th(kd)}$ for small kd, verifying that the weak dispersive effects are indeed properly accounted for by the last term.

Analysis of the Korteweg–de Vries equation shows that two types of solutions are possible, a periodic solution and a solitary wave solution. The periodic solution is called *cnoidal wave* because it is expressed in terms of elliptic functions $cn(x)$. The other possible solution of (2.15) involves only a single hump and is called a *solitary wave (soliton)*. Its profile is given by

$$\xi = \xi_0 \cdot ch^{-2}\left(\left(\frac{3\xi_0}{4d^3}\right)^{1/2}(x - ct)\right), \qquad (2.16)$$

where the speed of propagation is $c = c_0 + c_0\xi_0/(2d)$ (the propagation velocity increases with the amplitude of the hump).

An isolated hump propagates at constant speed with unchanging form. Solitons have been observed to exist not only at surface waves but also as internal waves in stratified fluid.

2.3. Waves at a Density Interface Between Liquids of Different Densities

Above we considered only waves at free surface of a liquid. However, waves can also exist at the interface between two immiscible liquids of different densities (the so-called *internal waves*). Such a sharp density gradient can, for example, be generated in the ocean by solar heating of the upper layer or in an estuary of river into which fresh water flows over sea water (which is more saline and consequently heavier). The situation can be idealized by considering a lighter fluid of density ρ_1 lying over a heavier fluid of density ρ_2.

2.3.1. *Internal waves between infinitely deep fluids*

We assume that the fluids are infinitely deep, so that only those solutions that decay exponentially from the interface are allowed. We have to solve the Laplace equations for the velocity potentials in both layers, subject to the continuity of p and v_2 at the interface ($z = 0$ instead of $z = \zeta$):

$$\frac{\partial^2 \phi_1^2}{\partial x^2} + \frac{\partial^2 \phi_1^2}{\partial z^2} = 0, \quad \frac{\partial^2 \phi_2^2}{\partial x^2} + \frac{\partial^2 \phi_2^2}{\partial z^2} = 0, \quad (2.17)$$

$$\phi_1 \to 0 \text{ at } z \to \infty, \quad \phi_2 \to 0 \text{ at } z \to -\infty, \quad (2.18)$$

$$\frac{\partial \phi_1}{\partial z} = \frac{\partial \phi_2}{\partial z} = \frac{\partial \zeta}{\partial t}, \quad \rho_1 \frac{\partial \phi_1}{\partial t} + \rho_1 g \zeta = \rho_2 \frac{\partial \phi_2}{\partial t} + \rho_2 g \zeta \quad \text{at } z = 0,$$
$$(2.19)$$

where $\zeta = \Re\left(\zeta_0 \exp(ikz - i\omega t)\right)$ is the interface displacement. Conditions (2.18) require that the solutions of equations (2.17) must be of the form of

$$\phi_1 = \phi_1^{(0)} \exp(-kz + ikx - i\omega t), \quad \phi_2 = \phi_2^{(0)} \exp(kz + ikx - i\omega t).$$

From the kinematic boundary conditions, we obtain $\phi_1^{(0)} = -\phi_2^{(0)} = i\omega\zeta_0/k$; the dynamic boundary condition (2.19) then gives the dispersion relation

$$\omega = \varrho_{12}\sqrt{gk}, \quad \varrho_{12} \equiv \sqrt{\frac{\rho_2 - \rho_1}{\rho_2 + \rho_1}}. \quad (2.20)$$

This equation shows that waves at the interface between two liquids of infinite thickness travel like deep water surface waves, with $\omega \propto \sqrt{g/\lambda}$, but at much reduced frequency. Therefore, internal waves have a smaller frequency (in general), and consequently a smaller phase velocity, than surface waves. If $\rho_1 = 0$, then (2.20) reduces to the expression for surface waves.

The horizontal velocity components in the two layers are

$$v_1 = \frac{\partial \phi_1}{\partial x} = -\omega\zeta_0 \exp(-kz + ikx - i\omega t),$$

$$v_2 = \frac{\partial \phi_2}{\partial x} = \omega\zeta_0 \exp(kz + ikx - i\omega t),$$

which show that the velocities in the two layers are oppositely directed. The interface is therefore a *vortex sheet* (the tangential velocity to this surface is discontinuous function). It can be expected that a continuously stratified medium, in which the density varies continuously as a function of z, will support internal waves whose vorticity is distributed throughout the flow. Consequently, internal waves in a continuously stratified fluid are not irrotational and do not satisfy the Laplace equation.

2.3.2. *Waves in a layer overlying on deep fluids*

Let us consider the case in which the upper layer is not infinitely thick but has a finite thickness; the lower layer is assumed to be infinitely thick. It is clear that the present configuration will allow two modes of oscillation, the first mode in which the free surface and the interface are in phase and the second mode in which they are oppositely directed.

Let d be the thickness of the upper layer, and let the origin be placed at the mean position of free surface. The equations for velocity potentials ϕ_1, ϕ_2 coincide with (2.17), but the boundary conditions are new:

$$\phi_2 \to 0 \quad \text{at } z \to -\infty, \tag{2.21}$$

$$\frac{\partial \phi_1}{\partial z} = \frac{\partial \xi}{\partial t}, \quad \frac{\partial \phi_1}{\partial t} + g\xi = 0 \quad \text{at } z = 0, \tag{2.22}$$

$$\frac{\partial \phi_1}{\partial z} = \frac{\partial \phi_2}{\partial z} = \frac{\partial \zeta}{\partial t}, \quad \rho_1 \frac{\partial \phi_1}{\partial t} + \rho_1 g\zeta = \rho_2 \frac{\partial \phi_2}{\partial t} + \rho_2 g\zeta \quad \text{at } z = -d. \tag{2.23}$$

We assume that a free surface displacement has the form $\xi = \Re(\xi_0 \exp(ikx - i\omega t))$, and an interface displacement has the form $\zeta = \Re(\zeta_0 \exp(ikx - i\omega t))$.

The velocity potentials in the layers must be of the form of

$$\phi_1 = (\phi_1^{(1)} \exp(kz) + \phi_1^{(2)} \exp(-kz)) \exp(ikx - i\omega t),$$

$$\phi_2 = \phi_2^{(0)} \exp(kz + ikx - i\omega t). \tag{2.24}$$

The boundary conditions at $z = 0$ and the first condition at $z = -d$ give the constants $\phi_1^{(1,2)}$ and $\phi_2^{(0)}$ in terms of the given amplitude ξ_0:

$$\phi_1^{(1)} = -\frac{i\xi_0}{2}\left(\frac{\omega}{k} + \frac{g}{\omega}\right), \quad \phi_1^{(2)} = \frac{i\xi_0}{2}\left(\frac{\omega}{k} - \frac{g}{\omega}\right),$$

$$\phi_2^{(0)} = -\frac{i\xi_0}{2}\left(\frac{\omega}{k} + \frac{g}{\omega}\right) - \frac{i\xi_0}{2}\left(\frac{\omega}{k} - \frac{g}{\omega}\right)\exp(2kd),$$

$$\zeta_0 = \frac{\xi_0}{2}\left(1 + \frac{gk}{\omega^2}\right)\exp(-kd) + \frac{\xi_0}{2}\left(1 - \frac{gk}{\omega^2}\right)\exp(kd). \qquad (2.25)$$

Substitution into the second boundary condition (2.23) gives the dispersion relation $\omega(k)$:

$$\left(\frac{\omega^2}{kg} - 1\right)\left(\frac{\omega^2}{kg}(\rho_1\text{sh}(kd) + \rho_2\text{ch}(kd)) - (\rho_2 - \rho_1)\text{sh}(kd)\right) = 0.$$

$$(2.26)$$

One possible root of equation (2.26) is $\omega^2 = kg$, which is the same as that for a deep water gravity wave. Equation (2.25) shows that in this case $\zeta_0 = \xi_0\exp(-kd)$, implying that the amplitude at the interface is reduced from that at the surface by the factor $\exp(kd)$; also, this equation shows that the motion of the interface and the free surface are locked in phase (they go up or down simultaneously). This mode is similar to a gravity wave propagating on the free surface of the upper liquid, in which the motion decays as $\exp(-kz)$ from free surface. It is called the *barotropic mode*, because the surfaces of constant pressure and density coincide in such a flow.

The other possible root of equation (2.26) is

$$\omega^2 = \frac{kg(\rho_2 - \rho_1)\,\text{sh}(kd)}{\rho_1\text{sh}(kd) + \rho_2\text{ch}(kd)}, \qquad (2.27)$$

which reduces to dispersive relation (2.20) if $kd \to \infty$. Substitution of (2.27) into (2.25) shows that

$$\xi = \zeta\left(1 - \frac{\rho_2}{\rho_1}\right)\exp(-kd).$$

Consequently, $\zeta(k, x, t)$ and $\xi(k, x, t)$ have opposite signs and the interface displacement is much larger than the surface displacement if the density difference is small.

This mode of behavior is called the *baroclinic* mode because the surfaces of constant pressure and density do not coincide. The horizontal velocity v_1 changes sign across the interface. The existence of a density difference has therefore generated a motion that is quite different from the barotropic behavior (for case, in which the fluids have infinite depth and no free surface, has only a baroclinic mode and no barotropic mode).

2.4. Summary

This chapter is devoted to the consideration of the so-called gravitational waves (surface and internal). These waves were characterized by the presence generally of the longitudinal displacements of particles of a liquid and the presence of external force (gravitation) as "restoring force".

In Section 2.1, the properties of the usual hyperbolic wave equation were considered, and also the concepts of phase/group velocities of a wave packet, potentials of velocity field ϕ were introduced. The possibility of various statements of boundary conditions for the equation on ϕ was shown.

Section 2.2 is devoted to the consideration of free surface waves for various cases of a ratio of characteristic parameters of a problem (the theory of waves on deep and shallow water, waves with finite amplitude).

In Section 2.3, we considered internal waves for various cases of a ratio of characteristic parameters of a problem (baroclinic and barotropic modes).

During the work on the material of this chapter, authors used the following books: [3, 6, 9, 13, 15]. We especially recommend for readers the fundamental monographs: R.K. Dodd, J.C. Eilbeck, J.D. Gibbon, H.C. Morris, *Solitons and Nonlinear Wave Equations* (Academic Press, New York, 1982) and J. Lighthill, *Waves in Fluids* (Cambridge University Press, Cambridge, 1978).

Chapter 3

Thermodynamics of Fluids

3.1. Equation of State

The *equation of state* is a necessary complement to the thermodynamic laws, which allow the latter to be applied to real substances. By "equation of state"(EOS), one usually means the relations which define the pressure p or (specific) internal energy ϵ_s of a physically homogeneous system in the thermodynamic equilibrium as functions of two arbitrary parameters that specify the state of matter. The equations of state cannot be derived solely from thermodynamic laws; they are determined experimentally or are calculated theoretically based on the representation about the structure of the matter provided by the methods of statistical mechanics.

3.1.1. *A perfect gas EOS*

The most simple and famous equations of state are ones for a perfect gas. The next step to physical reality is the van der Waals equation for the interacted particles, considered as elastic rigid bodies.

A perfect (ideal) gas has the following properties: (1) the pressure exerted by a gas at constant temperature is inversely proportional to its volume V (*Boyle's law*); (2) the pressure exerted by a fixed volume of gas is directly proportional to its thermodynamic temperature T (*Charles' law*); (3) the temperature T of a gas (with fixed mass in the container which can change geometrical parameters) is inversely proportional to its volume V (*Gay–Lussac's law*). Consequently, we can describe the properties of a perfect gas by *Mendeleev–Clapeyron's law*:

$$pV = \frac{m}{\mu}RT \quad \text{or, equivalently,} \quad pV = \nu RT, \qquad (3.1)$$

where m is the mass of the gas, μ is the molar (atomic) mass of the gas, $R = 8.3145$ J/(K· mol) is the *universal gas constant*, and $\nu = m/\mu$ is the number of moles of gas. This equality is called *equation of state* (EOS) for a *perfect gas*. It connects three thermodynamical parameters (p, V, T) by the algebraic way.

We can rewrite this EOS in the following forms: (1) $p = nk_BT = k_BT/v_s$, where $n = \nu N_A/V$ is the number density of particles per unit volume, $k_B = R/N_A$ is Boltzmann's constant, $v_s = 1/n$ is the volume per unit mass (*specific volume*), $N_A = 6.0225 \times 10^{23}$ is the number of particles in a mole of gas (*Avogadro's number*); (2) $p = \rho RT/\mu$, where $\rho = m/V$ is the mass density of the gas.

The *state* of a gas is described by *state variables* such as ρ, p, T. All thermodynamic properties of a gas are specified when the values of any two variables are given (it is a consequence of thermodynamic laws).

3.1.2. *The van der Waals and the Kamerlingh Onnes EOS*

Equation (3.1) provides a good approximation to the behavior of dilute (low-density) gases; more accurate expressions for imperfect real gases (and liquids) have more complicated forms.

It is impossible to derive equations of state from the laws of thermodynamics only; in order to obtain an EOS, one needs a microscopic description of the system. The first attempt to describe the behavior of a real matter by using its molecular structure was made by J.D. van der Waals. He argued that in the case of high densities and pressures, the ideal gas law needs to be modified to allow for two effects: the finite size of the molecules and the attraction between different molecules. The finite size of the molecules means that the volume actually available to the molecule v_s is reduced to $v_s - b$, where b is the molecular volume. Including this effect into the ideal gas law, we obtain $p(N_A v_s - N_A b_1) = \nu RT$. For $\nu = 1$, this equation has the form $p(V - b) = RT$, $b \equiv b_1 N_A$. The attraction between the molecules causes them to decelerate before they hit the container wall. This results in a reduction in the pressure. The force on an individual molecule is proportional to n and since the number of molecules striking the boundary per unit time is also proportional to n, we can expect the decrease in pressure to be proportional to the square of number density n^2. This leads to *van der Waals equation of state*:

$$p = \frac{RT}{V - b} - \frac{a}{V^2}. \tag{3.2}$$

We may rewrite this equation for any number of moles of matter ν:

$$\left(p + \frac{a\nu^2}{V^2}\right)(V - \nu b) = \nu RT. \tag{3.3}$$

The parameters $a \ll pV^2$, $b \ll V$ can be determined from a fit of the EOS to experimental values for temperatures above $T_c = 8a/(27Rb)$, where T_c is the *critical temperature*:

$$\left(\frac{\partial p}{\partial V}\right)_{T=T_c} = 0, \quad \left(\frac{\partial^2 p}{\partial V^2}\right)_{T=T_c} = 0, \quad V_c = 3b, \quad p_c = \frac{a}{27b^2}.$$

Here, p_c, V_c, T_c characterize the *critical point* of the *critical van der Waals isotherm*, when roots of the cubic equation (3.2) satisfy $V_1 = V_2 = V_3 = V_c$.

The equation of state for any substance can be written in the form of $pV = Vp(T, V) = f(T, V)$. Let's consider $\rho \sim 1/V$ as an argument of this EOS, then we have $pV = F(T, \rho)$; the right-hand side $F(T, \rho)$ can be expressed as a power series

$$pV = F(T, 0) + \frac{\partial F(T, 0)}{\partial \rho}\rho + \frac{1}{2}\frac{\partial^2 F(T, 0)}{\partial \rho^2}\rho^2 + \cdots. \tag{3.4}$$

The first term of the series $F(T, 0)$ gives EOS of a perfect gas (if $\rho \to 0$); the second, third, etc., coefficients are functions of the temperature T only. Therefore, equation (3.4) can be rewritten as

$$pV = RT\left(1 + \frac{B_2}{V} + \frac{B_3}{V^2} + \cdots\right), \tag{3.5}$$

where B_i are *virial coefficients* $(i = 2, 3, \ldots)$. The relation (3.5) is called the *Kamerlingh Onnes equation*. The van der Waals EOS can be transformed to this form, if one can expand the first term in the right-hand side of the equation

$$pV = RT/(1 - 4\zeta) - a/V, \quad \zeta = \frac{b}{4V},$$

which is equivalent to (3.2). We obtain

$$pV = RT(1 + 4\zeta + 16\zeta^2 + 64\zeta^3 + \cdots) - a/V,$$

or

$$pV = RT \left(1 + \frac{b - a/RT}{V} + \frac{b^2}{V^2} + \cdots \right).$$

Equations of state for liquids are based on the similarity of thermodynamic properties of ones and strongly compressed gases. Besides, there exists "quasi-crystal" or "hole" theory of liquids, in which the EOS has the following form (*Frenkel's equation*):

$$V - V_0 = N_A \exp \left(-\frac{u}{k_B T} \right) \Delta V,$$

where V is the molar volume of liquid with temperature T, V_0 is the minimal volume, which the liquid can possess, u is the energy of hole arising, and ΔV is the minimal volume of hole (defect in the quasi-crystal structure).

3.1.3. The EOS based on the quantum statistical models

Besides the equation of state in the form of $p = p(V, t)$, we can consider the so-called *caloric EOS*, i.e., equation for the *internal energy* $\epsilon_s = \epsilon_s(\rho, T)$. For the model of an ideal gas, the following conditions are true:

(1) a mean kinetic energy of one molecule of gas is defined as

$$\langle W_k \rangle = \frac{1}{N} \sum_{i=1}^{N} \frac{m_0 v_i^2}{2} = m_0 \langle v^2 \rangle,$$

where N is the number of molecules of gas and m_0 is the mass of one molecule;

(2) absolute temperature T for a monatomic gas is the measure of the mean kinetic energy of its molecules motion:

$$\frac{3}{2}k_B T = \langle W_k \rangle.$$

Therefore, the internal energy (for monatomic gas) is $\epsilon_s = (3/2)nk_B T$ $(n = N/V = \rho N/m)$.

Obtaining the EOS for matter with high density or temperature and other extreme characteristics is based on quantum-statistical models [27]. For instance, we may consider the so-called *Thomas–Fermi model*; one that is based on the use of Fermi–Dirac statistics and semiclassical approximation for electrons, i.e., one assumes that the electrons of atoms are continuously distributed in the phase space according to the Fermi–Dirac statistics.

The electron pressure is readily found if we use the Thomas–Fermi statistics and calculate the value of the pressure on the boundary

$$r = r_0 = \frac{m_e e^2}{\hbar^2} \left(\frac{3}{4\pi} \frac{A}{\rho N_A} \right)$$

of the *average atom* cell, where $V(r)|_{r=r_0} = 0$, $(dV/dr)|_{r=r_0} = 0$, $V(r)$ is the potential generated by the electrons and the atomic nuclei, e is the charge of electron, $\rho = nA/N_A$ is the density of matter, A is atomic weight of matter. If we "place" a rigid wall at the boundary of the atom cell, then the change in the density of the flux of electron momenta per unit of time during their reflection in the wall gives the electron pressure p_e. Let us choose the z-axis drawn from the center of the atom cell to be perpendicular to the wall. Then for electrons with momentum $\vec{p}$ ranging in the element of phase-space volume $d\vec{p} = dp_x dp_y dp_z = p^2\, dp \sin\vartheta d\vartheta\, d\varphi$, the density of the flux of momentum changes at the rate of $N(\vec{p})d\vec{p} \cdot p\cos\vartheta \cdot 2p\cos\vartheta$,

$$N(\vec{p})d\vec{p} = \frac{2}{(2\pi)^3} \frac{d\vec{p}}{1 + \exp\left((p^2/2 - V(r))/(k_B T) + \eta\right)}\bigg|_{r=r_0}$$

$$= \frac{2}{(2\pi)^3} \frac{p^2\, dp \sin\vartheta d\vartheta d\varphi}{1 + \exp\left(p^2/(2k_B T) + \eta\right)}.$$

Integration with respect to the variables φ $(0 < \varphi < 2\pi)$, ϑ $(0 < \vartheta < \pi/2)$ and p $(0 < p < \infty)$ yields

$$p_e = \frac{2}{(2\pi)^3} \int_0^\infty dp \int_0^{\pi/2} d\vartheta \int_0^{2\pi} d\varphi \frac{p^2 \sin\vartheta \cdot 2p^2 \cos^2\vartheta}{1 + \exp\left(p^2/(2k_B T) + \eta\right)}$$

$$= \frac{1}{3\pi^2} \int_0^\infty \frac{p^4 dp}{1 + \exp\left(p^2/(2k_B T) + \eta\right)} = \frac{(2k_B T)^{5/2}}{6\pi^2} I_{3/2}(-\eta), \quad (3.6)$$

$I_k(x) \equiv \int_0^\infty \frac{y^k dy}{1 + \exp(y - x)}$ is the Fermi–Dirac function, where $\eta = -\mu/(k_B T)$ and μ is the chemical potential. One can arrive at the same expression if we calculate the pressure as the average momentum transported by electrons per unit of time across a unit of surface:

$$p_e = \frac{2}{(2\pi)^3} \int_0^\infty dp \int_0^\pi d\vartheta \int_0^{2\pi} d\varphi \frac{p^2 \sin\vartheta \cdot p^2 \cos^2\vartheta}{1 + \exp\left(p^2/(2k_B T) + \eta\right)}$$

$$= \frac{(2\theta)^{5/2}}{6\pi^2} I_{3/2}(-\eta).$$

The electron density (concentration) on the boundary of the atom cell is given by

$$n(r_0) = n_e = \frac{(2k_B T)^{3/2}}{2\pi^2} I_{1/2}(-\eta). \quad (3.7)$$

Therefore, using the asymptotics of the function $I_k(x)$, we conclude that, at high temperatures, when $\eta \gg 1$ and the average kinetic energy of the electrons in the continuous spectrum is larger than the potential energy, $p_e = n_e k_B T$, i.e., the well-known formula for the pressure of an ideal gas. For large negative values of η, i.e., for large densities and low temperatures, formulas (3.6) and (3.7) yield

$$p_e = \frac{1}{5}(3\pi^2)^{2/3} n_e^{5/3}.$$

To find the total pressure, we must take into account the pressure of ion skeletons. At high temperatures, nuclei are usually regarded as an ideal gas. In this case, the total pressure is given by the formula

$p = 2.942 \cdot 10^4 (p_e + k_B T / v_c)$ (GPa), $v_c = 4\pi r_0^3 / 3$ is the volume of the atomic cell.

There exist other quantum statistical models: Hartree–Fock, Hartree–Fock–Slater, etc. These are used to construct the EOS for substances and mixtures too. For bosons, one can obtain equations of state by similar manner.

3.2. Thermodynamical Properties of Fluids and Gases

The basic aim of thermodynamics is the investigation of laws and properties of a *heat form of the matter motion* and connected with them physical phenomena. A heat form of the matter motion is a chaotic motion of particles (atoms, molecules, etc.) of the macroscopic bodies. The properties of substances from this viewpoint can be described with the help of the four laws of thermodynamics and their consequences. From these laws, we can deduce the function $\mathcal{F}$ of a number of independent macroscopic variables, which determines the behavior of a thermodynamic system (i.e., we construct the EOS: $\mathcal{F} = p(V, T)$ or $\mathcal{F} = \epsilon(\rho, T)$, etc.). These independent variables characterize the *equilibrium state* of the system.

3.2.1. *The laws of thermodynamics*

The laws of thermodynamics are as follows:

- (The *zeroth law*) If two systems are in thermal equilibrium with a third system, then they are in thermal equilibrium with each other.
- (The *first law*) There exists a function of state ϵ (the *internal energy*), such that an amount of energy E is supplied to an otherwise isolated system, bringing it from an equilibrium state $(\bar{1})$ to an equilibrium state $(\bar{2})$, then $E = \epsilon(\bar{1}) - \epsilon(\bar{2})$, irrespective of the way in which this energy was supplied.
- (The *second law*)
  (1) There exists no thermodynamic transformations of which the only result is to convert heat from a heat reservoir to work (*Kelvin's formulation*).

(2) There exist no thermodynamic transformations of which the only result is the transfer of heat from a colder heat reservoir to a hotter heat reservoir (*Clausius' formulation*).

- (The *third law* or *Nernst's theorem*) The change of entropy of a system S during a process between two equilibrium states at the same temperature, but characterized by different values of another thermodynamic parameter $\mathcal{T}$, tends to zero as the temperature approaches absolute zero.

Let us consider the very important consequence of zeroth law. If the system is in a thermodynamic equilibrium state, then all other systems in thermal equilibrium with it have a property in common, namely the property of being in thermal equilibrium with one another. This property is called *temperature T*. In other words, systems in *thermal equilibrium* are said to have the *same temperature*; systems not in thermal equilibrium are said to have different temperatures.

The *absolute temperature* of a system (measured in Kelvins) by definition is given by $T = \frac{pV}{\nu R}$, where p, V, ν are the pressure, volume and number of moles of a sufficiently dilute (perfect) gas brought into thermal contact with the system of interest.

3.2.2. *The first law of thermodynamics*

Let us consider the first law not as an axiom, but as an extension of the law of conservation of energy including thermal processes. The mechanical work performed in compressing the gas is given by

$$A = -\int_{(\bar{1})}^{(\bar{2})} pdV, \tag{3.8}$$

where p is the pressure and the integral is performed along the path in state space from state $(\bar{1})$ to state $(\bar{2})$ (this integral depends on the path taken in general). Pressure is defined as the force per unit area $(F = pS)$. If we consider a container (with the gas) with piston which is lowered by a small amount dx, then the work performed is given by $\delta W = F\delta z = -p\delta V$, where the symbol δx does not mean "a change of" x, but simply an infinitesimal amount of x. Because the final temperature depends only on amount of energy supplied, one can introduce an *internal energy* function $\epsilon(T)$. From experiments, we know ϵ is proportional to T: $\epsilon(T) = C_V T$, where *heat capacity* is in fact only weakly dependent on the type of gas. For gases, it is given as $C_V = i\nu R/2$, where $i = 3$ for monatomic gases, $i = 5$ for diatomic gases and $i = 7$ for N-atomic gases $(N > 2)$; for solids, at sufficiently high temperature, $C_V = 3\nu R$ (the *law of Dulong an Petit*).

If the system is not thermally isolated, heat can flow through the walls of the container and the total energy supplied is the sum of the heat Q supplied to the system and the work A performed on the system. Then we have $A + Q = \epsilon(\overline{2}) - \epsilon(\overline{1})$ and, consequently, $C_V = (\partial \epsilon / \partial T)_V$ (the subscript V indicates that the volume is kept constant). Note that the state space only applies to equilibrium states of the system. The processes of change of system parameters are well defined if they occur very slowly so that the system is in equilibrium during the entire period of time when the processes take place; this type of processes is called *quasi-static*. As a system is nearly in equilibrium during quasi-static process, such a process is *reversible*; that is, reversing the direction of the process in state space results in another valid one. If the system is thermally insulated so that no heat flows through the walls, then the quasi-static process is called *adiabatic*. For a small change δV in volume, the work performed on the gas is $\delta A = -p \delta V$ and $\delta \epsilon = C_V \delta T$. It follows that

$$\frac{dV}{dT} = -\frac{c_V}{p} = -\frac{C_V V}{\nu RT}.$$

If we denote $\gamma = 1 + \nu R/C_V$, then the last equation can be rewritten as $TV^{\gamma-1} = $ const and the EOS for the ideal gas is $pV^{\gamma} = $ const $(\gamma = 5/3$ for monatomic ideal gas, $\gamma = 7/5$ for diatomic gas).

But by the first law, $\int_{(\overline{1})}^{(\overline{2})} d\epsilon = \epsilon(\overline{2}) - \epsilon(\overline{1})$ is irrespective of the path $(\overline{1}) - (\overline{2})$. Consequently, $\delta \epsilon = d\epsilon$ and we can write the law in infinitesimal form:

$$\delta Q + \delta A = d\epsilon. \tag{3.9}$$

If $\epsilon = \epsilon(T, V)$, we have

$$d\epsilon = \left(\frac{\partial \epsilon}{\partial T}\right)_V dT + \left(\frac{\partial \epsilon}{\partial V}\right)_T dV = C_V dT + \left(\frac{\partial \epsilon}{\partial V}\right)_T dV, \quad (3.10)$$

where the subscript on partial derivatives indicates the variable to be kept fixed. Using equation (3.9), we have for the heat differential

$$\delta Q = C_V dT + \left(p + \left(\frac{\partial \epsilon}{\partial V}\right)_T\right) dV. \quad (3.11)$$

If we consider ρ, T as independent variables, then

$$d\epsilon = \left(\frac{\partial \epsilon}{\partial T}\right)_p dT + \left(\frac{\partial \epsilon}{\partial p}\right)_T dp \equiv C_p dT + \left(\frac{\partial \epsilon}{\partial p}\right)_T dp, \quad (3.12)$$

where $C_p = (\delta Q)_p / dT$ and

$$\delta Q = C_p dT + \left(p + \left(\frac{\partial \epsilon}{\partial V}\right)_T\right)\left(\frac{\partial V}{\partial p}\right)_T dp, \quad (3.13)$$

$$C_p = C_V + \left(p + \left(\frac{\partial \epsilon}{\partial V}\right)_T\right)\left(\frac{\partial V}{\partial p}\right)_p, \quad (3.14)$$

where C_p is the *heat capacity at constant pressure*. For ideal gas, we have $\epsilon = C_V T$, $(\partial V / \partial T)_p = \nu R / p$, $(\partial V / \partial p)_T = -V/p$, since

$$\delta Q = C_p dT - V dp, \quad C_p = C_V + \nu R.$$

3.2.3. *The properties of entropy*

The *entropy* concept is exceedingly important in both thermodynamics and kinetic theory of gases. In this section, we consider the properties of entropy in detail as an example of similar analysis for other thermodynamic functions.

The simplest example of a system to which it can be applied is a single homogeneous substance. In this simplest case, a complete description of its thermodynamic state requires a specification of its content, i.e., amount of each chemical substance contained, and further a specification of two other quantities such as, for example, volume and pressure. If the system is not homogeneous, then in order to describe its thermodynamic state, we have to consider it as composed of a number of homogeneous parts called *phases* each of which is described by specifying its content and a sufficient number of other properties.

A system of fixed material content is called a *closed system* and a system of variable content is called an *open system*. Similarly, a phase of fixed content is called a *closed phase* and a phase of variable content is called an *open phase*.

With the help of these definitions, we introduce the entropy of system (phase).

There exists a function S of the state of a system called *entropy* of the system having the following properties.

(1) The entropy S_Σ of the system Σ is the sum of entropies of its parts $\Sigma_1, \Sigma_2, \ldots, \Sigma_\tau$ so that $S_\Sigma = \sum_{k=\tau}^{\tau} S_{\Sigma_k}$ ($\tau \geq 1$). In this respect the entropy is *extensive* parameter similar to mass, volume and internal energy.

(2) The entropy S_{Σ_j} of a closed phase Σ_j is determined by the internal energy ϵ_{Σ_j} and the volume V_{Σ_j} of the phase so that

$$dS_{\Sigma_j} = \left(\frac{\partial S_{\Sigma_j}}{\partial \epsilon_{\Sigma_j}}\right)_{V_{\Sigma_j}} d\epsilon_{\Sigma_j} + \left(\frac{\partial S_{\Sigma_j}}{\partial V_{\Sigma_j}}\right)_{\epsilon_{\Sigma_j}} dV_{\Sigma_j}.$$

(3) The quantity $(\partial S_{\Sigma_j}/\partial \epsilon_{\Sigma_j})_{V_{\Sigma_j}}$ is always positive.
(4) The entropy of an insulated closed system Σ increases in any natural change, and its maximum at equilibrium: $dS_\Sigma \geq 0$.
(5) In any reversible adiabatic process, the entropy remains constant: $dS_\Sigma = 0$. (For *specific entropy* or entropy per unit mass of fluid, we use the notation s.)

Let us consider the thermally insulated system Σ composed of two closed phases, Σ_1, Σ_2, each being maintained at constant volume and in thermal contact with each other. We have

$$dV_{\Sigma_1} = dV_{\Sigma_2} = 0, \tag{3.15}$$

$$d\epsilon_{\Sigma_1} = d\epsilon_{\Sigma_1} + d\epsilon_{\Sigma_2} = 0, \tag{3.16}$$

$$dS_\Sigma = dS_{\Sigma_1} + dS_{\Sigma_2} \geq 0. \tag{3.17}$$

By virtue of (3.15), we may rewrite (3.16) as

$$\left(\frac{\partial S_{\Sigma_1}}{\partial \epsilon_{\Sigma_1}}\right)_{V_{\Sigma_1}} d\epsilon_{\Sigma_1} + \left(\frac{\partial S_{\Sigma_2}}{\partial \epsilon_{\Sigma_2}}\right)_{V_{\Sigma_2}} d\epsilon_{\Sigma_2} \geq 0 \qquad (3.18)$$

and by virtue of (3.16), this becomes

$$\left(\left(\frac{\partial S_{\Sigma_1}}{\partial \epsilon_{\Sigma_1}}\right)_{V_{\Sigma_1}} - \left(\frac{\partial S_{\Sigma_2}}{\partial \epsilon_{\Sigma_2}}\right)_{V_{\Sigma_2}}\right) d\epsilon_{\Sigma_1} \geq 0. \qquad (3.19)$$

We now define a positive quantity T by

$$T = \left(\frac{\partial \epsilon}{\partial S}\right)_V = \left(\left(\frac{\partial S}{\partial \epsilon}\right)_V\right)^{-1} \qquad (3.20)$$

and rewrite (3.19) as

$$(T_{\Sigma_2} - T_{\Sigma_1}) d\epsilon_{\Sigma_1} \geq 0. \qquad (3.21)$$

Therefore, the quantity $d\epsilon_{\Sigma_1} = -d\epsilon_{\Sigma_2}$ has the same sign as $T_{\Sigma_2} - T_{\Sigma_1}$. The property of T (*thermodynamic temperature*) expressed by formula (3.20) is obviously that of an "absolute" temperature defined in the consequence from zeroth law of thermodynamics.

For a single closed phase Σ_1, we have

$$d\epsilon_{\Sigma_1} = \left(\frac{\partial \epsilon_{\Sigma_1}}{\partial S_{\Sigma_1}}\right)_{V_{\Sigma_1}} dS_{\Sigma_1} + \left(\frac{\partial \epsilon_{\Sigma_1}}{\partial V_{\Sigma_1}}\right)_{S_{\Sigma_1}} dV_{\Sigma_1}$$

$$= T_{\Sigma_1} dS_{\Sigma_1} + \left(\frac{\partial \epsilon_{\Sigma_1}}{\partial V_{\Sigma_1}}\right)_{S_{\Sigma_1}} dV_{\Sigma_1}. \qquad (3.22)$$

For a reversible adiabatic change, S_{Σ_1} remains constant and consequently

$$d\epsilon_{\Sigma_1} = \left(\frac{\partial \epsilon_{\Sigma_1}}{\partial V_{\Sigma_1}}\right)_{S_{\Sigma_1}} dV_{\Sigma_1} = \delta A = -p_{\Sigma_1} dV_{\sigma_1}, \qquad (3.23)$$

where p_{Σ_1} denotes the pressure of the phase Σ_1. Therefore,

$$\left(\frac{\partial \epsilon_{\Sigma_1}}{\partial V_{\Sigma_1}}\right)_{S_{\Sigma_1}} = -p_{\Sigma_1} \qquad (3.24)$$

and substituting (3.24) into (3.22), we obtain

$$d\epsilon_{\Sigma_1} = T_{\Sigma_1} dS_{\Sigma_1} - p_{\Sigma_1} dV_{\Sigma_1}. \tag{3.25}$$

Comparing this with the statement of the first law (for fixed phase Σ_1) $d\epsilon = \delta Q + \delta A$, we see that for a reversible process $TdS = \delta Q$.

Let us consider the change in entropy when the system is neither thermally insulated nor in complete internal equilibrium. Let the system Σ be composed of phases $\Sigma_1, \Sigma_2, \ldots$ each in internal equilibrium. If two or more parts of the system have the same composition but different temperatures, these are to be regarded as different phases. Now, consider an infinitesimal change in Σ in which the quantities of heat gained by phases $\Sigma_1, \Sigma_2, \ldots$ are $\delta Q_{\Sigma_1}, \delta Q_{\Sigma_2}, \ldots$. Evidently, the changes inside the system Σ are independent of where the heat δQ_{Σ_k} comes from or where the heat δQ_{Σ_k} $(k = 1, 2, \ldots)$ goes to. We may therefore without affecting the changes inside Σ arrange for Σ_1 to exchange heat only with a system $\Sigma_1^\dagger$ which is in internal equilibrium and has the same temperature T_{Σ_1} as Σ_1; and similarly for $\Sigma_2, \ldots$. We also arrange for the composite system $\Sigma + \Sigma_1^\dagger + \Sigma_2^\dagger + \cdots$ to be thermally insulated. We accordingly have

$$dS_\Sigma + dS_{\Sigma_1^\dagger} + dS_{\Sigma_2^\dagger} \geq 0. \tag{3.26}$$

We also have

$$dS_{\Sigma_1^\dagger} = \frac{\delta Q_{\Sigma_1^\dagger}}{T_{\Sigma_1^\dagger}} = -\frac{\delta Q_{\Sigma_1}}{T_{\Sigma_1}}, \quad dS_{\Sigma_2^\dagger} = \frac{\delta Q_{\Sigma_2^\dagger}}{T_{\Sigma_2^\dagger}} = -\frac{\delta Q_{\Sigma_2}}{T_{\Sigma_2}}, \ldots \tag{3.27}$$

Substituting (3.27) into (3.26), we obtain

$$dS_{\Sigma_1} \geq \frac{\delta Q_{\Sigma_1}}{T_{\Sigma_1}} \tag{3.28}$$

the inequality relating to a change and the equality to a reversible process. The property of entropy described by (3.28) may be expressed as follows:

$$dS_{\Sigma_1} = d_e S_{\Sigma_1} + d_i S_{\Sigma_1}, \quad d_e S_{\Sigma_1} = \delta Q_{\Sigma_1}/T_{\Sigma_1}, \quad d_i S_{\Sigma_1} \geq 0,$$

where $d_e S$ denotes the increase in S_{Σ_1} associated with interaction of Σ_1 with its surrounding and $d_i S_{\Sigma_1}$ denotes the increase in S_{Σ_1} associated with a change occurring inside Σ_1.

3.2.4. *The third law of thermodynamics*

M. Planck in 1906 reformulated the third law of thermodynamics as follows: the entropy of all crystals at the absolute zero of temperature is the same and can therefore be normalized to zero.

There is considerable experimental evidence for this rule. Moreover, it can be justified on theoretical grounds with the help of quantum mechanics. It is important to realize that the crystal (as well as more general systems) must be in equilibrium for its law to hold.

An important consequence of the third law is that the absolute zero of temperature cannot be reached with a finite number of quasi-static processes: in order to cool a material, one has to extract heat from it. This is achieved by changing its state, i.e., by changing one of its thermodynamic parameters. However, as the absolute zero of temperature is approached, a given change in this parameter will result in even smaller changes of entropy and hence a smaller amount of heat being extracted. Eventually, the heat loss due to leakage will equal the amount of heat extracted in the process and the temperature stabilizes.

3.2.5. *The thermodynamic functions*

Let us consider some consequences of the first and second laws of thermodynamics. The specific *enthalpy* of fluid (i.e., the enthalpy of unit mass of fluid) is defined as $h = \epsilon_s + pV_s$, $V_s \equiv 1/\rho$. A small change in the parameters of state corresponds to small changes in the functions h, ϵ_s and s related by

$$dh = d\epsilon_s + p\,dV_s + V_s dp = T\,ds + V_s dp,$$

and from the first law (in the form of $d\epsilon_s = \delta q - d(1/\rho)$), we have

$$\delta q = dh - \frac{dp}{\rho},$$

when δq is the heat input per unit mass. We then see that for the process at constant pressure ($\delta q = c_p\,dT$):

$$c_p = \left(\frac{\partial h}{\partial T}\right)_p.$$

Comparing the last formula with definition of $c_V = (\partial \epsilon_s / \partial T)_\rho$, we see that enthalpy plays the same role in isobaric processes that internal energy does in processes at constant volume. For ideal gas, $\epsilon_s = c_V T$ and $c_p = c_V + p/(\rho T) = c_V + R$, since $h = c_V T + RT = c_p T$; the specific entropy $ds = c_V (dT/T) - R(d\rho/\rho) = c_p (dT/T) - R(dp/\rho)$, since $s = s_0 + c_V \ln T - R \ln \rho$ or $s = s_0^\dagger + c_p \ln T - R \ln p$.

The quantity $G = \epsilon - TS + pV$ is called *Gibbs function*; the quantity $F = \epsilon - TS$ is called *Helmholtz function* (or free energy). Besides, there are functions $J = S - \epsilon/T$, $Y = S - \epsilon/T - pV/T$ (*Massieu functions*). All these *characteristic functions* are used widely in thermodynamics and chemistry.

3.3. Summary

In this Chapter, the brief review of some concepts of thermodynamics which are directly used in the theory of the hydrodynamical and kinetic equations is given. The basic objects of study were the equation of a condition of substance (EOS) and laws of thermodynamics.

In Section 3.1, the concept of the equation of a condition of substance has been introduced and the most known equations were considered: for ideal gas, van der Waals and Kamelings Onnes EOS were considered. It was shown that there is a possibility to construct various equations of state using different quantum-statistical models. We obtained the caloric EOS by using the Thomas–Fermi model for finite temperature of the medium.

In Section 3.2, laws of thermodynamics and their basic consequences were considered. Definitions were given and properties of the basic thermodynamic functions (entropy, enthalpy, free energy, etc.) considered as well.

The material of this chapter is standard for courses of classical thermodynamics and therefore we did not consider it in detail. During the writing of the chapter, we used the following books: [6, 27], E.A. Guggenheim, *Thermodynamics: An Advanced Treatment for Chemists and Physicists* (Elsevier Science Publ., Amsterdam, 1967) and S. Flugge, *Handbuch der Physik. Band XII. Thermodynamik der Gase* (Springer-Verlag, Berlin, 1958). We recommend for readers the books: D. Chandler, *Introduction to Modern Statistical Mechanics* (Oxford University Press, Oxford, 1987), K. Huang, *Statistical Mechanics* (John Wiley & Sons, New York, 1987) and R. Kubo, M. Toda, N. Hashitsume, *Statistical Physics* (Springer-Verlag, Berlin 1985); in these works, the modern state of statistical physics and thermodynamics is considered.

Chapter 4

Shock Waves

4.1. One-Dimensional Fluidodynamics and Rankine–Hugoniot Conditions

When mass velocity of fluid can be greater than sound speed, the motion of a fluid possesses very interesting properties (in physical sense). Investigation of supersonic flows is possible in the simple one-dimensional case. These flows are not model only, but have the obvious physical applications. Besides, in the one-dimensional case, it is simple to consider the "genetic" connection between solutions of equations of hydrodynamics and the Boltzmann equation.

4.1.1. *Acoustic approximation in hydrodynamics*

For the one-dimensional isentropic flow ($\vec{v} = \{v, 0, 0\}$), the Euler equations (for compressible fluid) are as follows:

$$\frac{\partial v}{\partial t} + v \frac{\partial v}{\partial x} + \frac{1}{\rho} \frac{\partial p}{\partial x} = 0, \tag{4.1}$$

$$\frac{\partial \rho}{\partial t} + v \frac{\partial \rho}{\partial x} + \rho \frac{\partial v}{\partial x} = 0. \tag{4.2}$$

The disturbances with small amplitude that propagate in a compressible medium are called the *acoustic waves*. In this case, we can write the variables p, ρ as

$$p = p_0 + p^\dagger, \quad \rho = \rho_0 + \rho^\dagger, \quad |p^\dagger| \ll p_0, \quad |\rho^\dagger| \ll \rho_0,$$
$$\vec{v}(\nabla \vec{v}) = vv'_x = 0,$$

where p_0 is the constant equilibrium pressure, and ρ_0 is the constant equilibrium density of fluid. Then we can rewrite (4.1) as $\partial v/\partial t + \rho_0^{-1}(\partial p^\dagger/\partial x) = 0$ or in the form of

$$\frac{\partial p^\dagger}{\partial t} + \rho_0 \left(\frac{\partial p}{\partial p_0} \right)_{S=\text{const}} \frac{\partial v}{\partial x} = 0, \qquad (4.3)$$

since for isentropic fluid,

$$p^\dagger = \left(\frac{\partial p}{\partial \rho_0} \right)_{S=\text{const}} \rho^\dagger. \qquad (4.4)$$

Here, equations (4.3) and (4.4) describe the sonic wave. Let us use the *velocity potential* ϕ ($\vec{v} = \text{grad}\,\phi$): from $\partial v/\partial t + \text{grad}\,p^\dagger/\rho_0 = 0$ (the Euler equation for $p^\dagger$), we have the relation $p^\dagger = -\rho \cdot \partial \phi/\partial t$ and equation (4.3) can be rewritten as

$$\frac{\partial^2 \phi}{\partial t^2} - c^2 \frac{\partial^2 \phi}{\partial x^2} = 0, \qquad c = \left(\sqrt{\frac{\partial p}{\partial \rho}} \right)_{S=\text{const}}.$$

This equation is the *wave equation* and c is the (local) *sound speed* in the fluid. We consider the case $(\partial p/\partial \rho)_S \geq 0$ only (the substance in which $(\partial p/\partial \rho)_S < 0$ is called *Chaplygin's gas* (see the Appendices)).

We return to the investigation of the system of isentropic equations (4.1)–(4.2). With the help of the definition of c, one may write $\rho^{-1}(\partial p/\partial x) = c^2 \partial(\ln \rho)/\partial x$. If $p = p_0 + p^\dagger$, $v = v^\dagger$ ($v_0 = 0$, i.e., fluid in the rest), we have (neglecting products of small quantities) the new form of the Euler's equations (4.1)–(4.2):

$$\frac{\partial v^\dagger}{\partial t} + v^\dagger \frac{\partial v^\dagger}{\partial x} + c^2 \rho^{-1} \frac{\partial \rho^\dagger}{\partial x}, \quad \frac{\partial \rho^\dagger}{\partial t} + v \frac{\partial \rho^\dagger}{\partial x} + \rho \frac{\partial v^\dagger}{\partial x} = 0. \qquad (4.5)$$

To first order (neglecting again squares of small quantities), these become the linearized equations:

$$\rho_0 \frac{\partial v^\dagger}{\partial t} + c^2 \frac{\partial \rho^\dagger}{\partial x} = 0, \quad \frac{\partial \rho^\dagger}{\partial t} + \rho_0 \frac{\partial v^\dagger}{\partial x} = 0. \qquad (4.6)$$

Eliminating $v^\dagger$ by differentiating the first equation in x and the second equation in t, we obtain the wave equation for variable $\rho^\dagger$:

$$\frac{\partial^2 \rho^\dagger}{\partial t^2} - c^2 \frac{\partial^2 \rho^\dagger}{\partial x^2} = 0, \qquad (4.7)$$

with the general solution $\rho^\dagger(x,t) = \Phi_1(x+ct) + \Phi_2(x-ct)$, where $\Phi_i \in C^2(\mathbb{R}^1)$, $i = 1, 2$, are any functions of its arguments.

Equations (4.1)–(4.2) are *quasi-linear hyperbolic equations* in the one-dimensional case. The general form of these equations is

$$\widehat{A}_1(x,t,\vec{f})\frac{\partial \vec{f}(x,t)}{\partial t} + \widehat{A}_2(x,t,\vec{f})\frac{\partial \vec{f}(x,t)}{\partial x} = \vec{B}(x,t,\vec{f}), \qquad (4.8)$$

where $\vec{f} = \{f_\ell\}_1^N$ is a vector function consisting of unknown functions (components) (in our case, $\vec{f} = (\rho, v)^\top$), $\widehat{A}_i = (a_i)_{k\ell}$ ($i = 1, 2$, $k, \ell = 1, \ldots, N$) are the matrices of coefficients and $\vec{B} = \{b_k\}_1^N$ is a vector of the right-hand sides. If $\widehat{A}_1$ and $\widehat{A}_2$ do not depend on $\vec{f}$, then the system (4.8) is called the *semilinear* one.

4.1.2. *Examples of quasi-linear hydrodynamic equations*

Let us consider the simple *Example I* of quasi-linear equation, namely, linear equation $\partial f(x,t)/\partial t - \zeta(\partial f(x,t)/\partial x) = 0$, $\zeta = \mathrm{const} > 0$. The solution of this equation with initial data $f(x,0) = f_0(x)$ is $f(x,t) = f_0(x+\zeta t)$. Thus, $f(x,t)$ is a wave, with shape f_0, traveling with speed ζ to the left on the x-axis. On the lines $x + \zeta t = \mathrm{const}$, f is constant. We can think of information from the initial data as being propagated along these lines which are called *characteristics* of our linear equation with the right-hand side equal to zero. If initial data are given on the curve C transverse to the characteristics (i.e., nowhere tangent to them), then our equation is solved by setting $f(x_0, t_0)$ equal to the value of the initial data on the curve C at the point where C intersects the characteristic through (x_0, t_0). If initial data are given on the curve not transverse to the characteristics, such as the characteristic itself, then the equation does not necessarily have solutions.

Now, we consider more complicated *Example II*: $\partial f/\partial t + \zeta(x,t)(\partial f/\partial x) = 0$, where ζ is the differentiable function of arguments x and t. The characteristics of this equation are curves $t(\tau)$ and $x(\tau)$, satisfying

$$\frac{dt}{d\tau} = 1, \qquad \frac{dx}{d\tau} = \zeta(x,t) \qquad (4.9)$$

(here, τ is the *natural parameter* along these curves). These curves locally exist and never intersect without being coincident by the existence and uniqueness theorem for ordinary differential equations (see Fig. 4.1) [8].

As for *Example III*, we consider the non-homogeneous semilinear equation $\partial f/\partial t + \zeta(x,t)(\partial f/\partial x) = b(x,t)$. Its characteristics are the same as that in *Example II*. In addition, we have $\frac{d}{d\tau}f(x(\tau), t(\tau)) = b(x(\tau), t(\tau))$ and the solution of the equation is

$$f(x(\tau), t(\tau)) = f(x(\tau_0), t(\tau_0)) + \int_{\tau_0}^{\tau} b(x(\tau_1), t(\tau_1))d\tau_1,$$

which determines f off any curve M if initial data are given on M and M is transverse to the characteristics (see Fig. 4.2).

At last, we analyze the case of nonlinear equation (*Example IV*): $\partial f(x,t)/\partial t + f \cdot (\partial f/\partial x) = 0$ (i.e., $\zeta = f$). Let us search for curves $x(\tau)$, $t(\tau)$, along which f is constant. By chain rule, we have

$$\frac{d}{d\tau}f(x(\tau), t(\tau)) = \frac{\partial f}{\partial x}\frac{dx}{d\tau} + \frac{\partial f}{\partial t}\frac{dt}{d\tau}.$$

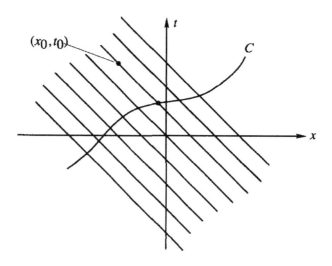

Fig. 4.1. Characteristics intersecting the curve C [8].

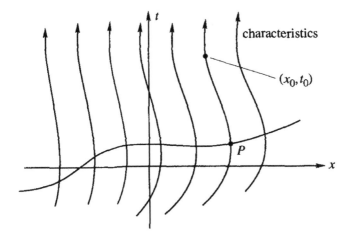

Fig. 4.2. Characteristics of a variable coefficient linear equation that are non-intersecting curves.

As a consequence, we obtain the following characteristic system: $dt/d\tau = 1$, $dx/d\tau = f$. Thus, the characteristics depend on f. Along the curve defined by $dt/d\tau = 1$ and $dx/d\tau = f$, we have $df/d\tau = 0$, i.e., $f = \text{const}$. The situation is similar to the one in the linear case, but characteristics can now intersect (this is easy to see directly: if $f = \text{const}$ on a characteristic, then $dt/d\tau = 1$, $dx/d\tau = f$ is a straight line, and such lines issuing from different points can intersect).

Therefore, we have a system of three ordinary differential equations. The solution of this system through (x_0, t_0, f_0) is still unique, but the characteristics in the (x, t) plane are the projections of the three-dimensional solutions of the system on the plane, and thus may indeed intersect. When such intersection occurs, our method of solution by characteristics breaks down and the solution is no longer uniquely determined (see Fig. 4.3).

We can consider the more general case of nonlinear equation and Cauchy problem for it:

$$\frac{\partial f(x,t)}{\partial t} + \varphi(x, t, f, \mathtt{J}) = 0, \quad \mathtt{J} = \frac{\partial f}{\partial x}, \quad f(x,0) = f_0(x),$$
$$\varphi \in C^2, \quad f_0(x) \in C^2.$$

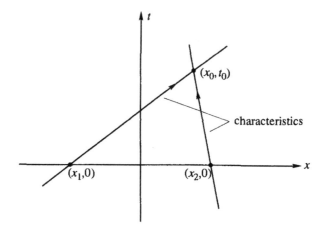

Fig. 4.3. The characteristics of a nonlinear equation may intersect.

If we differentiate this equation by x, then we obtain the Cauchy problem for two quasi-linear equations:

$$\frac{\partial \daleth}{\partial t} + \left(\frac{\partial}{\partial \daleth} \varphi(x, t, f, \daleth) \right) \frac{\partial \daleth}{\partial x} = -\frac{\partial \varphi}{\partial f} \daleth - \frac{\partial \varphi}{\partial x},$$

$$\frac{\partial f}{\partial t} = -\varphi(x, t, f, \daleth),$$

$$f(x, t = 0) = f_0(x), \quad \daleth(x, t = 0) = \frac{df_0(x)}{dx}.$$

This system can be reduced to the (characteristic) system of three ordinary differential equations:

$$\frac{dx}{dt} = \frac{\partial}{\partial \daleth} \varphi(x, t, f, \daleth), \quad \frac{df}{dt} = \daleth \left(\frac{\partial}{\partial \daleth} \varphi(x, t, f, \daleth) \right) - \varphi,$$

$$\frac{d\daleth}{dt} = -\daleth \left(\frac{\partial}{\partial f} \varphi(x, t, f, \daleth) \right) - \left(\frac{\partial}{\partial x} \varphi(x, t, f, \daleth) \right),$$

with initial data

$$x(x_0, t = 0, f_0, \daleth_0) = x_0, \quad f(x_0, t = 0, f_0, \daleth_0) = f_0,$$

$$\daleth(x_0, t = 0, f_0, \daleth_0) = \daleth_0.$$

Therefore, the solution of our nonlinear equation is

$$f(x,t) = f\left(x_0(x,t), t, f_0(x_0(x,t)), df_0(x_0(x,t))/dx_0\right),$$

where $f(x_0(x,t), t, f_0(x_0(x,t)), df_0(x_0(x,t))/dx_0)$ is the solution of the characteristic system with initial data.

4.1.3. *The characteristics and Riemann invariants*

Now we return to the system (4.8) and define its characteristics. Let $\widehat{A}_1 = \widehat{I}$, $\vec{B} = 0$ for simplicity, then the rate of change of $\vec{f}$ along the curve $(x,t)(\tau)$ is

$$\frac{d\vec{f}}{d\tau} = \frac{\partial \vec{f}}{\partial t}\frac{dt}{d\tau} + \frac{\partial \vec{f}}{\partial x}\frac{dx}{d\tau}$$

$$= \left[-\widehat{A}_2(x,t,\vec{f})\frac{dt}{d\tau} + \frac{dx}{d\tau}\widehat{I}\right]\frac{\partial f}{\partial x} \equiv \widehat{C}(x,t,\vec{f})\frac{\partial f}{\partial x}. \quad (4.10)$$

We may define a *characteristic* of (4.8) to be a curve with the following property: if data are given on that curve, the differential equation does not determine the solution at any point not on the characteristic. If the characteristic is not parallel to the x-axis, this means we cannot determine $\partial \vec{f}/\partial x$ from knowledge of $\vec{f}$ on the curve. From (4.10), this will happen if the matrix $\widehat{C}(x,t,\vec{f})$ is a singular one, or

$$\frac{dt}{d\tau} = 1, \quad \frac{dx}{d\tau} = \lambda(x,t,\vec{f}),$$

where $\lambda(x,t,\vec{f})$ is an eigenvalue of $\widehat{A}_2(x,t,\vec{f})$. Note that there are N characteristics through each point and that the characteristics depend on $\vec{f}$.

If M is a curve transverse to all the characteristics, then the $\widehat{C}(x,t,\vec{f})$ is invertible, and $\partial \vec{f}/\partial x = \widehat{C}^{-1}\partial \vec{f}/\partial s$. Thus, the initial data on M uniquely determine a solution near M (we say "near" because, as in *Example IV*, the characteristics for the same eigenvalue can intersect). In this general case, we can't expect $\vec{f}$ to be constant along the characteristics. However, we might seek functions $r_1, \ldots, r_N$ of $\vec{f}$ that are constant along the characteristics associated with the eigenvalues $\lambda_1, \ldots, \lambda_N$. Indeed, suppose we can find function

r associated with the eigenvalue λ with the property that

$$(\widehat{A}_2)^T \frac{\partial r}{\partial \vec{f}} = \lambda \frac{\partial r}{\partial \vec{f}},$$

i.e., $\partial r / \partial v f$ is an eigenvalue of the transpose of $\widehat{A}_2$. We claim that r is constant along the corresponding characteristic. Such functions r are called the *Riemann invariants* of the equation.

Let us apply these ideas for equations (4.1)–(4.2), written in the following form:

$$\frac{\partial \vec{f}}{\partial t} + \widehat{A}\frac{\partial \vec{f}}{\partial x} = 0, \quad \vec{f} = \begin{pmatrix} \rho \\ v \end{pmatrix}, \quad \widehat{A} = \begin{pmatrix} v & \rho \\ c^2 \rho^{-1} & v \end{pmatrix}. \tag{4.11}$$

The eigenvalues of matrix $\widehat{A}$ are $\lambda_1 = v + c$ and $\lambda_2 = v - c$. The characteristics are the curves

$$C_+: \quad \frac{dx}{dt} = \lambda_1 = v + c,$$

$$C_-: \quad \frac{dx}{dt} = \lambda_2 = v - c. \tag{4.12}$$

From the definition of Riemann invariants, we have $\partial r / \partial \vec{f} = \vec{h}_\pm \equiv (h_\pm^{(1)}, h_\pm^{(2)})^T$ and

$$\begin{pmatrix} v & c^2 \rho^{-1} \\ \rho & v \end{pmatrix} \begin{pmatrix} h_\pm^{(1)} \\ h_\pm^{(2)} \end{pmatrix} = (v \pm c) \begin{pmatrix} h_\pm^{(1)} \\ h_\pm^{(2)} \end{pmatrix}.$$

The components of the eigenvector $\vec{h}_\pm$ are $h_\pm^{(1)} = \pm c \rho^{-1}$ and $h_\pm^{(2)} = 1$. For the Riemann invariants $r_\pm$, we have the system of equations:

$$\frac{\partial r_\pm}{\partial \rho} = h_\pm^{(1)}, \quad \frac{\partial r_\pm}{\partial v} = h_\pm^{(2)}.$$

The functions $r_\pm = v \pm \int \rho^{-1} c(\rho) d\rho$ are constants along the $\pm$ characteristics, respectively. It is necessary to note that r_+ needs to be constant on C_- and r_- does not need to be constant on C_+. Of course, if equation is nonlinear as in *Example IV*, there may be difficulties with two C_+ or two C_- characteristics crossing.

To obtain Riemann invariants, one may use the so-called *hodograph transformation* [28].

4.1.4. *The shock waves and Rankine–Hugoniot relations*

Now, we consider the waves in fluid. Earlier, we have obtained the equation for the velocity potential ϕ in the one-dimensional case:

$$\frac{\partial^2 \phi}{\partial t^2} - c^2 \frac{\partial^2 \phi}{\partial x^2} = 0, \quad v = \nabla \phi = \frac{d\phi}{dx}, \tag{4.13}$$

which has the general solution $\phi = \Phi_1(x - ct) + \Phi_2(x + ct)$, where $\Phi_{1,2}$ are arbitrary functions of their arguments. This solution represents a superposition of two *traveling plane waves*, where $c = (\partial p/\partial \rho)^2_{S=\text{const}}$ is the adiabatic speed of sound.

For $p^\dagger$, $\rho^\dagger$, we can obtain equations similar to (4.13); so, choosing $\phi = \Phi(x - ct)$, we have (for $v = v_0 + v^\dagger$) $v^\dagger = (\partial \phi/\partial x) = \Phi'$, and

$$\rho^\dagger = -(\rho_0/c^2)(\partial \phi/\partial t) = (\rho_0/c)\Phi' = (v^\dagger/c)\rho_0, \quad p^\dagger = c^2 \rho^\dagger = c\rho_0 v^\dagger.$$

Therefore, in acoustic waves, $p^\dagger$, $\rho^\dagger$, $v^\dagger$ are all in phase.

Acoustic waves (disturbances with small amplitude) are solutions of linear fluidodynamic equations. When nonlinear phenomena become important, the character of wave solutions of these equations becomes principally new. In particular, in an acoustic disturbance, a region of compression tends to overrun a rarefaction that precedes it; thus, as an acoustic wave propagates, the leading part of the profile progressively steepens, eventually becoming a near discontinuity, which one can identify as a *shock wave* (or shock simply).

Once a shock forms, it moves through the fluid supersonically and therefore outruns preshock acoustic disturbances by which adjustments in local fluid properties might otherwise take place. It can therefore persist as a distinct entity in the flow until it is damped by dissipative mechanisms. The material behind a shock wave is hotter, denser, and has a higher pressure and entropy than the material in front of it. The higher the velocity of the shock, the more pronounced the change in material properties across the discontinuity. The rise in entropy across a shock front implies that wave energy has been dissipated irreversibly (this process damps the propagation shock).

We now consider the shock waves and its properties with the help of the one-dimensional nonlinear equation of fluidodynamic type in the form of

$$\frac{\partial v}{\partial t} + \frac{\partial}{\partial x}(F(v)) = 0 \quad \text{or, equivalently,}$$

$$\nabla_{x,t}\vec{\mathcal{F}} = 0, \quad \vec{\mathcal{F}} = \begin{pmatrix} F(v) \\ v \end{pmatrix}, \tag{4.14}$$

where $\nabla_{x,t}(f,g)^T \equiv \partial f/\partial x + \partial g/\partial t$, $\vec{\mathcal{F}}$ is the vector function of variable v.

Let's introduce the *weak form* of (4.14):

$$\int \chi \cdot \nabla_{x,t}\vec{\mathcal{F}}dxdt = \int \nabla_{x,t}\chi \cdot \vec{\mathcal{F}}dxdt = 0,$$

$$\nabla_{x,t}\chi = (\partial\chi/\partial x, \partial\chi/\partial t)^T, \tag{4.15}$$

where $\chi \in C^\infty$, $\text{supp}\,\chi = \mathcal{T} \subset (x,t)$. The *weak solution* of (4.14) is the function $v(x,t)$ that satisfies (4.15) for all smooth $\chi(x,t)$ with compact support. Now, we have the possibility to investigate the properties of the solution (in weak sense) of equation (4.14) near a jump discontinuity. Let v be a weak solution with discontinuity across a smooth curve $\tilde{\tau} \subset \mathcal{T}$, where $\mathcal{T} \subset (x,t)$ and

$$\chi \in C^\infty(\mathcal{T}), \quad \text{supp}\,\chi = \mathcal{T}, \quad \tilde{\tau} = \mathcal{T}_1 \cap \mathcal{T}_2, \quad \mathcal{T} = \mathcal{T}_1 \cup \mathcal{T}_2.$$

Therefore, we have

$$\int \nabla_{x,t}\chi \cdot \vec{\mathcal{F}}dxdt = \left(\int_{\tilde{\tau}} \chi\vec{\mathcal{F}} \cdot \vec{n}dSdt - \int_{\mathcal{T}_1} \chi\nabla_{x,t}\vec{\mathcal{F}}dxdt \right)$$

$$- \left(\int_{\tilde{\tau}} \chi\vec{\mathcal{F}} \cdot \vec{n}dSdt - \int_{\mathcal{T}_2} \chi\nabla_{x,t}\vec{\mathcal{F}}dxdt \right), \tag{4.16}$$

where $\vec{n}$ is outward normal vector to $\partial\mathcal{T}_1$ and $\nabla_{x,t}\vec{\mathcal{F}} = 0$. Consequently, (4.16) can be written as

$$\int_\tau \chi(\vec{\mathcal{F}}|_{\mathcal{T}_1} - \vec{\mathcal{F}}|_{\mathcal{T}_2}) \cdot \vec{n}dS = 0, \tag{4.17}$$

or, equivalently, $\vec{\mathcal{F}}|_{\mathcal{T}_1} \cdot \vec{n} - \vec{\mathcal{F}}|_{\mathcal{T}_2} \cdot \vec{n} \equiv [\vec{\mathcal{F}}\vec{n}] = 0$ on $\tilde{\tau}$ (because (4.17) is valid for all χ). The symbol $[\vec{\mathcal{F}}\vec{n}]$ denotes the *jump in* $\vec{\mathcal{F}}\vec{n}$ *across* $\tilde{\tau}$.

If curve $\tilde{\tau}$ is parameterized by $x = x(t)$ and $\vec{n} = (1, -\dot{x})/\sqrt{1 + \dot{x}^2}$, then (4.17) becomes $\dot{x}[v] - [\mathcal{F}] = 0$ (on $\tilde{\tau}$). Here, $\dot{x} = dx/dt$ is the *speed of the discontinuity*.

The *system of conservation laws* in the one-dimensional case

$$\frac{\partial v_i}{\partial t} + \frac{\partial}{\partial x} F_i(v_1, \ldots, v_N) = 0, \quad i = 1, \ldots, N, \qquad (4.18)$$

for unknown vector $\vec{v} = (v_1, \ldots, v_N)$ can be considered in the same manner as that of (4.14). In particular, the Euler system of equations

$$\frac{\partial \rho}{\partial t} + \frac{\partial (\rho v)}{\partial x} = 0, \quad \frac{\partial v}{\partial t} + v \frac{\partial v}{\partial x} + \rho^{-1} \frac{\partial p}{\partial x} = 0,$$

$$\frac{\partial e_s}{\partial t} + \mathrm{div}((e_s + p)v) = 0, \quad e_s = \frac{\rho v^2}{2} + \rho \epsilon_s, \qquad (4.19)$$

where e_s is specific total energy of fluid and may be rewritten in the following form:

$$\frac{\partial \rho}{\partial t} + \frac{\partial}{\partial x} (\rho v) = 0, \quad \frac{\partial}{\partial t} (\rho v) + \frac{\partial}{\partial x} (\rho^{-1} (\rho v)^2 + p) = 0, \qquad (4.20)$$

$$\frac{\partial e_s}{\partial t} + \frac{\partial}{\partial x} ((e_s + p)(\rho v)\rho^{-1}) = 0. \qquad (4.21)$$

For the isentropic fluid, we have $p = A\rho^\gamma$, $A = \mathrm{const}$, $\gamma = c_p/c_V = \mathrm{const} \geq 1$; then, the specific enthalpy h can be written as $h = \int^\rho \gamma A s^{\gamma-2} ds = \gamma/(\gamma - 1) A\rho^{\gamma-1}$ and for the specific internal energy, we obtain the relation:

$$\epsilon_s = \frac{A\rho^{\gamma-1}}{\gamma - 1} = \frac{p}{\rho(\gamma - 1)}. \qquad (4.22)$$

Therefore, jump relations across a discontinuity $\tilde{\tau}$ (with velocity $\dot{x}$) in (x, t) plane for the Euler system have the form of

$$\dot{x}[\rho] = [\rho v], \quad \dot{x}[\rho v] = [\rho^{-1}(\rho v)^2 + p], \quad \dot{x}[e_s] = [(e_s + p)v]. \quad (4.23)$$

These are called the *Rankine–Hugoniot relations* (RHR). In the coordinate system moved with the velocity $\dot{x}$ along x-axis, the RHR (at origin) become

$$\rho_0 v_0 = \rho_1 v_1, \quad \rho_0 v_0^2 + p_0 = \rho_1 v_1^2 + p_1,$$
$$((e_s)_0 + p_0)v_0 = ((e_s)_1 + p_1)v_1, \tag{4.24}$$

where the subscripts (0) and (1) indicate states "pre" and "post" of the discontinuity.

Let us define $j = \rho_0 v_0 = \rho_1 v_1$. If $j = 0$, then the discontinuity is called a *contact* one; because $v_0 = v_1 = 0$, this discontinuity moves with the fluid and $p_0 = p_1$ (but $\rho_0 \neq \rho_1$ in general). If $j \neq 0$, then discontinuity is called a *shock*. Because $v_0 \neq 0$, $v_1 \neq 0$, fluid is crossing the shock. The side that consists of gas that has not crossed the shock is called the *front* of the shock (state "pre", subscript (0)); the other side is called *back* (state "post", subscript (1)).

From RHR, we have $j = (p_1 - p_0)/(v_0 - v_1)$ and $j^2 = (p_1 - p_0)/(V_0^{(s)} - V_1^{(s)})$ (since $v_i = j/\rho_i = jV_i^{(s)}$, $V_i^{(s)} = \rho_i^{-1}$ are specific volumes, $i = 0, 1$); consequently, $v_0 - v_1 = \sqrt{(p_1 - p_0)(V_0^{(s)} - V_1^{(s)})}$. These identities are consequences of the *mechanical jump relations* only (the first and second equations of the system (4.24)). To bring in the energy, combine the first and third equations of RHR to give $(e_0)_s V_0^{(s)} - (e_1)_s V_1^{(s)} = p_1 V_1^{(s)} - p_0 V_0^{(s)}$. However,

$$(e_0)_s V_0^{(s)} - (e_1)_s V_1^{(s)} = (\epsilon_s)_0 - (\epsilon_s)_1 - (p_0 - p_1)(V_0^{(s)} + V_1^{(s)})/2.$$

Thus, we have

$$(\epsilon_s)_1 - (\epsilon_s)_0 + \frac{p_0 + p_1}{2}(V_1^{(s)} - V_0^{(s)}) = 0. \tag{4.25}$$

This relation is called *Hugoniot equation* for the shock. We can introduce the so-called *Hugoniot function*,

$$H(p, V^{(s)}) = \epsilon_s(p, V^{(s)}) - \epsilon_s(p_0, V_0^{(s)}) + \frac{p_0 + p}{2}(V^{(s)} - V_0^{(s)}),$$
$$H(p_1, V_1^{(s)}) = 0 \tag{4.26}$$

(it is the hyperbola in $(p, V^{(s)})$-plane).

4.1.5. *The general structure of shock waves*

The shocks can be essentially different in the structure. Let us consider nonlinear equation $\partial v/\partial t + v\partial v/\partial x = \partial v/\partial t + \partial(v^2/2)/\partial x = 0$ with initial data $v(x, t = 0) = 0$ if $x \geq 0$ and $v(x, t = 0) = 1$ if $x \leq 0$. Its characteristics are straight lines and $v = \text{const}$ along them; these straight lines intersect (see Fig. 4.4(a)). To keep the characteristics from crossing, we introduce a shock with propagation speed $\dot{x} = [v^2/2]/[v] = 1/2$; thus, we get a globally defined weak solution by letting $v = 1$ behind the shock and $v = 0$ in front of the shock (see Fig. 4.4(b)). If we consider the initial data $v(x, t = 0) = 1$, if $x \geq 0$ and $v(x, t = 0) = 0$, if $x \leq 0$, then the characteristics do not fill out the (x, t) plane.

Through every point on the path of the shock in the (x, t) plane, one can draw two characteristics, one on each side of the shock; either both of them can be traced back to the initial line or both of them can be traced upward to larger t. In either case, the shock is needed to avoid characteristics intersecting and creating a multivalued solution. One may say that this shock *separates* the characteristics. The shock is said to obey the *entropy condition* if two characteristics that intersect on each point can be traced backward to the initial line (as in Fig. 4.4(b)). A shock that does not obey the entropy condition is called a *rarefaction shock* (it violates basic thermodynamic principles in general case). The properties of these waves

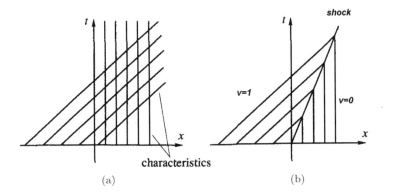

Fig. 4.4. Characteristics for the initial data $v = 0$ if $x \geq 0$ and $v = 1$ if $x < 0$ [8].

are exotic: (1) if a rarefaction shock is allowed, then the problem will not have a unique solution; (2) a solution that includes rarefaction shocks need not depend continuously on the initial data. As in the fluid dynamics, *compressive shocks* are only investigated, they satisfy the entropy conditions. Thus, we exclude solutions such as those in Fig. 4.5(b). This restriction will make the weak solution of the problem unique.

A shock *separate characteristics of a given family*, if (1) it satisfies the jump conditions; (2) through every point of the trajectory of the shock in (x, t) plane, one can draw two characteristics of the family, one on each side of the shock; (3) either both characteristics can be traced back to the initial line or they can both be traced upward to increasing t. The *entropy condition* for our system is as follows: a shock satisfies the entropy condition if, when it separates

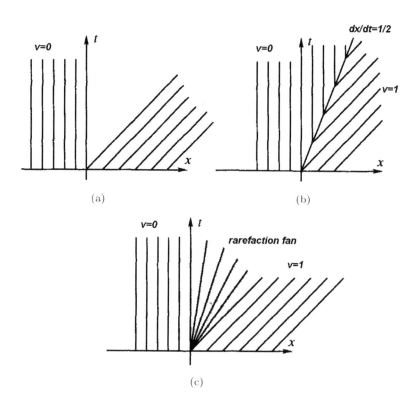

Fig. 4.5. Characteristics for the initial data $v = 1$ if $x \geq 0$ and $v = 0$ if $x < 0$ [8].

characteristics of one family, the characteristics on each side can be traced back to the initial data. A shock is called *compressive* if the pressure behind the shock is larger than the pressure in front of the shock.

Let us consider the entropy condition for the systems of conservation laws.

Consider the system $\partial \vec{v}/\partial t + \widehat{A}(\vec{v})\partial(\vec{v})/\partial x = 0$, where the matrix $\widehat{A}$ has N real eigenvalues λ_i ($i = 1, \ldots, N$) (see (4.11)). Correspondingly, we have N families of characteristics given by the curves $dx_i/dt = \lambda_i$. In the two-dimensional case of isentropic flow, there are two families of characteristics, C_+ and C_-, associated with $\lambda_1 = v + c$, $\lambda_2 = v - c$; in the 3×3 system of gas dynamics,

$$\lambda_1 = v + c, \quad \lambda_2 = v - c, \quad \lambda_0 = v, \qquad (4.27)$$

and we have three corresponding families C_+, C_-, C_0.

Let us investigate the three-dimensional case in detail on the example of the system (4.20)–(4.22). Consider a shock facing to the right, i.e., its front is on its right, its back is on its left, and the fluid crosses it from the right to the left. The shock cannot separate C_0 characteristics. Indeed, by our conventions, the velocity v_0 in front of the shock is negative. By the jump relation $\rho_0 v_0 - \rho_1 v_1$, we obtain that $u_1 < 0$. Therefore, the configuration of the C_0 characteristics is such that on the left (labeled with a (1)), the characteristics go to the future and on the right (labeled with a (0)) they go to the past, and so the shock does not separate them (see Fig. 4.6).

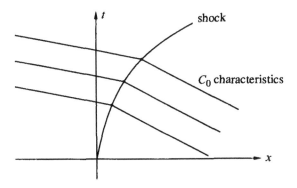

Fig. 4.6.   The appearance of C_0 characteristics separated by a shock [8].

The right-facing shock cannot separate C_- characteristics either. Indeed, as $v_0 < 0$, we have $v_1 < 0$ as well and $v_0 < c$, $v_1 < c$. Thus, one has a picture qualitatively similar to that shown in Fig. 4.6 for C_0 characteristics. Consequently, right-facing shock can only separate C_+ characteristics.

Let us rewrite (4.25) as

$$\left(\frac{\gamma - 1}{\gamma + 1}\right) v_0^2 + \left(1 - \left(\frac{\gamma - 1}{\gamma + 1}\right)\right) c_0^2$$

$$= \left(\frac{\gamma - 1}{\gamma + 1}\right) v_1^2 + \left(1 - \left(\frac{\gamma - 1}{\gamma + 1}\right)\right) c_1^2,$$

$$\epsilon = \frac{p}{\rho(\gamma - 1)}, \quad c^2 = \gamma \frac{p}{\rho}.$$

We denote the common value of both sides as c_*^2. Since $\gamma > 1$, we have $(\gamma - 1)/(\gamma + 1) < 1$ and

$$(|v_0| > c_0) \Leftrightarrow |v_0| > c_*, \quad (|v_1| > c_1) \Leftrightarrow |v_1| > c_*.$$

Rewrite the expression for c_*^2 as follows:

$$c_*^2 = \left(\left(\frac{\gamma - 1}{\gamma + 1}\right) v_1^2 - \left(1 - \left(\frac{\gamma - 1}{\gamma + 1}\right)\right) \frac{\gamma p_1}{\rho_1}\right)$$

$$= \left(\frac{\gamma - 1}{\gamma + 1}\right) v_1^2 + \left(1 + \left(\frac{\gamma - 1}{\gamma + 1}\right)\right) \frac{p_1}{\rho_1}.$$

However, $\rho_0 v_0^2 + p_0 = \rho_1 v_1^2 + p_1$ and we have

$$\rho_1 c_*^2 = \left(\frac{\gamma - 1}{\gamma + 1}\right) (\rho_1 v_1^2 + p_1) + p_1,$$

$$\rho_0 c_*^2 = \left(\frac{\gamma - 1}{\gamma + 1}\right) (\rho_0 v_0^2 + p_0) + p_0.$$

Therefore, $c_*^2 = (p_1 - p_0)/(\rho_1 - \rho_0)$; but we know that the left-hand side of this relation is equal to $j^2 = (\rho_0 v_0)(\rho_1 v_1)$. Consequently, $v_0 v_1 = c_*^2$. It follows that $|v_1| > c_*$ implies $|v_0| < c_*$ and $|v_0| > c_*$ implies $|v_1| < c_*$. Consequently, the velocity of a (compressive) shock is supersonic on one side and is subsonic on the other.

A shock is compressive if and only if $\rho_0 < \rho_1$. This fact follows the relation $c_*^2 = (p_1 - p_0)/(\rho_1 - \rho_0) > 0$ and the definition of compressive shock (in this shock $p_1 > p_0$).

A compressive shock obeys the entropy condition; we know that $|v_0| > c(\rho_0)$ and $|v_1| < c(\rho_1)$, consequently, $\dot{x} = v_1 + c(\rho_1) > 0$ in the "post" region of (x, t) plane and $\dot{x} = v_0 + c(\rho_0) < 0$ in the "pre" region. Non-compressive shock (when $\rho_1 < \rho_0$) violates the entropy condition: from the jump condition $\rho_0 v_0 = \rho_1 v_1$, we have $v_1 < v_0 < 0$; $c(\rho) = \sqrt{\gamma p / \rho}$ is the increasing function of ρ, then $v_1 + c(\rho_1) < v_0 + c(\rho_0)$. However, $|v_1| > c(\rho_1)$, $|v_0| < c(\rho_0)$, i.e., $v_1 + c(\rho_1) < 0$ and $v_0 + c(\rho_0) > 0$. Consequently, our proposition is true.

4.2. Weak Shocks and the Boltzmann Equation

The problem of investigation of shock waves has very important physical applications and we ought to analyze it with the help of kinetic equations formalism. The main aim of this consideration is to find the connection between solutions of special type of NSE (i.e., system of quasi-linear equations) on the macroscopic level and the solution of integro-differential Boltzmann equation described the microscopic motion of molecules of the fluid. Also, we need to verify the origin of shock for supersonic flows and obtain the exact profile of the shock front. The optimal mathematical apparatus for investigation of properties of the Boltzmann equation near $M = 1$ (where M is the Mach number) is the theory of bifurcation of solutions of nonlinear equations and the Lyapunov–Schmidt construction of the "branching equation".

4.2.1. *The properties of the Boltzmann collisional operator*

We will consider the one-dimensional steady state or "stationary" shock. For this case, the Boltzmann equation for particles (molecules, atoms) of fluid without external force has the form of

$$\widehat{G}(f) \equiv v_x \cdot \frac{\partial f(x, \vec{v})}{\partial x} - \mathcal{J}(f, f) = 0, \quad \vec{v} = \{v_x; v_y; v_z\}, \quad \vec{g} = |\vec{v} - \vec{v}_1|,$$

$$(4.28)$$

$$\mathcal{J}(f, f) = \int (f(x, (\vec{v})^\dagger) f(x, (\vec{v}_1)^\dagger)$$
$$- f(x, \vec{v}) f(x, \vec{v}_1)) \cdot B(g, \vartheta) d\varphi d\vartheta d\vec{v}_1,$$

where $\vec{v}$, $\vec{v}_1$ are velocities of the particles before their impact, $(\vec{v})^\dagger$, $(\vec{v}_1)^\dagger$ are velocities after the impact, $B(g, \vartheta)$ is the dispersion function, $v_y = v_z = 0$. We will consider the simplest case of "hard spheres", when $B(g, \vartheta) = B_0 \sin(2\vartheta)$, $B_0 = \text{const} > 0$. For $x \to \pm\infty$, the asymptotic solutions of this equation are (local) Maxwellians:

$$f_{(\pm)}^{(0)}(\vec{v}) = \frac{\rho_{(\pm)}}{(2\pi RT_{(\pm)})^{3/2}} \exp\left(-\frac{(v_x - u_{(\pm)})^2 + v_y^2 + v_z^2}{2RT_{(\pm)}}\right), \quad (4.29)$$

where $\rho_{(\pm)}$, $T_{(\pm)}$, $u_{(\pm)}$ are the density, temperature and mass velocity, respectively, for $x = \pm\infty$, and R is the universal gas constant. We will consider the case when $u_{(-)}/c = M \geq 1$, where c is the speed of sound ($c = c_{(-)} = \sqrt{5RT_{(-)}/3}$) in the system. Let's introduce the new variables $\vec{w} = \{w_x = v_x - u_{(-)}; w_y = v_y; w_z = v_z\}$ and $\widetilde{f}(x, \vec{v}) = f(x, \vec{v})/f_{(-)}^{(0)}(\vec{v}) - 1$ (*Oseen's linearization*). Then the Boltzmann equation can be rewritten as

$$\widehat{G}(\widetilde{f}; u_{(-)}) \equiv (w_x + u_{(-)}) \cdot \frac{\partial \widetilde{f}(x, \vec{w})}{\partial x} - (\widehat{L}_1(\widetilde{f}) + \widehat{L}_2(\widetilde{f}, \widetilde{f})) = 0,$$

$$\widehat{L}_1(\widetilde{f}) \equiv \left(\frac{1}{f^{(0)}} \frac{\partial \mathcal{J}(0, 0)}{\partial \widetilde{f}}\right)(\widetilde{f})$$

$$= \int (\widetilde{f}(\vec{w}^\dagger) + \widetilde{f}(\vec{w}_1^\dagger) - \widetilde{f}(\vec{w})$$

$$- \widetilde{f}(\vec{w}_1)) f_{-}^{(0)}(\vec{w}_1) B(g, \vartheta) d\vartheta d\varphi d\vec{w}_1,$$

$$\widehat{L}_2(\widetilde{f}, \widetilde{f}) \equiv \left(\frac{1}{f^{(0)}} \frac{\partial^2 \mathcal{J}(0, 0)}{\partial \widetilde{f}^2}\right)(\widetilde{f}, \widetilde{f})$$

$$= \int (\widetilde{f}(\vec{w}^\dagger) \widetilde{f}(\vec{w}_1^\dagger) - \widetilde{f}(\vec{w}) \widetilde{f}(\vec{w}_1)) f_{-}^{(0)}(\vec{w}_1) B(g, \vartheta) d\vartheta d\varphi d\vec{w}_1,$$

$$(4.30)$$

where $\widehat{L}_1(\widetilde{f})$ is the first-order *Fréchet derivative* of the nonlinear operator $\mathcal{J}$ in the vicinity of the function $\widetilde{f} = 0$ (i.e., linearized Boltzmann operator of collisions) and $\widehat{L}_2(\widetilde{f}, \widetilde{f})$ is the second-order Fréchet derivative of $\mathcal{J}$ in the same vicinity. We can see that $\widehat{L}_{1,2} \neq \widehat{L}_{1,2}(u_{(-)})$ and $\widetilde{f}(x = -\infty; \vec{w}) = 0$.

We introduced earlier (see (1.78) and (1.79)) the bilinear form:

Let's consider the bilinear form

$$\mathcal{I}(k, s) = \frac{1}{2} \int (k^\dagger s_1^\dagger + k_1^\dagger s^\dagger - k\widetilde{s}_1 - k_1 \widetilde{s}) B(g, \vartheta) d\vartheta d\varphi d\vec{w}_1,$$

where $k = k(\vec{w})$, $k^\dagger = k(\vec{w}^\dagger)$, etc. H. Grad in his work [29] proved that

$$\mathcal{I}(k, s) = \mathcal{I}_\Sigma \equiv \frac{1}{8} \int (k^\dagger s_1^\dagger + k_1^\dagger s^\dagger - ks_1 - k_1 s)$$

$$\times (\phi(\vec{w}) + \phi(\vec{w}_1) - \phi(\vec{w}^\dagger) - \phi(\vec{w}_1^\dagger)) B(g, \vartheta) d\vartheta d\varphi d\vec{w}_1$$

$$(4.31)$$

for any proper functions $\phi(\vec{w})$. If we assume $k = f_{(-)}^{(0)}$, $s = f_{(-)}^{(0)} h(\vec{w})$ and $\phi = s$, from this relation, one may obtain

$$\int f_-^{(0)}(\vec{w}) s \widehat{L}_1(h) d\vec{w} = -\frac{1}{4} \int f_-^{(0)}(\vec{w}) f_-^{(0)}(\vec{w}_1)(h^\dagger + h_1^\dagger - h_1 - h_2)$$

$$\times (s^\dagger + s_1^\dagger - s - s_1) B(g, \vartheta) d\vec{w} d\vec{w}_1 d\vartheta d\varphi.$$

It is easy to see that the change $s \to h$, $h \to s$ has no influence on the value of the integral in the left-hand side of the last relation. Therefore, one can introduce the *Hilbert space H* with internal multiplication rule $\langle k(\vec{w}), s(\vec{w}) \rangle_H = \int f_-^{(0)}(\vec{w}) k(\vec{w}) s(\vec{w}) d\vec{w}$. The linear operator $\widehat{L}_1$ is self-adjoint: $\langle g, \widehat{L}_1(h) \rangle_H = \langle h, \widehat{L}_1(g) \rangle_H$, $g, h \in \mathrm{dom}(\widehat{L}_1)$; its kernel is three dimensional:

$$(\widehat{L}_1(h) = 0) \Leftrightarrow (h \in \{\psi_1 = 1; \psi_2 = w_x/\sqrt{RT_-};$$

$$\psi_3 = \sqrt{1/6}(w_x^2/(RT_- - 3))\}),$$

where ψ_i $(i = 1, 2, 3)$ are orthonormal summational invariants in reduced form.

For any $h \in H$, we have $\langle \psi_i, \widehat{L}_1(h) \rangle_H = 0$ (it is the consequence of self-adjointness of $\widehat{L}_1$) and $\zeta_0 \equiv \langle \psi_i, \widehat{L}_2(h, h) \rangle_H = 0$

$(i = 1, 2, 3)$ (this property follows that of (4.31) if $h = s = f_-^{(0)}\widetilde{f}$, $\phi = \psi_i$ $(i = 1, 2, 3)$, since $\zeta_0 = \int \psi_i \mathcal{I}(f_-^{(0)}\widetilde{f}, f_-^{(0)}\widetilde{f})d\vec{v}$). Therefore, we obtain from (4.28) the system of hydrodynamic conservation laws $\Psi_i \equiv \partial\langle\psi_i(\vec{v}), v_x\widetilde{f}(x, \vec{v})\rangle_H/\partial x = 0$ $(i = 1, 2, 3)$. For $x = \pm\infty$, we obtain the Rankine–Hugoniot relations (4.24).

The conditions $\Psi_i = 0$ mean that the values $\mu_H \equiv \langle\psi_i, v_x\widetilde{f}\rangle_H$ $(i = 1, 2, 3)$ are not functions of variable x and

$$\mu_H = \int v_x\psi_i\widetilde{f}(x, \vec{v})f_-^{(0)}(\vec{v})d\vec{v} = \int v_x\psi_i f(x, \vec{v})d\vec{v} - \int v_x\psi_i f^{(0)}d\vec{v}$$

$$= \int v_x\psi_i f(x, \vec{v})d\vec{v} - \int v_x\psi_i f(-\infty, \vec{v})d\vec{v} = 0.$$

Consequently, there is the subspace $H_1 \subset H$, such that $\dim \operatorname{Ker} \widehat{L}_1 = 1$ and H_1 can be defined as $H_1 = H \ominus \operatorname{span}\{v_x\psi_i\}$, where span $\{v_x\psi_i\}$ is linear manifold spanned on the vectors $v_x\psi_i(\vec{v})$ $(i = 1, 2, 3)$. The base element of the $\widehat{L}_1$ kernel is the function $\psi_0 = \sqrt{3/10}\sum_{i=1}^{3}\psi_i^{(0)}\psi_i$, where $\{\psi_i^{(0)}\}_{i=1}^{3} = \{1; \sqrt{5/3}; \sqrt{2/3}\}$ is the null vector of the matrix $H_{ik} \equiv ((w_x + c/\sqrt{RT_-})\psi_i, \psi_k)_H$, $i, k = 1, 2, 3$ (we note that $(\psi_0, \psi_0)_H = 0$).

The operator $\widehat{L}_1$ can be represented as

$$\widehat{L}_1(\widetilde{f}) = \widehat{K}_B(\widetilde{f}) - \nu_B \cdot \widetilde{f}(\vec{w}),$$

$$\widehat{K}_B(\widetilde{f}) = 2\pi \int B(g, \vartheta)(\widetilde{f}(\vec{w}^\dagger) + \widetilde{f}(\vec{w}_1^\dagger) - \widetilde{f}(\vec{w}_1))d\vec{w}_1 d\vartheta,$$

$$\nu_B = 2\pi \int B(g, \vartheta)f_-^{(0)}(\vec{w}_1)d\vec{w}_1 d\vartheta$$

where $\nu_B > 0$ (see [29]) and the operator $\widehat{K}_B$ is self-adjoint compact in the Hilbert space H. Therefore, the mapping $\widehat{L}_1(u_{(-)})$ ($u_{(-)}$ is the parameter) is the *Fredholm operator* and

$$\operatorname{ind}(\widehat{L}_1(u_{(-)})) = \dim \operatorname{Ker}(\widehat{L}_1(u_{(-)})) - \dim \operatorname{CoKer}(\widehat{L}_1(u_{(-)})) = 0$$

(any Fredholm operator may be represented as the sum of the compact and continuously invertible operators).

Therefore, we obtain, as the consequence of Theorem 1.7 of [30], that the solution of the equation $\mathcal{J}(\tilde{f}, \tilde{f}) = 0$ has the *bifurcation point* $(u_{(-)} = c, \tilde{f} = 0)$, that is, for $u_{(-)} < c$, the equation $\mathcal{J}(f, f) = 0$ has the solution $f = f_-^{(0)}$ ($\tilde{f} = 0$) and for $u_{(-)} \geq c$, this equation has new solution $f = f_+^{(0)}$ ($\tilde{f} = f_+^{(0)}/f_-^{(0)} - 1$).

To investigate equation (4.30), we consider the properties of solutions of $\partial \widehat{G}(\tilde{f}, u_{(-)})/\partial \tilde{f} = 0$. The essential generalized $(w_x + u_{(-)})\partial/\partial x$-spectrum of the operator $\widehat{L}_1(u_{(-)})$ is the same as that of essential $\beta(u_{(-)}) \cdot (w_x + u_{(-)})\widehat{\text{Lap}_2}$-spectrum of $\widehat{\text{Lap}_2}\widehat{L}_1(u_{(-)})$, where

$$\widehat{\text{Lap}_2}(F(x)) \equiv F^{(\text{Lap}_2)}(\beta) = \beta \int_{-\infty}^{+\infty} \exp(-\beta x)F(x)dx,$$

$$(\widehat{\text{Lap}_2})^{-1}(F^{(\text{Lap}_2)}(\beta)) \equiv F(x) = \frac{1}{2\pi i} \int_{\eta_0-i\infty}^{\eta_0+i\infty} \exp(\beta x)F^{(\text{Lap}_2)}(\beta)\frac{d\beta}{\beta},$$

$$\beta \in \mathbb{C}, \quad \beta_{(-)} < \text{Re}\,(\beta) < \beta_{(+)}, \quad \beta_{(-)} < \eta_0 < \beta_{(+)}, \quad \eta_0 \in \mathbb{R}^1,$$

i.e., $\widehat{\text{Lap}_2}$ is the two-sided Laplace transform (by variable x) operator. By the local stability of the Fredholm properties of operator $\widehat{L}_1$ for bounded perturbations (see [31]), we can conclude that there exists the point $(\beta_0(u_{(-)}), f_-^{(0)})$, in some vicinity of which there exists the unique non-trivial branch of solution of (4.30) $\Omega(x, w_x)$ such that $\beta_0(u_{(-)}) \in \mathbb{R}_+^1$ for $u_{(-)} \geq c$ and $\beta_0(u_{(-)}) \in \mathbb{R}_-^1$ for $u_{(-)} \leq c$. This non-trivial solution of $\widehat{G}(\tilde{f}; u_{(-)})$ for $u_{(-)} \geq c$ is $\tilde{f} = \Omega(x, w_x)$ and it can be identified with the shock wave, which we considered earlier (for Navier–Stokes equations).

4.2.2. *The profile of weak shock*

There exists the possibility to obtain the important information about the structure of the shock on the base of kinetic approach. We consider the generalized eigenvalues problem:

$$\widehat{S}(\beta, u_{(-)}, \tilde{f}^{(\text{Lap}_2)}) \equiv \beta \cdot \widehat{A}(u_{(-)})(\tilde{f}^{(\text{Lap}_2)}) - \widehat{L}_1(\tilde{f}^{(\text{Lap}_2)}) = 0,$$

$$\beta \in \mathbb{C}, \quad \widehat{A}(u_{(-)})(\cdot) \equiv \widehat{A}(\cdot) = (w_x + u) \cdot (\cdot). \tag{4.32}$$

The generalized $\widehat{A}$-spectrum of $\widehat{L}_1$ may be decomposed as

$$\widehat{A}\sigma(\widehat{L}_1) = C(\widehat{A}\sigma(\widehat{L}_1)) \oplus P(\widehat{A}\sigma(\widehat{L}_1)),$$

where $C(\widehat{A}\sigma(\widehat{L}_1))$ is the generalized continuous $\widehat{A}$-spectrum of the operator $\widehat{L}_1$, $P(\widehat{A}\sigma(\widehat{L}_1)) = \{\beta_0\}$ is the generalized $\widehat{A}$-point spectrum of $\widehat{L}_1$ including the *unique generalized $\widehat{A}$-eigenvalue* $\beta_0 = \beta_0(u_{(-)}) \in \mathbb{C}$, which is connected with the *generalized $\widehat{A}$-eigenfunction* $\widetilde{\psi}_0$. We introduce two families of projective operators of special type (see [32]), associated with $\beta_0(u_{(0)})$:

$$\widehat{P}_2(\beta_0(u_{(-)})) = \frac{1}{2\pi i} \int_{c(\beta_0)} \widehat{A}\widehat{R}_\beta(\widehat{A}, \widehat{L}_1)d\beta, \quad \widehat{P}_2(\beta_0(u_{(-)})): Y \to Y,$$

$$\widehat{P}_1(\beta_0(u_{(-)})) = \frac{1}{2\pi i} \int_{c(\beta_0)} \widehat{R}_\beta(\widehat{A}, \widehat{L}_1)\widehat{A}d\beta, \quad \widehat{P}_1(\beta_0(u_{(-)})): X \to X,$$

where $\widehat{R}_\beta(\widehat{A}, \widehat{L}_1) = (\beta\widehat{A} - \widehat{L}_1)^{-1}$ is the generalized resolvent of $\widehat{L}_1$, such that the *Hilbert identity* has the form of

$$\widehat{R}_{\beta_1}(\widehat{A}, \widehat{L}_1) - \widehat{R}_{\beta_2}(\widehat{A}, \widehat{L}_1) = (\beta_2 - \beta_1)\widehat{R}_{\beta_1}(\widehat{A}, \widehat{L}_1)\widehat{A}\widehat{R}_{\beta_2}(\widehat{A}, \widehat{L}_1).$$

$c(\beta_0)$ is the closed contour in the complex β-plane, in which the point $\beta_0(u_{(-)})$ is placed, X, Y are Banach spaces over $\mathbb{R}^3$ such that $L_1(f): X \to Y$, $f = f(\vec{v}) \in X$.

Let us apply the projectors $\widehat{P}_1(\beta_0)$ and $\widehat{E} - \widehat{P}_1(\beta_0)$ to $f(x, \vec{v}) \in Z_1(\mathbb{R}^1; X(\mathbb{R}^3))$:

$$f(x, \vec{v}) = a(x) \cdot \widetilde{\psi}_0(u_{(-)}) + b(x, u_{(-)}),$$

$$\widehat{P}_1(u_{(-)})f(x, \vec{v}) = a(x)\widetilde{\psi}_0, \quad (\widehat{E} - \widehat{P}_1(u_{(-)}))f(x, \vec{v}) = b(x, u_{(-)}).$$

Here $\widehat{E}: X \to X$ is identity operator in the Banach space X; $Z_1 \equiv AC^1$, where AC is the space of absolutely continuous vector-valued integrable functions, differentiable by the first argument, with the norm $\|f(x, \vec{v})\|_{AC} = \int \|\partial f(x, \vec{v})/\partial x\|_X dx)$ and $f \in AC^1$ is equivalent to $f \in AC$, $\partial f/\partial x \in AC$.

Now, let us apply the projectors $\widehat{E} - \widehat{P}_2(\beta_0)$ and $\widehat{P}_2(\beta_0)$ to non-linear equation $\widehat{G}(f, u_{(-)}) = \widehat{A}(u_{(-)})\partial f/\partial x - \mathcal{J}(f, f) = 0$ (the stationary Boltzmann equation (4.30) in which we noted $f = \widetilde{f}$ for simplicity):

$$(\widehat{E} - \widehat{P}_2(\beta_0))\widehat{G}(f, u_{(0)})$$

$$\equiv \widehat{A}\frac{\partial b}{\partial x} - \widetilde{L}_1(b) - (\widehat{E} - \widehat{P}_2(\beta_0))L_2(a\widetilde{\psi}_0 + b, a\widetilde{\psi}_0 + b) = 0,$$

(4.33)

$$\widehat{A}\frac{\partial a}{\partial x}\widetilde{\psi}_0 - \widetilde{L}_1(a\widetilde{\psi}_0) = \widehat{P}_2(u_{(-)})L_2(a\widetilde{\psi}_0 + b, a\widetilde{\psi}_0 + b).$$

(4.34)

Equations (4.33) and (4.34) are called the *Lyapunov–Schmidt branching equations* for the nonlinear problem $\widehat{G}(f, u_{(-)}) = 0$, $f(x = \pm\infty, w_x) = f_{\pm}^{(0)}$. For information about a general Lyapunov–Schmidt approach to nonlinear equations, the readers may refer to [33, 34].

There is the possibility to find the explicit forms of projective operators $\widehat{P}_1$, $\widehat{P}_2$ with the help of Banach–Hahn theorem (see [35]):

$$\widehat{P}_1(u_{(-)})f = \langle \varsigma_1^*, f \rangle \widetilde{\psi}_0(u_{(-)}), \quad \widehat{P}_2(u_{(-)})f = \langle \varsigma_2^*, f \rangle \widehat{A}\widetilde{\psi}_0(u_{(-)}),$$

$$\langle \varsigma_1^*, \widetilde{\psi}_0 \rangle = 1,$$

where linear forms $\varsigma_k^* \in W_k^*$, $W_1 = X$, $W_2 = Y$ and W_k^* are topological adjoint to W_k spaces $(k = 1, 2)$. Besides, we have the following spaces decompositions:

$$X = W_1 = W_1^{(+)} \oplus W_1^{(-)}, \quad W_1^{(-)} = \{f \mid f \in X : \langle \varsigma_1^*, f \rangle = 0\},$$

$$W_1^{(+)} = \operatorname{span}\{\psi_0(u_{(-)})\} = \operatorname{Ker}(\beta\widehat{A} - \widehat{L}_1), \quad \beta \in c(\beta_0) \in \mathbb{C},$$

$$Y = W_2 = W_2^{(+)} \oplus \operatorname{Ran}(\beta\widehat{A} - \widehat{L}_1),$$

$$\operatorname{Ran}(\beta\widehat{A} - \widehat{L}_1) = \{f \mid f \in Y : \langle f, \varsigma_2^* \rangle = 0\},$$

$$W_2^{(+)} = \operatorname{span}\{w_2\} = \widehat{A}W_1^{(+)}, \quad w_2 = \widehat{A}\psi_0.$$

Therefore, the second Lyapunov–Schmidt equation (4.34) can be rewritten as

$$\frac{da}{dx} = \beta_0(u_{(-)}) \cdot a(x) + \beta_1(u_{(-)}) \cdot a^2 + \widehat{\beta}_2(a, b),$$

$$\beta_1(u_{(-)}) = \langle \varsigma_2^*, \widehat{L}_2(\widetilde{\psi}_0, \widetilde{\psi}_0) \rangle,$$

(4.35)

$$\widehat{\beta}_2(a, b) = \langle \varsigma_2^*, \widehat{L}_2((a\widetilde{\psi}_0 + b)^2) \rangle - \langle \varsigma_2^*, \widehat{L}_2(a\widetilde{\psi}_0, a\widetilde{\psi}_0) \rangle,$$

$$\widehat{\beta}_2 : AC^1(\mathbb{R}^1) \times AC(\mathbb{R}^1; (\widehat{E} - \widehat{P}_1)X) \to AC(\mathbb{R}^1).$$

To obtain $b = b(a)$, we apply the *implicit abstract function theorem* [33] to equation (4.33), which may be written in the following form:

$$\widehat{\Upsilon}(u_{(-)})(b_1) = (\widehat{E} - \widehat{P}_2(u_{(-)}))(a_1^2(x) \cdot \widehat{L}_2(\tilde{\psi}_0, \tilde{\psi}_0)$$
$$+ 2\tau a_1(x) \cdot \widehat{L}_2(\tilde{\psi}_0, b_1) + \tau^2 \cdot \widehat{L}_2(b_1, b_1)),$$

where τ, a_1, b_1 are defined by the relations $a(x) = \tau a_1(x)$, $b(x, u_{(-)}) = \tau^2 b_1(x, u_{(-)})$. We can consider the *pseudo-inverse operator* $(\widehat{\Upsilon}(u_{(-)}))^{-1}$ (see, for example, [33]) in the vicinity $\tilde{O}_\varepsilon(\tau = 0)$ of the point $(a_1 = a_1^{(0)}; b_1 = b_1^{(0)}; s = 0)$, where $a_1^{(0)}$ is the vector with the bounded norm such that the vector

$$b_1^{(0)}(x, u_{(-)}) = (\widehat{\Upsilon}(u_{(-)}))^{-1}((a_1^{(0)})^2 \cdot (\widehat{E} - \widehat{P}_2(u_{(-)}))\widehat{L}_2(\tilde{\psi}_0, \tilde{\psi}_0))$$

has the bounded norm too. In this vicinity,

$$\tilde{O}_\varepsilon(\tau = 0) = \{a_1^{(0)} : \|a_1^{(0)}(x) - a_1(x)\|_{AC^1} < \varepsilon;$$
$$b_1^{(0)} : \|b_1^{(0)}(x) - b_1(x)\|_{AC} < \varepsilon; \ \tau : |\tau| < \varepsilon; \ \varepsilon \in \mathbb{R}_+^1\}$$
$$\in AC^1(\mathbb{R}^1) \times AC(\mathbb{R}^1; (\widehat{E} - \widehat{P}_1)X) \times \mathbb{R}^1,$$

there exists the unique Fréchet-differentiable mapping $b_1(x, u_{(-)}) = b_1(a(x), \tau)$, $b_1(0, \tau) = 0$.

Therefore, equation (4.35) can be rewritten in closed form (since we obtain $b = b(a)$):

$$\frac{da_1(x_1)}{dx_1} = a_1(x_1) - \beta_1[\beta_0] \cdot a_1^2(x_1) + \beta_2(a_1; b_1(a_1(x_1); [\beta_0]),$$

$$\beta_0 = x_1/x = \tau,$$

$$\beta_2(a_1; b_1; [\beta_0]) = 2\beta_0 a_1 \cdot \langle \varsigma_2^*, \widehat{L}_2(\tilde{\psi}_0, b_1)\rangle + \beta_0^2 \langle \varsigma_2^*, \widehat{L}_2(b_1, b_1)\rangle.$$

$$(4.36)$$

One may consider the following *proposition*: estimates $\beta(u_{(-)}) = O(u_{(-)} - c)$, $\|a(x)\|_{AC^1} = O(u_{(-)} - c)$ are true in the vicinity $\tilde{O}(M = 1)$, where c is the speed of sound [36, 37]. As a consequence, if we exclude from consideration in (4.36) the terms proportional to

$O(u_{(-)} - c)^k$, $k \geq 3$, we obtain the Riccatti's ordinary differential equation:

$$\frac{da_1(x_1)}{dx_1} = a_1(x_1) - \beta_1[\beta_0]a_1^2(x_1). \tag{4.37}$$

This equation has the base (non-trivial) solution: $a_1^{(0)} = (\beta_1)^{-1}$ $\exp(x_1)/(\exp(x_1) + 1)$. We will consider the solution of (4.36) close in some sense to this base solution:

$$a_1(x_1) = a_1^{(0)}(x_1) \cdot (1 + \lambda(x_1)), \quad \lim_{\beta_0 \to 0} \lambda(x_1) \to 0.$$

This solution describes the *profile of weak shock*, that is, the shock has "pre" and "post" conditions in $x = -\infty$ and $x = +\infty$, respectively; if we consider as the (generalized) base solution of Riccatti's equation the discontinuous function, we obtain other profile a_1. But in physical sense, the weak shock is very interesting because real shock waves can be considered as weak in some approximation.

Therefore, the second Lyapunov–Schmidt equation becomes

$$\frac{d\lambda(x_1)}{dx_1} + \frac{\lambda(x_1)e^{x_1}}{e^{x_1} + 1} = -\frac{\lambda^2(x_1)e^{x_1}}{e^{x_1} + 1} + \beta_0 \cdot \beta_1[\beta_0] \cdot (1 + e^{-x_1})$$

$$\times \beta_2(s_1^{(0)}(1 + \lambda); b(a_1^{(0)}(1 + \lambda)). \tag{4.38}$$

Now, we consider operator $\widehat{N} : \mathbb{R}^1 \times AC^1(\mathbb{R}^1) \to AC(\mathbb{R}^1)$ associated with this differential equation; it is bounded in some vicinity $\widetilde{O}_{\delta c, \delta a_1}$ $(c; a_1^{(0)})$ of the point $(c; a_1^{(0)})$. Operator $\widehat{N}$ is Fréchet-differentiable by λ in this vicinity, and

$$\widehat{N}_\lambda \equiv \frac{\partial \widehat{N}(\beta_0 = 0, \lambda = 0)}{\partial \lambda} = \frac{d}{dx_1} + \frac{e^{x_1}}{e^{x_1} + 1}.$$

But $\widehat{N}_\lambda$ is not a homeomorphism $AC^1(\mathbb{R}^1)$ on $AC(\mathbb{R}^1)$ because

$$(\widehat{N}_\lambda)^{-1}(\cdot) = \int_{-\infty}^{x_1} \frac{e^\xi + 1}{e^{x_1} + 1}(\cdot)d\xi$$

is not defined as bounded operator for all $x_1 \in \mathbb{R}^1$ on the set of absolutely continuous functions. Therefore, we can consider for (4.38) the bifurcation problem, that is, the problem of arising non-trivial solution of this equation; the trivial solution is $a_1(x_1) \equiv 0$ or $\lambda = \lambda^{(0)} \equiv -1$.

Now, we investigate the structure of spectrum of $\widehat{N}_\lambda$. The eigenfunctions (bounded in $AC^1(\mathbb{R}^1)$) of $\widehat{N}_\lambda$ are

$$\lambda = \lambda(\gamma \,|\, x_1) \equiv \frac{\exp(\gamma x_1)}{\exp(x_1) + 1}, \quad \gamma \in [0, \, 1].$$

For $\gamma \in \Gamma_\lambda \equiv (-\infty; \, 0) \cup (1; \, +\infty)$, the generalized eigenfunctions $\lambda(\gamma \,|\, x_1)$ are unbounded, that is, $\Gamma_\lambda = C\sigma(\widehat{N}_\lambda)$. For investigation of the branching of the solution $a_1^{(0)}$ of equation (4.36), let us consider the point $\gamma = 1 \in P\sigma(\widehat{N}_\lambda)$; its eigenfunction is $\lambda(\gamma = 1 \,|\, x_1) = e^{x_1}/(e^{x_1} + 1)$. The projective operator for Lyapunov–Schmidt construction with the range span$\{\lambda(1 \,|\, x_1)\}$ may be presented as $\widehat{\pi}_{(1)}(\gamma) = \widetilde{\pi}(\gamma)|_{\gamma=1} \cdot \lambda(1 \,|\, x_1)$, where $\widetilde{\pi}(\gamma)|_{\gamma=1}$ is the value of some linear functional $\widetilde{\pi}(\gamma)$ in the point $\gamma = 1$ ($\widehat{\pi}_{(1)}\lambda(1 \,|\, x_1 = \lambda(1 \,|\, x_1))$; the existence of this functional is guaranteed by the consequence of the Banach–Hahn theorem. Commutative properties of $\widehat{\pi}_{(1)}$ are $\widehat{\pi}_{(1)}\widehat{N}_\lambda = \widehat{N}_\lambda\widehat{\pi}_{(1)} = \widehat{\pi}_{(1)}$.

The pair of Lyapunov–Schmidt equations are as follows:

$$\widehat{\pi}_{(1)}\widehat{N}_\lambda(\lambda) = \lambda(+\infty) \cdot \frac{e^{x_1}}{e^{x_1} + 1} = -\frac{e^{x_1}}{e^{x_1} + 1}\left(\frac{\lambda(+\infty)e^{x_1}}{e^{x_1} + 1}\right)^2 |_{x_1 \to +\infty}$$

$$+ \beta_0\beta_1((1 + e^{-x_1})\beta_2(\lambda))|_{x_1 \to \infty}, \qquad (4.39)$$

or, equivalently,

$$\lambda(+\infty) = -\lambda^2(+\infty) + \beta_0\beta_1\beta_2(\lambda(x_1))|_{x_1 \to +\infty},$$
$$(\widehat{E} - \widehat{\pi}_{(1)})\widehat{N}_\lambda(\lambda) = -\lambda(1 \,|\, x_1) \cdot \lambda^2(x_1) + \beta_0\beta_1(1 + e^{-x_1})\beta_2(\lambda)$$
$$+ \lambda(1 \,|\, x_1)\lambda^2(+\infty) - \beta_0\beta_1\lambda(1 \,|\, x_1)$$
$$\cdot \beta_2(\lambda(x_1))|_{x_1 \to +\infty}. \qquad (4.40)$$

Consequently, we have the system of algebraic equations, solutions of which define the function $\lambda(x_1)$.

Therefore, we can formulate the following result: in the vicinity of the point $(\beta_0 = 0) \in \mathbb{R}^1$, there exists the unique mapping $\lambda(\beta_0; \lambda(+\infty)) : \mathbb{R}^1 \times \mathbb{R}^1 \to AC^1(\mathbb{R}^1)$ such that $\lambda(\beta_0; \lambda(+\infty))$ is the unique solution of equation (4.38), with the property $\lambda(0,0) = 0$.

Consequently, the exact profile $a_1(x_1)$ of the shock (see (4.36)) is unique in this vicinity.

Therefore, the consideration of weak shock based on the kinetic approach (i.e., with the use of Boltzmann equation) allows the possibility to investigate the structure of shock front and obtain its characteristics.

4.3. Summary

This chapter is devoted to the consideration of the basic properties of one-dimensional shock waves and their properties on the basis of the research of structure of the equations of hydrodynamics and the kinetic Boltzmann equation.

In Section 4.1, the properties of linear, quasi-linear and non-linear first-order differential equations were analyzed, concepts of characteristics and Riemann invariants were introduced. Rankine–Hugoniot relations on the discontinuity were obtained, and cases when shock waves satisfy physical entropy condition were investigated. In Section 4.2, the physical substantiation of the origin of a weak shock wave at a kinetic level was considered. With the use of methods of the nonlinear functional analysis, the phenomenon of bifurcation of solutions of the stationary Boltzmann equation, resulting in the transition from local Maxwell distributions to the global solution with complicated structure was investigated. The secondary branching of solutions was considered (this phenomenon may be interpreted as the physical base of fluid turbulization).

During the work on the material of this chapter, authors used the following books and papers: [8–10, 21–25, 36], etc. We especially recommend for readers Refs. [33–35] for a detailed study of problems mentioned in this chapter.

Part II
Stability and Instabilities in Fluidodynamics

Chapter 5

Instabilities in Fluidodynamics and
Its Consequences

On the example of shock, we saw that nonlinear effects in hydrodynamics may play a very important (critical in the some cases)
role. Now, we consider other examples of influence of nonlinearity
and special structure of fluidodynamic equations on the real physical
properties of liquid and gas. We begin the investigation from the consideration of *convection* in fluid dynamics, which is the most simple
and widely known physical effect in hydrodynamics connected with
the critical behavior of the fluid. More complicated types of instability are Saffman, Rayleigh–Taylor, Kelvin–Helmholtz, Richtmyer–
Meshkov ones; we will consider these examples too.

5.1. An Introduction to Thermal Convection

5.1.1. *Schwarzschild's criterion for instability*

Let us find the criterion for the convective instability of a homogeneous fluid.

We consider the heating of fluid in gravitational field, which leads
to the vertical temperature stratification. Earth's atmosphere forms
such an example, since the Sun's rays heat the ground, which reradi-
ate infrared radiation that warms the overlying air. Since the warm
fluid is buoyant relative to the cool one, when does such an adverse

temperature gradient become unstable with respect to the tendency to develop overturning motions (thermal convection)? Let us consider an isolating small blob of gas immersed in the ambient medium. We assume the blob to be initially in thermal and mechanical equilibrium with its surroundings, and we wish to analyze whether a small displacement of the blob would lead to forces that restore the blob to its original position (stability), or that accelerate the blob further away from equilibrium (instability). If the blob is small, the sound-travel time across it to establish pressure equilibrium with the surroundings will be shorter than the corresponding time required for heat to diffuse across it. In other words, if the blob were to move by a slight amount downward (to make dp positive), say, it would maintain pressure equilibrium by compressing adiabatically (constant s) in the new ambient medium of higher p. Thus, the change in the internal density of the blob is related to the change of its (and the external medium's) pressure via

$$d\rho_{\text{blob}} = (\partial\rho/\partial p)_s dp. \tag{5.1}$$

The corresponding change in the density of the ambient medium equals

$$(d\rho)_{\text{medium}} = (\partial\rho/\partial p)_s dp + (\partial\rho/\partial s)_p ds, \tag{5.2}$$

where dp and ds are the differences of the pressure and specific entropy of the medium at the new position compared to the old that arise because of the constraints of mechanical and thermal balance. The blob will continue to move downward if it suffers a buoyancy deficit relative to its new surroundings, that is, if its density increase exceeds that of the ambient medium: $(d\rho)_{\text{blob}} > (d\rho)_{\text{medium}}$. Since the blob is small, the two dp's in equations (5.1) and (5.2) are considered to be equal. Thus, *convective instability* will arise if the ambient medium satisfies

$$\left(\frac{\partial\rho}{\partial s}\right)_p ds < 0 \tag{5.3}$$

for a downward displacement.

The quantity $(\partial\rho/\partial s)_p$ does not lend itself to immediate thermodynamic interpretation. Let us transform it to a more simple form by invoking the relation $(\partial(\rho^{-1})/\partial s)_p = (\partial T/\partial p)_s$. All thermodynamically stable substances have temperatures that increase upon adiabatic compression, i.e., $(\partial T/\partial p)_s > 0$. Consequently, $(\partial \rho/\partial s)_p < 0$, and equation (5.3) becomes *Schwarzschild's criterion* for convective instability:

$$ds > 0 \quad \text{in the direction of gravity.} \tag{5.4}$$

We can simplify this criterion in concrete cases. For example, we consider convection in heated liquid in the gravitational field of the Earth (the vector of acceleration of gravity is $\vec{g} = -\vec{e}_z g$). Then, the condition of convection instability is $ds/dz < 0$. Expanding the derivative ds/dz, we have

$$\frac{ds}{dz} = \left(\frac{\partial s}{\partial T}\right)_p \frac{dT}{dz} + \left(\frac{\partial s}{\partial p}\right)_T \frac{dp}{dz} = \frac{c_p}{T}\frac{dT}{dz} - \left(\frac{\partial \rho^{-1}}{\partial T}\right)_p \frac{dp}{dz} < 0.$$

Since $dp/dz = -\rho g$, we obtain

$$-\frac{dT}{dz} > g\beta\frac{T}{c_p}, \quad \beta = \rho\left(\frac{\partial \rho^{-1}}{\partial T}\right)_p, \tag{5.5}$$

where β is the thermal expansion coefficient; for a thermodynamically perfect gas in equilibrium, $\beta T = 1$ and the last relation becomes $-dT/dz > g/c_p$. In general, thermal convection in the considered problem occurs if temperature decreases upward with a gradient whose magnitude exceeds the value $g\beta T/c_p$.

5.1.2. *The equations describing thermal convection*

Let us derive the equations describing thermal convection. We will suppose the fluid to be incompressible. This means that the pressure is supposed to vary only slightly through the fluid, so that the density change due to changes in pressure may be neglected. For example, in the atmosphere, where the pressure varies with the height, this assumption means that we will not consider columns of air of great height, in which the density varies considerably over the height of the column. The density change due to non-uniform heating of the fluid cannot be neglected; it results in the forces which bring about the convection. We write $T = T_0 + T^\dagger$, $\rho = \rho_0 + \rho^\dagger$, where T_0 and

ρ_0 are some constant mean temperature and density, respectively; $T_0 \gg |T^\dagger|$, $\rho_0 \gg |\rho^\dagger|$. In the pressure $p = p_0 + p^\dagger$, p_0 is not constant. It is pressure corresponding to mechanical equilibrium, when the temperature and density are constant and equal to T_0 and ρ_0, respectively. It varies with height according to the hydrostatic equation $p_0 = \rho_0 \vec{f}_{\mathrm{ex}} \cdot \vec{r} + \mathrm{const} = -\rho_0 g z + \mathrm{const}$. In a fluid column with height h, the hydrostatic pressure drop is $\rho_0 g h$. This causes a density change $\sim \rho g h / c^2$, where c is the speed of sound. According to the condition stated, this change must be negligible not only in comparison with the density itself but also in comparison with the thermal change $\rho^\dagger = (\partial \rho_0 / \partial T)_p T^\dagger = -\rho_0 \beta T^\dagger$ $(\beta = -\rho^{-1} \partial \rho / \partial T)$. That is, we must have $g h / c^2 \ll \beta \delta T$, where δT is a characteristic temperature difference.

Let us substitute in the Navier–Stokes equation (with the term including the external field)

$$\frac{\partial \vec{v}}{\partial t} + (\vec{v} \cdot \nabla)\vec{v} = -\frac{1}{\rho}\nabla p + \nu \triangle \vec{v} + \vec{f}_{\mathrm{ex}}, \tag{5.6}$$

the above-mentioned relations for p, ρ, T:

$$\frac{\partial \vec{v}}{\partial t} + (\vec{v}\nabla)\vec{v} = -\nabla \left(\frac{p^\dagger}{\rho_0}\right) + \nu \triangle \vec{v} - \beta T^\dagger \vec{f}_{\mathrm{ex}}. \tag{5.7}$$

In the thermal conduction equation, the viscosity term can be shown to be small in free convection compared with other terms, and may therefore be omitted. We thus obtain

$$\frac{\partial T^\dagger}{\partial t} + \vec{v} \cdot \nabla T^\dagger = \xi \triangle T^\dagger, \quad \xi = \kappa / \rho_0 c_p. \tag{5.8}$$

Equations (5.7), (5.8) and the equation of continuity div $\vec{v} = 0$ form a *Boussinesq system* or a complete system of equations governing thermal (free) convection.

This system of five equations for the unknown functions $\vec{v}$, $p^\dagger / \rho_0$, $T^\dagger$ contains three parameters, ξ, ν and βg $(g = |\vec{f}_{\mathrm{ex}}|)$. Moreover, the solution will involve the characteristic length $\mathcal{L}$ and temperature difference δT. There is no characteristic velocity, since there is no flow due to external forces, and the whole motion of the fluid is due to its non-uniform heating. From these quantities, one can form two independent dimensionless combinations: *Prandtl number* $Pr = \nu/\xi$ (it depends only on the properties of the fluid) and the *Rayleigh number* $Ra = \beta g \delta T \mathcal{L}^3/(\nu\xi)$, which is the chief characteristic of the convection as such. Convective flow may be either laminar or turbulent. The onset of turbulence is governed by the Rayleigh number: the convection becomes turbulent when $Ra \gg 1$.

5.2. Linear Analysis and Normal Modes

Now, we consider the general methods of linear stability analysis. The investigation in detail of thermal convection is very important both for physical applications and for mathematical formalism, applied in other problems of fluid dynamics such as diffusion and general theory of turbulence. This is because we shall develop a theory of stability on the base of examples of tasks connected with convection.

5.2.1. *Stability of basic flow and normal modes*

We specify the *basic flow* of interest (a solution of the governing equations of motion, whose stability we wish to investigate) by the velocity field $\vec{v}_*(\vec{r}, t)$ and pressure $p_*(\vec{r}, t)$ of an incompressible viscous fluid in a given region V with boundary ∂V. This fluid is governed by the Navier–Stokes equations without external forces in the dimensionless form (see Section 1.2 in Chapter 1):

$$\frac{\partial \vec{v}_*}{\partial t} + (\vec{v}_* \cdot \nabla)\vec{v}_* = -\nabla p_* + Re^{-1}\triangle\vec{v}_*, \quad \nabla \cdot \vec{v}_* = 0 \qquad (5.9)$$

in the given domain V; the boundary conditions are $(\vec{v}_*)|_{\partial V} = ((\vec{v}_*)^{(B)})|_{\partial V}$ (for example, $\vec{v}_* = (\vec{v}_*)_0$ on the part of ∂V, $\vec{v}_*$ is periodic on the rest of ∂V).

Now, for general initial values $\vec{v}_{(\text{tot})}(\vec{r}, 0)$ of the velocity and $p_{(\text{tot})}(\vec{r}, 0)$ of the pressure, there is a *total flow* with velocity $\vec{v}_{(\text{tot})}(\vec{r}, t)$ and pressure $p_{(\text{tot})}(\vec{r}, t)$ for $t > 0$, such that

$$\frac{\partial \vec{v}_{(\text{tot})}}{\partial t} + (\vec{v}_{(\text{tot})} \cdot \nabla)\vec{v}_{(\text{tot})} = -\nabla p_{(\text{tot})} + Re^{-1}\triangle\vec{v}_{(\text{tot})}, \quad \nabla\vec{v}_{(\text{tot})} = 0,$$

$\vec{v}_{(\text{tot})} = (\vec{v}_*)_0$ and so forth on ∂V. We may define the *perturbation quantities* $(\vec{v})^\dagger = \vec{v}_{(\text{tot})} - \vec{v}_*$ and $p^\dagger = p_{(\text{tot})} - p_*$. Then the Navier–Stokes equations become

$$\frac{\partial(\vec{v})^\dagger}{\partial t} + (\vec{v})^\dagger \cdot \nabla \vec{v}_* + \vec{v}_* \cdot \nabla(\vec{v})^\dagger + ((\vec{v})^\dagger \cdot \nabla)(\vec{v})^\dagger = -\nabla p^\dagger + Re^{-1}\triangle(\vec{v})^\dagger,$$

$$\nabla(\vec{v})^\dagger = 0 \quad (\text{in } V), \quad (\vec{v})^\dagger|_{\partial V} = ((\vec{v})^\dagger)^{(B)})|_{\partial V}. \qquad (5.10)$$

We shall say that a given basic *flow is stable* if all perturbations which are small initially remain small for all the time, and it is *unstable* if at least one perturbation which is small initially grows so much that it ceases to remain small after some time. One may reformulate this definition in terms of *Lyapunov's stability theory.*

A basic flow is *stable* if for all $\varepsilon > 0$ there exists $\delta(\varepsilon)$ such that if $\|(\vec{v})^\dagger(\vec{r}, t = 0)\| < \delta$ and $\|p^\dagger(\vec{r}, t = 0)\| < \delta$, then $\|(\vec{v})^\dagger(\vec{r}, t)\| < \varepsilon$ and $\|p^\dagger(\vec{r}, t)\| < \varepsilon$ for all $t > 0$.

The flow is called *asymptotically stable* if it is stable and $\|(\vec{v})^\dagger(\vec{r}, t)\| \to 0$, $\|p^\dagger(\vec{r}, t)\| \to 0$ as $t \to \infty$. In the theory of dynamical systems, an asymptotically stable solution, whether it be steady or unsteady, is called an *attractor.*

The definition of stability crucially concerns the evolution of *small* perturbations with time, so it is plausible that stability may be investigated by neglecting products of the perturbed quantities in the equations of motion and boundary conditions. This gives *linearized problem.*

From (5.10), we can obtain linearized problem

$$\frac{\partial(\vec{v})^\dagger}{\partial t} + \vec{v}_* \cdot \nabla(\vec{v})^\dagger + (\vec{v})^\dagger \cdot \nabla \vec{v}_* = -\nabla p^\dagger + Re^{-1}\triangle(\vec{v})^\dagger,$$

$$\nabla(\vec{v})^\dagger = 0 \text{ in } V, \qquad (5.11)$$

and $(\vec{v})^\dagger = 0$ or is periodic on the boundary ∂V.

If the basic flow is steady ($\vec{v}_*$ is independent of t), then the linearized problem has coefficients independent of t. It follows plausibly that we may separate the variables, so that the general solution of an initial-value problem is a linear superposition of *normal modes,* each having the form of

$$(\vec{v})^\dagger(\vec{r}, t) = \exp(\sigma t)\vec{v}_\ddagger(\vec{r}), \quad p^\dagger(\vec{r}, t) = \exp(\sigma t)p_\ddagger(\vec{r}),$$

$$\sigma = \Re(\sigma) + i\Im(\sigma) \in \mathbb{C},$$

where the eigenvalue σ and corresponding eigenfunctions $\vec{v}_\ddagger$, $p_\ddagger$ can be found in principle by solving the resultant equations and boundary conditions, namely,

$$\sigma\vec{v}_\ddagger + \vec{v}_* \cdot \nabla\vec{v}_\ddagger + \vec{v}_\ddagger \cdot \nabla\vec{v}_* = -\nabla p_\ddagger + Re^{-1}\triangle\vec{v}_\ddagger, \quad \nabla\vec{v}_\ddagger = 0 \text{ in } V,$$

and $\vec{v}_\ddagger = 0$ or is periodic on ∂V. The eigenvalues σ of this real problem are real or occur in complex conjugate pairs. If V is bounded, then there is a countable infinity of discrete eigenvalues. The basic flow is stable if $\Re(\sigma) < 0$ for all σ and unstable if $\Re(\sigma) > 0$ for at least one eigenvalue σ. We may define the *critical Reynolds number* Re_c such that if $Re \leq Re_c$, then $\Re(\sigma) \leq 0$ for all eigenvalues and that $\Re(\sigma) > 0$ for at least one mode and for at least one value of Re in any neighborhood of Re_c. Then we say that the basic flow is *marginally stable* when $Re = Re_c$, because the flow is stable if $Re < Re_c$ and unstable for all small enough positive values of $Re - Re_c$; for *neutral stable states*, we have $\Re(\sigma) = 0$.

5.2.2. *The Benard problem in hydrodynamics*

In this section, we shall investigate the *Benard convection* (thermal convection in a layer of fluid) and consider its stability.

Consider a layer ($z \in [0; d]$) confined between two isothermal walls, in which the lower wall is maintained at a higher temperature. We have rewritten the Boussinesq system of equations (5.7), (5.8) as

$$\frac{\partial v_k}{\partial t} + v_j\frac{\partial v_k}{\partial r_j} = -\frac{1}{\rho_0}\frac{\partial p^\dagger}{\partial r_k} + \nu\triangle v_k - g(1 - \alpha(T - T_0))\delta_{k,3}, \quad (5.12)$$

$$\frac{\partial T}{\partial t} + v_k\frac{\partial T}{\partial r_k} = \xi\triangle T, \quad \frac{\partial v_k}{\partial r_k} = 0, \quad k = 1, 2, 3, \quad (5.13)$$

where $\{r\}_{k=1}^3 = \{x, y, z\}$ and $\{v_k\}_{k=1}^3 = \{v_x, v_y, v_z\}$; the density is given by the equation of state $\rho = \rho_0(1 - \beta(T - T_0))$ (ρ_0 representing the reference density at the reference temperature T_0).

The total flow variables (background plus perturbation) are $T = T_{(bg)}(z) + T^\dagger(\vec{r}, t)$, $p = p_0(z) + p(\vec{r}, t)$, $v_k = v_k(\vec{r}, t)$. The basic state satisfies

$$0 = -\frac{1}{\rho_0}\frac{\partial p_0}{\partial r_k} - g(1 - \beta(T_{(bg)} - T_0))\delta_{k,3}, \quad k = 1, 2, 3,$$

$$0 = \xi\frac{d^2 T_{(bg)}}{dz^2}. \tag{5.14}$$

The last equation of heat transfer gives the linear temperature distribution by z: $T_{(bg)}(z) = T_0 - \Lambda(z + d/2)$, where $\Lambda \equiv \Delta T/d$ is the magnitude of the vertical temperature gradient and T_0 is the temperature of lower wall. We can rewrite the Boussinesq equations (5.12), (5.13) as

$$\frac{\partial v_k}{\partial t} + v_j\frac{\partial v_k}{\partial r_j} = -\frac{1}{\rho_0}\frac{\partial}{\partial x_k}(p_0(z) + p(\vec{r}, t))$$

$$- g(1 - \beta(T_{(bg)} + T^\dagger - T_0))\delta_{k,3} + \nu\triangle v_k,$$

$$\beta = -\rho^{-1}\left(\frac{\partial \rho}{\partial T}\right)_p > 0, \tag{5.15}$$

$$\frac{\partial T^\dagger}{\partial t} + v_j\frac{\partial}{\partial v_j}(T_{(bg)} + T^\dagger) = \xi\triangle(T_{(bg)} + T^\dagger). \tag{5.16}$$

Subtracting the mean state equation (5.14) from the perturbed equation (5.15), and neglecting squares of perturbations, we have

$$\frac{\partial v_k}{\partial t} = -\frac{1}{\rho_0}\frac{\partial p}{\partial r_k} + g\beta T^\dagger\delta_{k,3} + \nu\triangle v_k, \tag{5.17}$$

$$\frac{\partial T^\dagger}{\partial t} - \Lambda v_z = \xi\triangle T^\dagger, \tag{5.18}$$

where v_z is the vertical component of velocity. The advection term in (5.18) results from

$$v_j\frac{\partial T_{(bg)}}{\partial r_j} = v_z\frac{dT_{(bg)}}{dz} = -\Lambda v_z.$$

Equations (5.17) and (5.18) govern the behavior of perturbations in the system.

At this point, it is useful to pause that the Rayleigh number $Ra = g\beta\Lambda d^4/(\xi\nu)$ is the ratio of buoyancy force to viscous force. From (5.18), the velocity scale is found by equating the advective and diffusion terms, giving

$$v_z \sim \frac{\xi T^\dagger}{\Lambda d^2} \sim \frac{\xi\Lambda}{d\Lambda} = \frac{\xi}{d}.$$

An examination of the last two terms in (5.17) shows that

$$\frac{\text{Buoyancy force}}{\text{Viscous force}} \sim \frac{g\beta T^\dagger}{\nu v_z/d^2} \sim \frac{g\beta\Lambda d}{\nu v_z/d^2} = \frac{\beta g\Lambda d^4}{\nu\xi},$$

which is the Rayleigh number.

Taking the Laplacian of the z-component ($k = 3$) of (5.17), we obtain

$$\frac{\partial}{\partial t}(\triangle v_z) = -\frac{1}{\rho_0}\triangle\frac{\partial p}{\partial z} + g\beta\triangle T^\dagger + \nu\triangle(\triangle v_z). \qquad (5.19)$$

The pressure term in the last equation can be eliminated by taking the divergence of equation (5.17) and using the continuity equation $\partial v_k/\partial r_k = 0$ ($k = 1, 2, 3$). This gives

$$-\frac{1}{\rho_0}\frac{\partial^2 p}{\partial x_k^2} + g\beta\frac{\partial T^\dagger}{\partial x_k}\delta_{k,3} = 0.$$

Differentiating with respect to z, we obtain

$$-\frac{1}{\rho_0}\triangle\frac{\partial p}{\partial z} + g\beta\frac{\partial T^\dagger}{\partial z^2} = 0.$$

Hence, equation (5.19) becomes

$$\frac{\partial}{\partial t}(\triangle v_z) = g\beta\left(\frac{\partial^2}{\partial x^2} + \frac{\partial^2}{\partial y^2}\right)T^\dagger + \nu\triangle(\triangle v_z). \qquad (5.20)$$

Equations (5.18) and (5.20) govern the development of perturbations in the system. The boundary conditions on the upper and lower walls are that the no-slip condition is satisfied and that the walls are maintained at constant temperatures. These conditions require

$v_k = T^\dagger = 0$ ($k = 1, 2, 3$) at $z = \pm d/2$. Because the conditions on v_x and v_y hold for all x and y, it follows from the continuity equation that $\partial v_z / \partial z = 0$ at the walls. The boundary conditions therefore can be written as

$$v_z = \frac{\partial w_z}{\partial z} = T^\dagger = 0 \quad \text{at } z = \pm \frac{d}{2}. \tag{5.21}$$

Let us introduce dimensionless variables $\tilde{t} = \xi t / d^2$, $\tilde{r}_k = \tilde{r}_k / d$, but in the following, we drop "tilde", so t, r_k will be dimensionless variables. Equations (5.17), (5.20) and boundary conditions (5.21) then become

$$\left(\frac{\partial}{\partial t} - \triangle \right) T^\dagger = \frac{\Lambda d^2}{\xi} v_z, \tag{5.22}$$

$$\left(Pr^{-1} \frac{\partial}{\partial t} - \triangle \right) \triangle v_z = \frac{g\beta d^2}{\nu} \left(\frac{\partial^2}{\partial x^2} + \frac{\partial^2}{\partial y^2} \right) T^\dagger, \tag{5.23}$$

$$v_z = \frac{\partial v_z}{\partial z} = T^\dagger = 0 \quad \text{at } z = \pm \frac{1}{2}, \tag{5.24}$$

where $Pr = \nu / \xi$ is the Prandtl number.

Because the coefficients of the equations (5.23) and (5.24) are independent of x, y, t, solutions exponential in these variables are allowed. We therefore assume normal modes having the form of

$$v_z = [(v_z)_\ddagger(z) \exp(ik_x x + ik_y y)] \cdot \exp(\sigma t),$$
$$T^\dagger = [T_\ddagger(z) \exp(ik_x x + ik_y y)] \cdot \exp(\sigma t).$$

The requirement that solutions remain bounded as $x, y \to \infty$ implies that the wavenumbers $k_x, k_y \in \mathbb{R}^1$. In other words, the *normal modes must be periodic in the directions of unboundness*. The growth rate σ is complex number. With this dependence, equations (5.23) and (5.24) become

$$\left(\sigma - \left(\frac{d^2}{dz^2} - (k_x^2 + k_y^2) \right) \right) T_\ddagger = \frac{\Lambda d^2}{\xi} (v_z)_\ddagger, \tag{5.25}$$

$$\left(\frac{\sigma}{Pr} - \left(\frac{d^2}{dz^2} - (k_x^2 + k_y^2) \right) \right) (v_z)_\ddagger = -\frac{g\beta d^2 (k_x^2 + k_y^2)}{\nu} T_\ddagger. \tag{5.26}$$

This system may be rewritten in compact form:

$$(\sigma - (\hat{\delta}^2 - \tilde{k}^2))T_{\ddagger} = \tilde{v}, \tag{5.27}$$

$$\left(\frac{\sigma}{Pr} - (\hat{\delta}^2 - \tilde{k}^2)\right)(\hat{\delta}^2 - \tilde{k}^2)\tilde{v} = -Ra \cdot \tilde{k}^2 T_{\ddagger}, \tag{5.28}$$

where $Ra = g\beta\Lambda d^4/(\xi\nu)$, $\tilde{k} \equiv \sqrt{k_x^2 + k_y^2}$ and $\tilde{v} \equiv \Lambda d^2(v_z)_{\ddagger}/\xi$. The boundary conditions now become $\tilde{v} = \hat{\delta}\tilde{v} = T_{\ddagger} = 0$ at $z = \pm 1/2$.

Now, we show that for Benard problem $\Im(\sigma) = 0$ (i.e., σ is real). We multiply (5.27) by $T_{\ddagger}^*$ (the complex conjugate of $T_{\ddagger}$), and integrate between $\pm 1/2$, using the boundary conditions. The various terms transform as follows:

$$\sigma \int T_{\ddagger}^* T_{\ddagger} dz = \sigma \int |T_{\ddagger}|^2 dz,$$

$$\int T_{\ddagger}^* \hat{\delta}^2 T_{\ddagger} dz = \left(T_{\ddagger}^* \hat{\delta}T_{\ddagger}\right)\Big|_{-1/2}^{1/2} - \int \hat{\delta}T_{\ddagger}^* \hat{\delta}T_{\ddagger} dz = -\int |\hat{\delta}T_{\ddagger}|^2 dz.$$

Equation (5.27) then becomes

$$\sigma \int |T_{\ddagger}|^2 dz + \int |\hat{\delta}T_{\ddagger}|^2 dz + \tilde{k}^2 \int |T_{\ddagger}|^2 dz = \int T_{\ddagger}^* \tilde{v} dz,$$

which can be written as

$$\sigma S_1 + S_2 = \int T_{\ddagger}^* \tilde{v} dz, \quad S_1 \equiv \int |T_{\ddagger}|^2 dz, \quad S_2 \equiv \int (|\hat{\delta}T_{\ddagger}|^2 + \tilde{k}^2 |T_{\ddagger}|^2) dz. \tag{5.29}$$

Similarly, multiply (5.28) by $\tilde{v}^*$ and integrate by parts. The first term in (5.28) gives

$$\frac{\sigma}{Pr} \int \tilde{v}^* (\hat{\delta}^2 - \tilde{k}^2) \tilde{v} dz$$

$$= \frac{\sigma}{Pr} \int \tilde{v}^* \hat{\delta}^2 \tilde{v} dz - \frac{\sigma \tilde{k}^2}{Pr} \int \tilde{v}^* \tilde{v} dz = -\frac{\sigma}{Pr} (|\hat{\delta}\tilde{v}|^2 + \tilde{k}^2 |\tilde{v}|^2) dz. \tag{5.30}$$

The second term in (5.28) gives

$$\int \tilde{v}^*(\hat{\delta}^2 - \tilde{k}^2)(\hat{\delta}^2 - \tilde{k}^2)\tilde{v}dz$$

$$= \int \tilde{v}^*\hat{\delta}^4\tilde{v}dz + \tilde{k}^4 \int \tilde{v}^*\tilde{v}dz - 2\tilde{k}^2 \int \tilde{v}^*\hat{\delta}^2\tilde{v}dz$$

$$= \int (|\hat{\delta}^2\tilde{v}|^2 + 2\tilde{k}^2|\hat{\delta}\tilde{v}|^2 + \tilde{k}^4|\tilde{v}|^2)dz. \tag{5.31}$$

Using equations (5.30) and (5.31), the integral of equation (5.28) becomes

$$\frac{\sigma}{Pr}\tilde{S}_1 + \tilde{S}_2 = Ra\,\tilde{k}^2 \int \tilde{v}^*T_{\ddagger}dz, \tag{5.32}$$

where

$$\tilde{S}_1 = \int (|\hat{\delta}\tilde{v}|^2 + \tilde{k}^2|\tilde{v}|^2)dz, \quad \tilde{S}_2 = \int (|\hat{\delta}^2\tilde{v}|^2 + 2\tilde{k}^2|\hat{\delta}\tilde{v}|^2 + \tilde{k}^4|\tilde{v}|^2)dz.$$

Integrals S_i, $\tilde{S}_i$ ($i = 1, 2$) are positive. The right-hand side of equation (5.32) is $Ra \cdot \tilde{k}^2$ times the complex conjugate of the right-hand side of equation (5.29). We can therefore eliminate the integral on the right-hand side of these equations by taking the complex conjugate of (5.29) and substituting into (5.32). This gives

$$\frac{\sigma}{Pr}\tilde{S}_1 + \tilde{S}_2 = Ra\,\tilde{k}^2(\sigma^* S_1 + S_2).$$

Equating imaginary parts gives $\Im(\sigma)(\tilde{S}_1/Pr + Ra\,\tilde{k}^2 S_1) = 0$. We consider only the case for which $Ra > 0$. The quantity $\tilde{S}_1/Pr + Ra\,\tilde{k}^2 S_1$ is then positive, and the preceding equation requires that $\Im(\sigma) = 0$.

We can rewrite equations (5.27) and (5.28) for the marginal state (in which $\sigma = 0$) in the form of

$$(\hat{\delta}^2 - \tilde{k}^2)T_{\ddagger} = -\tilde{v}, \quad (\hat{\delta}^2 - \tilde{k}^2)^2\tilde{v} = Ra\,\tilde{k}^2 T_{\ddagger}. \tag{5.33}$$

Eliminating $T_{\ddagger}$, we obtain

$$(\hat{\delta}^2 - \tilde{k}^2)^3\tilde{v} = -Ra\,\tilde{k}^2\tilde{v}. \tag{5.34}$$

The boundary conditions become $\tilde{v} = \hat{\delta}\tilde{v} = (\hat{\delta}^2 - \tilde{k}^2)^2\tilde{v} = 0$ at $z = \pm 1/2$. We have sixth-order homogeneous differential equation with six homogeneous boundary conditions. Non-zero solutions for such a system can only exist for a particular value of parameter Ra (for a given $\widetilde{k}$). It is therefore an eigenvalue problem. Note that the Prandtl number has dropped out of the marginal state.

Because the coefficients of the equations (5.34) are independent of z, the general solution can be expressed as a superposition of solutions of the form $\widetilde{v} = \exp(\mu z)$, where the six roots of μ are given by $(\mu^2 - \widetilde{k}^2)^3 = -Ra\,\widetilde{k}^2$. The three roots of this equation are

$$\mu^2 = -\widetilde{k}^2 \left(\left(\frac{Ra}{\widetilde{k}^4} \right)^{1/3} - 1 \right), \quad \mu^2 = \widetilde{k}^2 \left(1 + \frac{1}{2} \left(\frac{Ra}{\widetilde{k}^4} \right)^{1/3} (1 \pm i\sqrt{3}) \right).$$

$$(5.35)$$

Taking square roots, the six roots finally become $\pm i\mu_0$, $\pm\mu_1$, $\pm\mu_1^*$, where

$$\mu_0 = \widetilde{k} \left(\left(\frac{Ra}{\widetilde{k}^4} \right)^{1/3} - 1 \right)^{1/2},$$

and μ_1 and its conjugate μ_1^* are given by the two roots of (5.35).

The even solution of equation (5.34) is therefore

$$\widetilde{v} = c_1 \cos\mu_0 z + c_2 \operatorname{ch}\mu_1 z + c_3 \operatorname{ch}\mu_1^* z,$$

where c_j ($j = 1, 2, 3$) are "constants of integration". To apply the boundary conditions on this solution, we find the following derivatives:

$$\widehat{\delta v} = -c_1 \mu_0 \sin\mu_0 z + c_2 \mu_1 \operatorname{sh}\mu_1 z + c_3 \mu_1^* \operatorname{sh}\mu_1^* z,$$

$$(\widehat{\delta}^2 - \widetilde{k}^2)^2 \widetilde{v} = c_1 (\mu_0^2 + \widetilde{k}^2)^2 \cos\mu_0 z$$

$$+ c_2 (\mu_1^2 - \widetilde{k}^2)^2 \operatorname{ch}\mu_1 z + c_3 ((\mu_1^*)^2 - \widetilde{k}^2)^2 \operatorname{ch}\mu_1^* z.$$

The boundary conditions $\widetilde{v} = \widehat{\delta v} = (\widehat{\delta}^2 - \widetilde{k}^2)^2 \widetilde{v} = 0$ at $z = \pm 1/2$ then require

$$\widehat{\mathcal{M}} \begin{pmatrix} c_1 \\ c_2 \\ c_3 \end{pmatrix} = 0,$$

$$\widehat{\mathcal{M}} = \begin{pmatrix} \mathrm{ch}\,\frac{\mu_0}{2} & \mathrm{ch}\,\frac{\mu_1}{2} & \mathrm{ch}\,\mu_1^*2 \\ -\mu_0\sin\frac{\mu_0}{2} & \mu_1\,\mathrm{sh}\,\frac{\mu_1}{2} & \mu_1^*\,\mathrm{sh}\,\frac{\mu_1^*}{2} \\ (\mu_0^2+\widetilde{k}^2)^2\cos\frac{\mu_0}{2} & (\mu_1^2-\widetilde{k}^2)^2\,\mathrm{ch}\,\frac{\mu_1}{2} & ((\mu_1^*)^2-\widetilde{k}^2)^2\,\mathrm{ch}\,\frac{\mu_1^*}{2} \end{pmatrix}.$$

Here, c_j ($j = 1, 2, 3$) cannot all be zero if we want to have a non-zero solution, which requires that $\det\widehat{\mathcal{M}}$ must vanish. This gives a relation between Ra and the corresponding eigenvalue $K \equiv \widetilde{k}$ (see Fig. 5.1). The curve $K(Ra)$ represents marginally stable states, which separate regions of stability and instability. The lowest value of Ra is found to be $Ra_c = 1708$, attained at $\widetilde{k}_c = 3.12$. As all values of $\widetilde{k}$ are allowed by the system, the flow first becomes unstable when the Rayleigh number reaches a value of $Ra_c = 1708$. The wavelength at the onset of instability is $\lambda_c = 2\pi d/\widetilde{k} \approx 2d$.

Laboratory experiments agree remarkably well with these predictions, and the solution of the Benard problem is considered as one of the major successes of the linear stability theory.

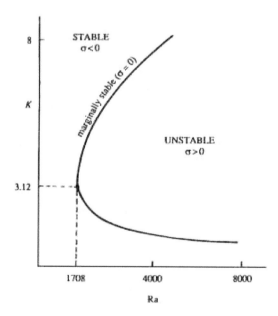

Fig. 5.1. Stable and unstable regions for Benard convection [13].

5.2.3. *Convection phenomena for a layer of fluid with stress-free surfaces*

Let us consider the solution of convection equations for a layer of fluid with stress-free surfaces. This case can be approximately realized in a laboratory experiment if a layer of liquid is floating on top of a somewhat heavier liquid. In this case, the boundary conditions are

$$v_z = T^\dagger = (\partial v_x/\partial z + \partial v_z/\partial x) = (\partial v_y/\partial z + \partial v_z/\partial y) = 0$$

at the surfaces, the latter two conditions resulting from zero stress. Because v_z vanishes (for all x and y) on the boundaries, it follows that the vanishing stress conditions require $\partial v_x/\partial z = \partial v_y/\partial z = 0$ at the boundaries. On differentiating the continuity equation with respect to z, it follows that $\partial^2 v_z/\partial z^2 = 0$ on these surfaces. In terms of the complex amplitudes, the eigenvalue problem is therefore

$$(\widehat{\delta}^2 - \widetilde{k}^2)^3 \widetilde{v}_z = -Ra\,\widetilde{k}^2 \widetilde{v}_z, \tag{5.36}$$

with $\widetilde{v}_z = (\widehat{\delta}^2 - \widetilde{k}^2)^2 \widetilde{v}_z = \widehat{\delta}^2 \widetilde{v}_z = 0$ or

$$\widetilde{v}_z = \widehat{\delta}^2 \widetilde{v}_z = \widehat{\delta}^4 \widetilde{v}_z = 0$$

at the surfaces $z = \pm 1/2$; the boundary conditions are not the same as that for the rigid boundaries:

$$\widetilde{v} = \widehat{\delta}\widetilde{v} = (\widehat{\delta}^2 - \widetilde{k}^2)^2 \widetilde{v} = 0 \quad \text{at } z = \pm 1/2$$

(see above). Successive differentiation of equation (5.36) shows that all even derivatives of $\widetilde{v}_z$ vanish on the boundaries. The eigenfunctions must therefore be $\widetilde{v}_z = c_0 \sin n\pi z$, $c_0 = \text{const}$, $n \in \mathbb{N}$. Substitution into (5.36) leads to the eigenvalue relation

$$Ra = \frac{1}{\widetilde{k}^2}\left(n^2\pi^2 + \widetilde{k}^2\right)^3,$$

which gives the Rayleigh parameter in the marginal state. For a given $\widetilde{k}^2$, the lowest value of Ra occurs when $n = 1$, which is the gravest mode. The critical Rayleigh number is obtained by finding the minimum value of Ra as $\widetilde{k}^2$ is varied, that is, by setting $d(Ra)/d(\widetilde{k}^2) = 0$.

This gives

$$\frac{d(Ra)}{d(\widetilde{k}^2)} = \frac{3(\pi^2 + \widetilde{k}^2)^2}{\widetilde{k}^2} - \frac{(\pi^2 + \widetilde{k}^2)^3}{\widetilde{k}^4} = 0,$$

which requires $\widetilde{k}_c^2 = \pi^2/2$. The corresponding critical value of Rayleigh parameter is $Ra_c = 27\pi^4/4 = 657$.

For a layer with free upper surface (where the stress is zero) and a rigid bottom wall, the solution of the eigenvalue problem gives $Ra_c = 1101$ and $\widetilde{k}_c = 2.68$.

5.3. Classical Types of Hydrodynamical Instabilities

With the help of above-considered linear analysis methods, we can investigate various types of instabilities which exist in hydrodynamics, astrophysics and plasma physics. In this section, we shall familiarize with several widespread kinds of hydrodynamical instability.

5.3.1. *Kelvin–Helmholtz instability*

Flows in which two layers of incompressible fluid move relative to each other, one "sliding" on the other, are unstable if the fluid is ideal; the surface of separation between these two fluid layers would be a *surface of tangential discontinuity*, on which the fluid velocity tangential to the surface is discontinuous. This instability is called the *Kelvin–Helmholtz instability*.

Let us consider the basic flow of two incompressible inviscid fluids in horizontal parallel infinite streams of different velocities and densities, one stream above the other:

$$\vec{v}_*(z) = v_*^{(2)} \cdot \vec{e}_x, \quad \rho_*(z) = \rho_*^{(2)}, \quad p_* = p_0 - \rho_*^{(2)} gz \quad \text{for } z > 0,$$

$$\tag{5.37}$$

$$\vec{v}_*(z) = v_*^{(1)} \cdot \vec{e}_x, \quad \rho_*(z) = \rho_*^{(1)}, \quad p_* = p_0 - \rho_*^{(1)} gz \quad \text{for } z < 0,$$

$$\tag{5.38}$$

where $v_*^{(1,2)}$ are the velocities of the two streams, $\rho_*^{(1,2)}$ are the densities, p_0 is the constant pressure, z is the height, and g is the acceleration due to gravity.

We assume the existence of a velocity potential ϕ on each side of the interface between the two streams with $\vec{w} = \nabla\phi$, where $\phi = \phi_2$ for $z > \xi$, $\phi = \phi_1$ for $z < \xi$, the interface having elevation $z = \xi(x, y, t)$ when the flow is disturbed. Then the equation of continuity and incompressibility gives $\nabla\vec{w} = 0$ and therefore

$$\Delta\phi_2 = 0 \quad \text{for } z > \xi, \quad \Delta\phi_1 = 0 \quad \text{for } z < \xi. \tag{5.39}$$

The boundary conditions are as follows:

(I) the initial disturbance may be supposed to occur in a finite region so that for all time $\nabla\phi \to \vec{v}_*$ as $z \to \pm\infty$;

(II) the fluid particles at the interface must move with the interface without the two fluids occupying the same point at the same time and without a cavity forming between the fluids. Therefore, the vertical velocity at the interface is given by

$$\frac{\partial\phi_k}{\partial z} = \frac{\partial\xi}{\partial t} + \frac{\partial\phi_k}{\partial x}\frac{\partial\xi}{\partial x} + \frac{\partial\phi_k}{\partial y}\frac{\partial\xi}{\partial y} \quad \text{at } z = \xi, \quad \text{for } k = 1, 2$$

(there are conditions of a discontinuity of tangential velocity at the interface);

(III) the normal stress of the fluid is continuous at the interface. This gives the dynamical condition that the pressure is continuous. Therefore,

$$\rho_1\left(c_1 - \frac{1}{2}(\nabla\phi_1)^2 - \frac{\partial\phi_1}{\partial t} - gz\right)$$
$$= \rho_2\left(c_2 - \frac{1}{2}(\nabla\phi_2)^2 - \frac{\partial\phi_2}{\partial t} - gz\right) \quad \text{at } z = \xi,$$

by Bernoulli's theorem for the irrotational flow, which is valid on each side of the surface $z = \xi$. In order that the basic flow satisfies this condition, the constants c_1, c_2 must be related so that $\rho_1(c_1 - (v_*^{(1)})^2/2) = \rho_2(c_2 - (v_*^{(2)})^2/2)$.

Equations (5.39) with boundary conditions (I), (II), (III) pose the full nonlinear problem for perturbations of the basic flow (5.37)–(5.38).

In linear approximation, we write $\phi_2 = v_*^{(2)} x + \delta\phi_2$, $\phi_1 = v_*^{(1)} x + \delta\phi_1$, and neglect products of small perturbations $\delta\phi_1$, $\delta\phi_1$, ξ. For $(\delta\phi_k)_{z=\xi}$ $(k = 1, 2)$, we have

$$(\delta\phi_k)_{z=\xi} = (\delta\phi_k)_{z=0} + \xi \cdot \left(\frac{\partial \delta\phi_k}{\partial z} \right)_{z=0} + \cdots ,$$

and linearization of (5.39) and equations of boundary conditions (I), (II), (III) give

$$\Delta(\delta\phi_2) = 0 \quad \text{for } z > 0, \quad \Delta(\delta\phi_1) = 0 \quad \text{for } z < 0; \tag{5.40}$$

$$\nabla(\delta\phi_k) \to 0 \quad \text{as } z \to \mp\infty \quad \text{for } k = 1, 2 \quad \text{respectively;} \tag{5.41}$$

$$\frac{\partial(\delta\phi_k)}{\partial z} = \frac{\partial \xi}{\partial t} + v_*^{(k)} \frac{\partial \xi}{\partial x} \quad \text{at } z = 0, \quad \text{for } k = 1, 2; \tag{5.42}$$

$$\rho_1 \left(v_*^{(1)} \frac{\partial(\delta\phi_1)}{\partial x} + \frac{\partial(\delta\phi_1)}{\partial t} + g\xi \right)$$
$$= \rho_2 \left(v_*^{(2)} \frac{\partial(\delta\phi_2)}{\partial x} + \frac{\partial(\delta\phi_2)}{\partial t} + g\xi \right) \quad \text{at } z = 0. \tag{5.43}$$

It can be seen that all coefficients of this linear partial differential system are constants and that the boundaries are horizontal. So, we use the method of normal modes, assuming that an arbitrary disturbance may be resolved into independent modes of the form

$$\xi = \xi_\ddagger \exp(ik_x x + ik_y y + \sigma t),$$
$$\delta\phi_k = (\delta\phi_k)_\ddagger \exp(ik_x x + ik_y y + \sigma t), \quad k = 1, 2. \tag{5.44}$$

This reduces the problem to an ordinary differential system with z as the dependent variable, where $(\delta\phi_k)_\ddagger$ are functions of z only and $\xi_\ddagger$ is a constant.

Equation (5.40) gives $(\delta\phi_2)_\ddagger = \varkappa_2^{(1)} \exp(-\tilde{k}z) + \varkappa_2^{(2)} \exp(\tilde{k}z)$, where $\varkappa_2^{(1,2)}$ are arbitrary constants and $\tilde{k} = \sqrt{k_x^2 + k_y^2}$. The boundary condition (5.41) at infinity implies that $\varkappa_2^{(2)} = 0$ and therefore

$(\delta\phi_2)_\ddagger(z) = \varkappa_2^{(1)} \exp(-\widetilde{k}z)$; similarly, we find that $(\delta\phi_1)_\ddagger(z) = \varkappa_1^{(1)} \exp(\widetilde{k}z)$. Equations (5.42) and (5.43) give three homogeneous linear algebraic equations for three unknown constants $\xi_\ddagger$, $\varkappa_1^{(1)}$, $\varkappa_2^{(1)}$:

$$\varkappa_2^{(1)} = -(\sigma + ik_x v_*^{(2)})\frac{\xi_\ddagger}{\widetilde{k}}, \quad \varkappa_1^{(1)} = (\sigma + ik_x v_*^{(1)})\frac{\xi_\ddagger}{\widetilde{k}}, \quad (5.45)$$

$$\rho_1\big(\widetilde{k}g + (\sigma + ik_x v_*^{(1)})^2\big) = \rho_2\big(\widetilde{k}g - (\sigma + ik_x v_*^{(2)})^2\big). \quad (5.46)$$

The solution of the equation (5.46) gives two modes with

$$\sigma = -ik_x\frac{\rho_1 v_*^{(1)} + \rho_2 v_*^{(2)}}{\rho_1 + \rho_2} \pm \left(\frac{k_x^2\rho_1\rho_2(v_*^{(1)} - v_*^{(2)})^2}{(\rho_1 + \rho_2)^2} - \frac{\widetilde{k}g(\rho_1 - \rho_2)}{\rho_1 + \rho_2}\right)^{1/2}.$$

$$(5.47)$$

Both are neutrally stable (i.e., $\Re(\sigma) = 0$) if $\widetilde{k}g(\rho_1^2 - \rho_2^2) \geq k_x^2\rho_1\rho_2$ $(v_*^{(1)} - v_*^{(2)})^2$, the equality gives marginal stability. One mode is asymptotically stable but the other unstable if $\widetilde{k}g(\rho_1^2 - \rho_2^2) <$ $k_x^2\rho_1\rho_2(v_*^{(1)} - v_*^{(2)})^2$. This is a necessary and sufficient condition for instability of the mode with wavenumbers k_x, k_y. Thus, the flow is always unstable (to modes with sufficiently large k_x, that is, to short waves) if $v_*^{(1)} \neq v_*^{(2)}$.

5.3.2. *Rayleigh–Taylor instability*

Let the setting of a problem be the same as that in the case of Kelvin–Helmholtz instability: there is the flow of two incompressible inviscid fluids in horizontal parallel infinite streams of different velocities and densities, one stream above the other. Thus, there is no tangential discontinuity.

When the basic flow is at rest $(v_*^{(1)} = v_*^{(2)} = 0)$, we find $\sigma = \pm(\widetilde{k}g(\rho_2 - \rho_1)/(\rho_1 + \rho_2))^{1/2}$. These eigenvalues can be interpreted differently if the whole fluid system has an upward vertical acceleration $\vec{g}_1$. Then, by the principle of equivalence in dynamics, we obtain $\sigma = \pm(\widetilde{k}g_2(\rho_2 - \rho_1)/(\rho_1 + \rho_2))^{1/2}$, where $g_2 = g_1 + g$ is the apparent gravitational, or net vertical, acceleration of the system. It follows that there is instability if and only if $g_2 < 0$, i.e., the net acceleration is directed from the lighter towards the heavier fluid. This is called the *Rayleigh–Taylor instability* (RTI). In the simplest case, this instability arises in the two-component fluid system at rest in gravitational field when a layer of fluid "2" with the density ρ_2 lies above the layer of fluid "1" with the density $\rho_1 < \rho_2$.

Thus, the Rayleigh–Taylor instability occurs when a heavy fluid rests on top of a light fluid in an effective gravitational field. Such an adverse arrangement would seem unlikely to exist as an initial state if $\vec{g}_2$ represents a "real" gravitational field, but the situation can arise naturally if $\vec{g}_2$ were actually associated with acceleration or deceleration of the entire region. For example, in astrophysics, RTI results from the effective reversal of gravity when a supernova shock wave runs through layers of the star with a radial stratification of heavy elements. Energetically, the heavy fluid prefers to be on the "bottom" (outward from the center), and the system wants to overturn. Since the light and heavy fluids are not interpenetrable freely, the overturning gets accomplished via "fingers" that drip past one another (see Fig. 5.2). The resulting mixing of heavy elements throughout the envelope of the supernova remnant can have observational consequences. Indeed, the simulation of time evolution of supernova SN 1987A was motivated by X-ray and gamma-ray observations of this supernova which indicated that radioactive cobalt is more thoroughly distributed among the explosive debris than was predicted by model calculations of thin-shell nucleosynthesis in the presupernova star.

The Kelvin–Helmholtz and Rayleigh–Taylor instabilities often appear together in the same problem. For example, Kelvin–Helmholtz instability can distort the opposed fingers of heavy and light fluid that try to slip past one another from the nonlinear development of a Rayleigh–Taylor instability. When we generalize the considerations from sudden jumps of fluid properties across sharp interfaces to more gentle gradients in extended regions, Rayleigh–Taylor instabilities fall in the same class of buoyancy-driven disturbances as thermal convection. Kelvin–Helmholtz instabilities then belong to the same category as the shear-flow disturbances.

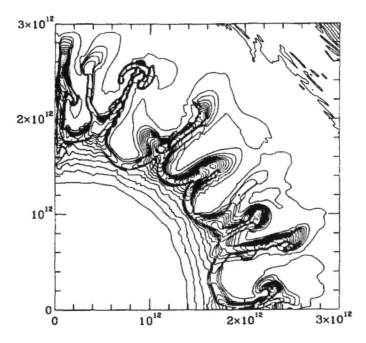

Fig. 5.2. Density contours in the hydrodynamical simulation of the passage of a supernova shock wave running through a model for SN1987A with a pre-explosion radius 3×10^{12} cm. The calculation assumes rotational symmetry about the vertical axis, and the shock front has propagated of the grid at time 9814 s after the initial explosion. The Rayleigh–Taylor instability (followed by Kelvin–Helmholtz instability) has led to the penetration of heavy elements into the helium mantle of the star [42].

Let us consider the Richardson criterion for the stability of an inviscid region containing a favorable buoyancy gradient but an adverse velocity gradient. Consider a gas that has a stable entropy gradient, $ds/dz < 0$, in a gravitational field g that points in the $-z$-direction. Suppose further that the gas flows in the x-direction with variable velocity $v(z)$, such that the shear dv/dz would give rise to Kelvin–Helmholtz instability, were it not for the stabilizing effect of the entropy gradient ds/dz. The *Richardson criterion* results from asking: how large can dv/dz get before ds/dz no longer provides a sufficient stabilizing influence?

The further derivation of the Richardson criterion proceeds via energy considerations. Consider first the buoyant work performed on a small blob that slowly moves a distance dz. We suppose that the size of the blob times $|dv/dz|$ is much less than the adiabatic sound speed, so that the blob maintains pressure equilibrium with the ambient medium when displaced. The change in the internal density of the blob is related to the change of its (and external medium's) pressure via $d\rho_{\text{blob}} = (\partial \rho/\partial p)_s dp$ and the corresponding change in the density of the ambient medium equals $d\rho_{\text{medium}} = (\partial \rho/\partial p)_s dp + (\partial \rho/\partial s)ds$, where dp and ds are the differences of the pressure and specific entropy of the medium at new position compared to the old that arise because of the constraint of mechanical and thermal balance. From these equations, we obtain that the buoyant force per unit volume at the end of the displacement has the expression

$$g((d\rho)_{\text{blob}} - (d\rho)_{\text{medium}}) = -g \left(\frac{\partial \rho}{\partial s} \right)_p ds,$$

where

$$\left(\frac{\partial \rho}{\partial s} \right)_p = \left(\frac{\partial \rho}{\partial T} \right)_p \cdot \left(\frac{\partial s}{\partial T} \right)_p^{-1} = \frac{T}{c_p} \left(\frac{\partial \rho}{\partial T} \right)_p < 0.$$

The average force per unit volume felt by the blob during the displacement equals half of the above value. Thus, the work required per unit mass to interchange two similar blobs, initially at z and $z + dz$, equals

$$(dA)_{\text{buoyancy}} = -\frac{g}{\rho} \left(\frac{\partial \rho}{\partial s} \right)_p ds dz. \tag{5.48}$$

With $(d\rho/ds)_p < 0$, $(dA)_{\text{buoyancy}}$ is positive for convectively stable regions, where $ds dz > 0$. In other words, it takes work to overturn gas in a region stable to convection.

Consider next the release of shear energy in the interchange of the same two blobs. Let the two blobs initially have speeds v and $v + dv$, and suppose that their mixing yields an average speed $v + dv/2$.

We then easily compute the difference in kinetic energies per unit mass between the final and initial states as

$$(dE_{(kin)})^{mix} = \frac{1}{2}(v^2 + (v + dv)^2) - \frac{1}{2}\left(2\left(v^2 + \frac{dv}{2}\right)^2\right) = \frac{1}{4}(dv)^2.$$

(5.49)

Consequently, $(dE_{(kin)})^{mix}$ is a positive-definite expression independent of the sign of dv. Thus, any departure from the uniform flow contains extra kinetic energy in the form of shear, which can be tapped, in principle, to drive dynamic mixing (Kelvin–Helmholtz instability). However, the work (5.48) needed to overcome the buoyancy excess will outweigh the extra kinetic (5.49) if $(dA)_{buoyancy} > (dE_{(kin)})^{mix}$. Thus, we have stability if

$$-\frac{g}{\rho}\left(\frac{\partial \rho}{\partial s}\right)_p ds dz > \frac{1}{4}(dv)^2.$$

Dividing by $(dv)^2$, we obtain the *Richardson criterion*

$$Ri \equiv -\frac{g}{\rho}\left(\frac{\partial \rho}{\partial s}\right)_p \cdot \left(\frac{ds}{dz}\right) \cdot \left(\frac{dv}{dz}\right)^{-2} > \frac{1}{4}$$

for stability.

5.3.3. *Rotational (Taylor–Couette) instability*

Let us consider the instability of the Couette flow between concentric rotating cylinders; the source of the instability is the adverse gradient of angular momentum.

This problem was first investigated by Rayleigh in 1888 (for inviscid liquid). Let $v_\theta(r)$ be the velocity at any radial distance; for inviscid flows, $v_\theta(r)$ can be any function, but only certain distributions can be stable. Imagine that two fluid rings of equal masses at radial distances $r = a_1$ and $r = a_2 > a_1$ are interchanged. As the motion is inviscid, W. Thomson's theorem requires that the circulation $\Gamma = 2\pi r v_\theta$ should remain constant during the interchange. That is, after the interchange, the fluid at $r = a_2$ will have the circulation (Γ_1) that it had at $r = a_1$ before the interchange.

Similarly, the fluid at $r = a_1$ will have the circulation (Γ_2) that it had at $r = a_2$ before the interchange. The conservation of circulation requires that the kinetic energy $E_k = v_\theta^2/2$ must change during the interchange: $(E_k)^{\text{initial}} = (1/8\pi^2)(\Gamma_1^2/a_1^2 + \Gamma_2^2/a_2^2)$, $(E_k)^{\text{final}} = (1/8\pi^2)(\Gamma_2^2/a_1^2 + \Gamma_1^2/a_2^2)$, so that the kinetic energy change per unit mass is $\Delta E_k = (E_k)^{\text{final}} - (E_k)^{\text{initial}} = (1/8\pi^2)(\Gamma_2^2 - \Gamma_1^2)(a_1^{-2} - a_2^{-2})$.

Because $a_2 > a_1$, a velocity distribution for which $\Gamma_2^2 > \Gamma_1^2$ would make $\Delta E_k > 0$, which implies that an external source of energy would be necessary to perform the interchange of the fluid rings. If Γ^2 decreases with r, then an interchange of rings will result in a release of energy; such a flow is unstable.

The (Rayleigh) criterion of instability of an inviscid Couette flow can therefore be formulated as follows: inviscid Couette flow is stable (to all axisymmetric perturbations) if $d\Gamma^2/dr > 0$ everywhere in the flow. This criterion is modified by Taylor's solution of the viscous problem (Fig. 5.3).

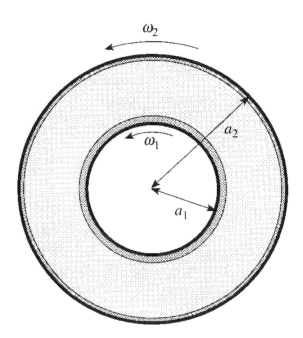

Fig. 5.3. Couette flow between rotating coaxial cylinders.

Let us define the *Taylor number* as follows:

$$Ta \equiv \frac{4\omega_1(a_2 - a_1)^4}{\nu^2}\left(\frac{\omega_1 a_1^2 - \omega_2 a_2^2}{a_2^2 - a_1^2}\right),$$

where ω_1, ω_2 are the angular speeds of the inner ($r = a_1$) and outer ($r = a_2$) cylinders, respectively, and ν is a viscosity coefficient of the liquid. Taylor's viscous solution indicates that the flow remains stable until a critical Taylor number is attained:

$$Ta_{\mathrm{cr}} = \frac{3416}{1 + \omega_2/\omega_1}.$$

The instability appears in the form of counter-rotating toroidal vortices (*Taylor vortices*). The streamlines are in the form of helixes, which axes wrapping around the annulus, somewhat like the stripes on a barber's pole. These vortices themselves become unstable at $Ta > Ta_{\mathrm{cr}}$ (Fig. 5.4).

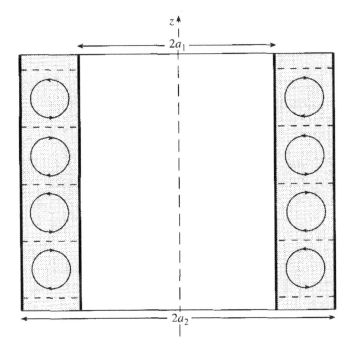

Fig. 5.4. The secondary circulation associated with the Couette flow between rotating coaxial cylinders when the Taylor number exceeds its critical value [15].

Phenomena analogous to the Taylor vortices are called *secondary flows* because they are superposed on a primary flow (such as the Couette flow in the present case).

5.3.4.   Rayleigh–Zel'dovich problem and large structures

Recently, special attention has been paid to the numerical modeling of the process of stabilization of the flow in different conditions both on the earth and in space [43, 44]. For example, investigations of the rotational motion are of principal fundamental value for the physics of accretion discs [45]. The problem of accretion of matter on compact sources is determined by the rate of loss of the angular moment in the accretion discs. Usually, it is assumed that the mechanism of the loss of the angular moment is associated with turbulent viscosity which exceeds molecular viscosity by many orders of magnitude. This results in the following problem: what is turbulent viscosity? and what is its physical meaning for accretion discs? We believe that it is necessary to separate the two physical processes. The first process is associated with the transfer (loss) of the angular moment. It is determined by large-scale turbulence structures. The second process is responsible for the transfer of the kinetic energy of turbulence into heat (dissipation). The problem of the relationship of these processes depends on many factors: the possible stationary nature of the process and the associated spectrum of turbulence, the geometrical special features of the flow, the possible effect of the magnetic field and other factors controlling the turbulence spectrum. Therefore, the problem of turbulence (including turbulent convection and turbulent mixing) becomes controlling for understanding the processes in astrophysical objects. The process of stabilization of the flow is of the non-stationary nature and is extremely complicated to investigate. However, advances have been made in understanding this process in recent years because of the application of the methods of mathematical modeling.

It is well known [44] that turbulence may be investigated by mathematical modeling or direct calculations of non-stationary movement of ordered and large-scale structures (solving the Euler equation) or by the application of specific type of hypotheses on the stochastic nature of the field of velocities and other physical quantities. In the second case, generally speaking, the physical reasons for the development of turbulence (supplying energy to chaotic motion) are not considered. We shall investigate the problem of formation of large structures. In this sense, the most representative picture is provided by the structure of the long-range wake during the flow around cylinder consisting of ordered vortices, mainly large-scale ones. However, small-scale vortices are also presented at the same time. The main problem of the theory of turbulence is to explain the development of such a spectrum of scales in the turbulent flow. Of course, it should be mentioned that the majority of investigations have been concerned with specific tasks in a narrow range of the variation of the physical parameters so that the problem could not be examined in sufficient detail.

However, more and more studies have appeared recently (for example, [46]) casting doubts on the problem of isotropic free turbulence. It is claimed (for example, [47]) that turbulence is of the alternating nature, i.e., the regions of turbulence are replaced by regions of laminar flow. Therefore, the author of [46] has suggested that it is more accurate to talk about the interaction in the flow (linear or nonlinear) and not about structures (laminar or turbulent). In this sense, the meaning of the Reynolds number, introduced for the characteristic of the problem, is difficult to understand because the Reynolds number starts to depend on the local structure of the flow. Therefore, instead of turbulent viscosity, it is necessary to introduce the concept of local dynamic viscosity and this has also been proposed in [46].

The development of turbulence is intimately allied with the evolution of instabilities in the fluid. As a simple example, we consider the problem of determination of the profile of velocity in the gap between the cylinders, starting with a highly non-equilibrium shear flow. The convective cells in the flow between cylinders appear at a Reynolds number of $Re \sim 60$ in the condition of incidence of the angular moment of the amount of motion outwards. At the same time, as for the plane-parallel flow, turbulence develops at $Re \sim 2000$. It was shown [48] that the increase of the angular moment outwards results in the stabilization of the laminar flow ($Re \sim 200000$ from the extrapolation of Taylor's results (1936)). Our investigations were directed to the examination of the nature of formation of the profile of velocity in the gap if in the initial stage the distribution of the shear flow was non-equilibrium. The aim of investigations was to examine the evolution of the structure of the flow (including vortices) on the model of the flow in the gap. Our concept was that at high Reynolds numbers, the inertia forces prevail over the viscous stresses and turbulization of the flow starts with the development of large scales — the formation of large vortices as a result of the shear flow (Kelvin–Helmholtz instability). In other words, the energy of chaotic motion in the turbulent flow is the energy of the shear flow itself.

We have considered the simplest case from the viewpoint of numerical modeling, i.e., the flow between two coaxial cylinders. As we already noted earlier, in the studies by Rayleigh, it was shown that the centrifugal force may stabilize the Taylor–Couette flow if the condition $d\Gamma^2/dr > 0$ is satisfied (see [49] for details).

The solution of the problem of the onset of turbulence is of great importance in astrophysical problems. For example, it is usually assumed that the accretion discs are characterized by the presence of conventional turbulent viscosity because the Reynolds numbers are very high. Nevertheless, in a number of investigations, these results are doubted and it is claimed that the flow of Kepler rotation may prove to be laminar. The main argument of these claims is justified by the stability of the differential rotation of the medium with the moment increasing with the radius. Zel'dovich carried out a number of analytical investigations of the flow between two rotating cylinders with different laws of the variation of moment in relation to radius. In particular, in [48], the dimensionless Taylor number was redefined:

$$Ta = \frac{d}{dr}(\omega r^2)^2 \cdot r^{-5} \left(\frac{d\omega}{dr}\right)^{-2},$$

which can be used to obtain the condition of stabilization of flow homogeneous with respect to density: $Ta > 0$. For example, if we consider the exponential dependence of the law of rotation $\omega \sim r^n$ which corresponds to the constant value of the Taylor number in the entire range $Ta = 8(n+2)/n^2$, the stabilization starts at $n > -2$; in solid-state rotation $Ta = \infty$ and the motion is stable. The criterion, introduced by Zel'dovich, is more convenient for astrophysical conditions in the evaluation of transition to the turbulent flow. These investigations are important because the movement of a homogeneous fluid with a rotating external cylinder is identical with the gravitationally stabilized shear flow.

The problem of evolution of the flow in the gap between cylinders is associated with the effect of different forces: the first is associated with inertia force $v\,dv/dr$, and the second force is determined by viscosity. In particular, the viscosity establishes the linear profile of the Couette plane-parallel flow between two moving sheets, if the Reynolds number is relatively small, although the problem of the value of the Reynolds number and the ratio of the forces require further examination. The point is that in the presence of a steep gradient of velocity and sufficiently high kinetic energy of the flow, the instabilities of the Kelvin–Helmholtz type can develop at the boundary of two flows with different velocities; in this case, the main role is played by the dynamic term in comparison with the viscosity term. In the development of the Kelvin–Helmholtz instability, viscosity is not included in the evaluation of the rate of growth. As shown in [50], the energy of the turbulent flows is associated with large scales (i.e., inertia forces) and these models can be investigated using Euler equations.

In this formulation of the problem, the inertia forces produce, as a result of the pair of forces, vortices (cyclones and anticyclones) which may result in the establishment of the profile of the flow in the gap. An important moment in this approach is the problem of breakup of the vortices. It is important to determine the forces resulting in the disintegration (viscous or dynamic) and the characteristic duration of these processes.

In the numerical experiments [51] based on the concept proposed previously, it was demonstrated that the kinetic energy of turbulent motion is associated with large-scale vortex structures. We believe that the high-frequency portion of the spectrum is generated in the course of nonlinear interactions of large-scale structures with one another and with the cylinder walls. The effect of nonlinear interactions is not restricted to the drastic increase in effective viscosity over molecular viscosity. Unlike molecular viscosity (which is scale-independent), the effective (dynamic) viscosity is sensitive to the size of flow structures (vortices). Larger structure is associated with higher viscosity. The dynamic viscosity characterizes the intensity of nonlinear interactions. The computations showed that nonlinear interactions taking place in different regions of the computational domain are characterized by different intensities. This observation suggests that dynamic viscosity has a local nature.

As noted by Zel'dovich [48] and confirmed by numerical modeling [51], the structure of the flow between cylinders is represented by alternating turbulence, i.e., the regions of almost laminar flow are replaced by regions of turbulence (large vortices).

5.3.5. *Other types of hydrodynamical instability*

There exist a very large number of various types of hydrodynamical instability (see, e.g., [14, 54] for details). As examples, we can mention the following types:

(1) *Saffman–Taylor instability, or Rayleigh–Taylor instability in a porous medium.* Let's consider the instability of a basic flow in which two incompressible viscous fluids move with a horizontal interface and uniform vertical velocity in a uniform porous medium. In this medium, the motion of a fluid is governed by *Darcy's law*, namely that $\vec{v} = \nabla\phi$, $\phi = -k(p + \rho g z)/\mu$, where ρ is the density of the fluid, μ is its dynamic viscosity, and k is the permeability of the medium to the fluid. Let the lower fluid have density ρ_1 and viscosity μ_1, and the upper fluid ρ_2 and μ_2; let the medium have permeability k_1 to the lower fluid and k_2 to the upper; and let the basic velocity of fluids be $u \cdot \vec{e}_z$. Then the flow is stable if and only if $(\mu_1/k_1 - \mu_2/k_2)u \geq g(\rho_2 - \rho_1)$.

(2) *An instability of an elliptic flow.* Let us consider the steady basic flow of uniform incompressible viscous fluid with

$$\vec{v}_* = \vec{u} = \frac{2\Omega}{a^2 + b^2}\{-a^2 y, \, b^2 x, \, 0\}, \quad p_* = \frac{2a^2 b^2 \rho \Omega^2 (x^2 + y^2)}{(a^2 + b^2)^2},$$

for positive constants a, b, ρ, Ω. Let's consider velocity perturbations of the form

$$\vec{u}^\dagger = \{aqz/c - ary/b, \, brx/a - bpz/c, \, cpy/b - cqx/a\}$$

for some functions p, q, r of t alone. We take the normal modes in the form of p, $q \propto \exp(\sigma t)$. Then the eigenvalue σ has the following form:

$$\sigma^2 = \frac{4\Omega^2 a^2 b^2 (c^2 - b^2)(a^2 - c^2)}{(b^2 + c^2)(a^2 + c^2)(a^2 + b^2)}.$$

Finally, we obtain that the flow is unstable if $a < c < b$ or $b < c < a$ but stable (to these perturbations) otherwise.

(3) *Jeans instability of self-gravitating gas.* The dynamics of self-gravitating gas are modeled by Poisson's equation $\Delta\phi = 4\pi G\rho$, as well as Euler's equations:

$$\frac{\partial \rho}{\partial t} + \nabla(\rho \vec{u}) = 0, \quad \rho\left(\frac{\partial \vec{u}}{\partial t} + \vec{u} \cdot \nabla \vec{u}\right) + \nabla p + \rho\nabla\phi = 0,$$

and an equation of state, $p = p(\rho)$, for a barotropic inviscid fluid, where ϕ is the gravitational potential and G is the gravitational constant.

Let us consider small perturbations $\phi^\dagger, \rho^\dagger, \vec{u}^\dagger$ of a basic state of rest with $\phi = \Phi(\vec{r})$, $\rho = R(\vec{r})$, $\vec{u} = 0$, and linearize the equations of motion by neglecting products of the perturbed quantities:

$$\Delta\phi^\dagger = 4\pi G\rho^\dagger, \quad R\frac{\partial \vec{u}^\dagger}{\partial t} = -c^2 \nabla\rho^\dagger - \rho^\dagger \nabla\Phi - R\nabla\phi^\dagger, \quad \frac{\partial \rho^\dagger}{\partial t} = -\nabla(R\vec{u}^\dagger),$$

where $c = \sqrt{dp(R)/dR}$ is the basic velocity of sound. We assume that the basic state is uniform with $R = \text{const}$, $\Phi = \text{const}$, and take normal modes with $\vec{u}^\dagger$, $\rho^\dagger$, $\phi^\dagger \propto \exp(i\vec{k} \cdot \vec{r} + \sigma t)$. Finally, we obtain that $\sigma^2 = 4\pi GR - c^2 \vec{k}^2$ (uniform gas is gravitationally unstable to long waves).

(4) *Richtmyer–Meshkov instability.* The Richtmyer–Meshkov instability is induced by the interaction of a traveling shock wave and a perturbed interface between two media of different densities. See the Appendices for details.

(5) *Chandrasekhar–Lebovitz instability.* This convective instability arises in gaseous homogeneous sphere for non-radial oscillations with frequencies of low orders and is described in [52].

5.4. Summary

In this Chapter, the basic properties of hydrodynamical instabilities were examined. The mathematical formalism, allowing to investigate instabilities development in fluid, was demonstrated. In Section 5.1, the phenomenon of thermal convection was considered; phenomenological reasons of Schwarzschild's criterion for convection were obtained. The system of Boussinesq equations also was considered. In Section 5.2, the formalism of analysis of occurrence and development of instability in hydrodynamical flow was considered. With the help of the mentioned mathematical apparatus, the investigation of Benard's convection and convection phenomena for a layer of fluid with free surface was carried out. Section 5.3 was devoted to the consideration of the wide spread in hydrodynamics and astrophysics instability types, in particular, instabilities of Kelvin–Helmholtz, Rayleigh–Taylor, Taylor–Couette, Jeans, etc. Criteria of instability occurrence, connected with parameters Richardson Ri and Taylor and Taylor–Zel'dovich Ta were introduced.

During the work on the material of this chapter, authors used the following the books and papers: [13, 16, 40, 43, 49, 52, 54], etc. We especially recommend for readers the Refs. [15, 65, 70] for a detailed study of problems mentioned in this chapter.

Chapter 6

Classical Turbulence in Fluid Dynamics

6.1. Introduction to Classical Turbulence

De facto, all flows encountered in engineering practice and in nature are turbulent. The boundary layer on an aircraft wing is turbulent, the atmosphere boundary layer over Earth's surface is turbulent, and major oceanic currents are turbulent. In this section, we consider certain elementary ideas about the dynamics of turbulent flows. We will see that such flows do not allow a strict analytical study, and one depends heavily on physical intuition and dimensional arguments.

In spite of our everyday experience with it, turbulence is not easy to define precisely. Some characteristics of turbulent flows are as follows:

- *Randomless*: Turbulent flows seem irregular, chaotic and unpredictable.
- *Nonlinearity*: Turbulent flows are nonlinear. It causes the relevant nonlinearity parameter (*Re*, *Ra*, etc.), to exceed a critical value. In unstable flows, small perturbations grow spontaneously and frequently equilibrate as finite amplitude disturbances. On further exceeding the stability criteria, the new state can become unstable to more complicated disturbances, and the flow eventually reaches a chaotic state.
- *Diffusivity*: Due to the macroscopic mixing of fluid particles, turbulent flows are characterized by a rapid rate of diffusion of momentum and heat.

- *Vorticity*: Turbulence is characterized by high levels of fluctuating vorticity. The identifiable structures in a turbulent flow are called *eddies*. A characteristic feature of turbulence is the existence of an enormous range of eddy sizes. The large eddies have a size of order of the width of the region of the turbulent flow; in a boundary layer, this is the thickness of the layer. The large eddies contain most of the energy. The energy is handed down from large to small eddies by nonlinear interactions, until it is dissipated by viscous diffusion in the smallest eddies.
- *Dissipation*: The vortex stretching mechanism transfers energy and vorticity to increasingly smaller scales, until the gradients become so large that they are dissipated by viscosity. Turbulent flows therefore require a continuous supply of energy to make up for the viscous losses.

The onset and evolution of turbulence in hydrodynamic flows can be described from various positions with the use of widely different mathematical tools. Since this problem is rather complicated, no complete theory for its description has been created until now, although first attempts were made as early as the late 19th century. Nevertheless, we can distinguish two basic approaches, which are, in a sense, opposite. They are based on the following antagonistic assumptions: (1) in turbulent flows, we deal with manifestations of stochastic processes (statistical turbulence); and (2) in turbulent flows, there are quasi-deterministic structures on macro- and mesoscales; their dissipation and decay give rise to a collection of unsteady small-scale turbulence processes (methods of statistical turbulence can be locally applied on small scales, although, in principle, there is an alternative, for example, based on the direct solution of the Boltzmann or Kac kinetic equations). The first approach goes back to Keller' and Friedmann's works (1925), who gave the general definition of Euler spatiotemporal correlation functions of thermal and hydrodynamic fields in turbulent flows and proposed a method for deriving dynamical equations for correlation functions from the Navier–Stokes equation. These equations are used to average physical flow characteristics, for example, the velocity and the size of the vortices. A mathematical description makes use of the assumption that perturbations on different scales are random and relies on probability theory (generalized to function spaces). This assumption underlies the derivation of simplified equations and is valid when the increments of flow instability development are estimated on different scales. However, observing the velocity variations at a point in turbulent flow, we can conclude that the velocity generally remains within its values in large vortices, since they are the basic elements in the turbulent flow structure. Fast fluctuations occur when small vortices go over large structures. It is this observation that underlies the second approach. Its basic ideas were formulated by O.M. Belotserkovskii (on the basis of numerical experiments). According to the results of computer simulation, we can believe that large vortices carry the basic energy of turbulent flows and determine the flow structure. The distribution of large vortices does not reflect the random nature of perturbations, but corresponds to the physical laws of fluid dynamics, when the inertial terms in the Navier–Stokes equation prevail over the stresses caused by viscosity (and the equation is reduced to the Euler one). Then the pair of forces arising from the pressure field and dynamical forces related to the velocity field gives rise to a vortex structure. The origin of large (non-classical "coherent") vortices and the flow structure have to be described using the theory of the Euler equation.

6.1.1. *The mechanism of origin of turbulence*

Let us consider the possible mechanism of the occurrence of small-scale turbulence in a liquid. The motion of viscous incompressible liquid is described by NSE (1.34) taking into account the condition $\nabla \vec{v}$. We investigate stability of stationary motion of a liquid in linear approximation, that is, we introduce the velocity field $\vec{v}$ as $\vec{v} = \vec{v}_* + \vec{v}^\dagger$, where $\vec{v}_*$ is the velocity of basic flow and $\vec{v}^\dagger$ is the velocity perturbation. We obtain for the considered system in the dimensionless form (see (5.10)):

$$\frac{\partial \vec{v}^\dagger}{\partial t} + (\vec{v}^\dagger \cdot \nabla)\vec{v}_* + (\vec{v}_* \cdot \nabla)\vec{v}^\dagger = -\nabla p^\dagger + Re^{-1}\Delta\vec{v}^\dagger,$$

$$\Delta\vec{v}^\dagger = 0, \tag{6.1}$$

where Re is a Reynolds number and we neglect the small term $(\vec{v}^{\,\dagger} \cdot \nabla)\vec{v}^{\,\dagger}$. The general solution of this equation is found as the sum terms of the form $\vec{v}^{\,\dagger}(\vec{r}, t) \propto \vec{v}^{\,\dagger}_{(1)}(\vec{r}) \times \exp(-i\omega t)$. If for all ω we have $\Im(\omega) < 0$, then perturbation $\vec{v}^{\,\dagger} \to 0$ for $t \to \infty$ (in mean), i.e., the flow is stable. As the flow for $Re \lesssim 1$ is stable, and for $Re \gg 1$ is unstable, then there exists Re_{cr} (critical Reynolds number) for which the stability is broken. Instability means increase of perturbation for $t \to \infty$; consequently, for $Re > Re_{\mathrm{cr}}$, there exists the frequency $\widetilde{\omega}$ such that $\widetilde{\omega} = \omega_1^{(1)} + i \cdot \omega_1^{(2)}$ and $\omega_1^{(1)} \in \mathbb{R}^1$, $\mathbb{R}^1 \ni \omega_1^{(2)} > 0$. We may assume that at enough small $\check{\delta} \equiv Re - Re_{\mathrm{cr}}$ inequality, $\omega_1^{(1)} \gg \omega_1^{(2)}$ is fulfilled, as at $Re = Re_{\mathrm{cr}}$ the occurrence of one frequency $\widetilde{\omega}$ with $\omega_1^{(2)} > 0$ precedes the value $\widetilde{\omega}$ with $\omega_1^{(2)} = 0$ and for other frequencies $\omega_1^{(2)} < 0$. Then the corresponding $\widetilde{\omega}$ value $\vec{v}^{\,\dagger}$ has the form of

$$\vec{v}^{\,\dagger} \propto \exp\left(\omega_1^{(2)}t - i\omega_1^{(1)}t\right) \cdot \vec{v}^{\,\dagger}_{(1)}(\vec{r}). \tag{6.2}$$

As a factor $\exp(\omega_1^{(2)}t)$ grows by time t, that assumption "$|\vec{v}^{\,\dagger}|$ is small" is broken in a short time, but for $\tau_1 < 1/\omega_1^{(2)}$ the motion of fluid (and the change of $\vec{v}^{\,\dagger}$), we can count (quasi-)periodic with period $2\pi/\omega_1^{(1)} \ll \tau_1$.

The amplitude of this motion can be written as $\mathrm{A}(t) = \Re\left(A_0 \exp(\omega_1^{(2)}t - i\omega_1^{(1)}t)\right)$. The averaging of value $d|A|^2/dt$ over time interval Δt $(2\pi/\omega_1^{(1)} \ll \Delta t \ll \tau_1)$ gives

$$\left\langle \frac{d|A|^2}{dt} \right\rangle = 2\omega_1^{(2)}|A|^2 - c_0|A|^4 \approx \frac{d|A|^2}{dt},$$

and at $c_0 > 0$ from this formula we obtain that with increasing time t, $|A|^2 \to |A|^2_{\mathrm{max}}$, where $|A|^2_{\mathrm{max}} = 2\omega_1^{(2)}/c_0$. For small enough value $\check{\delta}$, the dependence $\omega_1^{(2)}(\check{\delta}) \propto \check{\delta}$, and for $|A|_{\mathrm{max}}$, we obtain $|A|_{\mathrm{max}} \propto \sqrt{\check{\delta}}$.

For $c_0 < 0$ on the interval $[Re_0; Re_{\mathrm{cr}}]$, the function $|A|^2_{\mathrm{max}}(Re)$ is a two-valued parabola (abscissa of the vertex of parabola is Re_0, and Re_{cr} is the point of intersection of parabola with Re-axis); for $c_0 > 0$, the abscissa of vertex is $Re_0 = Re_{\mathrm{cr}}$. The analysis of behavior of value $d|A|^2/dt$ at $c_0 < 0$ shows that for $Re = Re_{\mathrm{cr}}$, the perturbation increases stepwise; on the interval $Re_0 < Re < Re_{\mathrm{cr}}$, the motion of fluid is stable for infinitesimal perturbations, but unstable for perturbations of finite amplitude.

Thus, for the perturbed flow, stable quasi-periodic (for $t \ll \tau_1$) motion of fluid arises with frequency $\omega_1^{(1)}$. At increase of Re, this motion remains stable only up to some value $Re = Re_{cr,2}$. From equations (6.1) with replacement $\vec{v}_* \to \vec{v}^\dagger$ and $\vec{v}^\dagger \to \vec{v}^{\dagger\dagger}$, an occurrence of one more period $(2\pi/\omega_2^{(1)})$ follows. After the bifurcation in the point $Re = Re_{cr,2}$, there exist two variants of motion of fluid: (1) if $\omega_1^{(1)}/\omega_2^{(1)}$ is an irrational number, then motion is quasi-periodic (trajectory is wound on the two-dimensional torus); (2) if $\omega_1^{(1)}/\omega_2^{(1)} \in \mathbb{Z}$, that motion is periodical but resonant.

As a result of the subsequent bifurcations, a chaotic state may arise in the flow of fluid. As one of the possible scenarios of transition of laminar flow in the turbulent one, we may consider the process of doubling of the period for *Poincaré mapping*.

As one-dimensional Poincaré mapping describing the above-mentioned process at increasing of Reynolds number, we consider the following mapping:

$$x_{k+1} = -\left(1 + (Re - Re_1)\right) \cdot x_k + x_k^2 + \mu x_k^3. \tag{6.3}$$

This mapping has a fixed point $\tilde{x} = 0$. If $Re - Re_1 < 0$ then $|dx_{k+1}/dx_k|_{\tilde{x}} < 1$, and the point $\tilde{x}$ is stable; if $Re - Re_1 > 0$, then $|dx_{k+1}/dx_k|_{\tilde{x}} > 1$, and the point $\tilde{x}$ is unstable. Consequently, at $Re = Re_1$, we get bifurcation of mapping. If we assume that x_k and $Re - Re_1$ are small enough, then the second iteration can be written in the form of

$$x_{k+2} = x_k + 2(Re - Re_1)x_k - 2(1 + \mu)x_k^3. \tag{6.4}$$

The point $\tilde{x}$ at $Re - Re_1 < 0$ is fixed and a stable one for mapping (6.4), but for $Re - Re_1 > 0$, this point is unstable. At the same time, two fixed points arise: $\tilde{x}_{1,2} = \pm\sqrt{(Re - Re_1)/(1 + \mu)}$. These points correspond to the motion with doubled period (2-cycle). The mapping (6.3) moves $\tilde{x}_1 \to \tilde{x}_2$, $\tilde{x}_2 \to \tilde{x}_1$, but the mapping (6.4) does not move these points. The recurrence of bifurcations leads to the origin of a *strange attractor* (an attracting set of unstable paths) [53] in the limit of increase Re: $Re_1, Re_2, \ldots \to Re_{cr}$. Thus, we obtain the

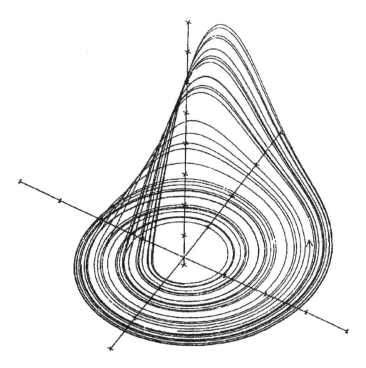

Fig. 6.1. Attractive set (attractor) for the model system of equations $dx/dt = -y(t) - z(t)$, $dy/dt = x(t) + c_1 y(t)$, $dz/dt = c_2 + z(t) \cdot (x(t) - c_3)$. The trajectory of system at first corresponds to skintight spiral orbits in a plane x, y, and then corresponds to considerably more rarefied orbits along an axis z, forming "jets" with the subsequent returning in an initial plane [55].

state of chaos in the system, i.e., a laminar flow becomes a turbulent one (Fig. 6.1).

The scenario of the turbulence origin described above is not unique. There are other scenarios analyzed; one of them is the scenario of the origin of turbulence via *intermittence* (see below).

6.1.2. *The homogeneous isotropic turbulence*

The turbulence is said to be *homogeneous* if all points in space are statistically equivalent; it is said to be *isotropic* if all directions are statistically equivalent. Homogeneity indicates that mean properties of fluid do not vary with absolute position in a particular direction.

The most important implication of this assumption is that the *velocity correlation tensor* $\varrho_{ij}(\vec{r}, \vec{r}^{\dagger}) = \varrho_{ij}(\vec{r} - \vec{r}^{\dagger}) \equiv \varrho_{ij}(\Delta\vec{r})$ (with analogous results for $\varrho_{ijk}, \ldots$), where

$$\varrho_{ij}(\vec{r}, \vec{r}^{\dagger}; t, t^{\dagger}) \equiv \langle \xi_i(\vec{r}, t)\xi_j(\vec{r}^{\dagger}, t^{\dagger}) \rangle,$$

$$\varrho_{ijk}(\vec{r}, \vec{r}^{\dagger}, \vec{r}^{\dagger\dagger}; t, t^{\dagger}, t^{\dagger\dagger}) \equiv \langle \xi_i(\vec{r}, t)\xi_j(\vec{r}^{\dagger}, t^{\dagger})\xi_k(\vec{r}^{\dagger\dagger}, t^{\dagger\dagger}) \rangle, \ldots,$$

$$i, j, k, \ldots = 1, 2, 3,$$

$$\vec{\xi}(\vec{r}, t) = \vec{v}(\vec{r}, t) - \vec{u}(\vec{r}, t), \quad \langle v_i(\vec{r}, t) \rangle = u_i(\vec{r}, t),$$

$$\langle \xi_i(\vec{r}, t) \rangle = 0$$

($\vec{v}$ is the velocity in the NSE (1.35)). The isotropy implies independence of direction as well as independence of position in the fluid; this means that all velocity moments are invariant under rotations of the coordinate frame and under reflections in coordinate planes; the principal implication is the additional symmetry requirement: $\varrho_{ij}(\Delta\vec{r}) = \varrho_{ji}(\Delta\vec{r})$.

Let us consider the infinitesimal material line element $\Delta\vec{r}(t)$ between two fluid particles P_1 and P_2, separated by the non-random distance $\Delta\vec{r}(0)$ at $t = 0$. Since the infinitesimal separation vector $\Delta\vec{r}(t)$ must be linearly related to $\Delta\vec{r}(0)$, it follows that there exists a matrix $\widehat{\alpha}_{ij}$ (which is a random function of time and the initial spatial position of particle P_1 but not of the direction or magnitude of $\Delta\vec{r}(0)$) such that $\Delta r_j(t) = \widehat{\alpha}_{ij}\Delta r_j(0)$. Therefore, $(\Delta r(t))^2 = \widehat{\beta}_{jk}\Delta r_j(0)\Delta r_k(0)$, $\widehat{\beta}_{jk} = \widehat{\alpha}_{ij}\widehat{\alpha}_{ik}$, where $\widehat{\beta}_{jk}$ is a real symmetric matrix. Since $\Delta\vec{r}(0)$ is not random, it follows that

$$\langle (\Delta\vec{r}(t))^2 \rangle = \langle \widehat{\beta}_{jk} \rangle \Delta r_j(0)\Delta r_k(0), \quad \langle h(t) \rangle \equiv \lim_{T \to \infty} \frac{1}{T} \int_{t_0}^{t_0+T} h(t)dt.$$

Isotropy and homogeneity imply that $\langle \widehat{\beta}_{jk} \rangle = \gamma(t)\delta_{jk}$ since no point $\vec{r}$ or direction can be preferred. Hence, $\langle (\Delta\vec{r}(t))^2 \rangle = \gamma(t)(\Delta\vec{r}(0))^2$. Now for fixed t, denote the (real) eigenvalues of $\widehat{\beta}$ by β_i and corresponding eigenvectors $\vec{b}_i$, $i = 1, 2, 3$. Since $\vec{b}_i$ form an orthogonal triad and the volume of the infinitesimal element about particle P_1 spanned by $(\delta \cdot \vec{b}_i)$ $(\delta \ll 1)$ is conserved in evolution from 0 to t (by incompressibility of fluid), it follows that $\beta_i > 0$ $(i = 1, 2, 3)$ and $\beta_1\beta_2\beta_3 = 1$ (after time evolution from 0 to t the vectors about P_1 resulting from $\delta \vec{b}_i$ at time $t = 0$ have lengths $\delta \beta_i$, and are still orthogonal). By the equality

$$\gamma = \frac{1}{3} \langle Sp \, \widehat{\beta} \rangle = \frac{1}{3} \left\langle \sum_{i=1}^{3} \beta_i \right\rangle \geq \left\langle \left(\prod_{i=1}^{3} \beta_i \right)^{1/3} \right\rangle = 1.$$

Hence, we have $\langle (\Delta \vec{r}(t))^2 \rangle \geq (\Delta \vec{r}(0))^2$, i.e., infinitesimal line elements are stretched (on the average) by homogeneous isotropic incompressible turbulence. The same proof applies to show that surface elements are stretched to show line stretching in an arbitrary number of dimensions.

It is necessary to note that one may generalize our analysis of homogeneous turbulence on the manifolds (Riemann and Weyl); in this case, the properties of turbulent fluid can differ from the considered ones essentially.

6.1.3. *Hydrodynamic equations with the fluctuating velocity*

Let us represent fluid velocity and pressure (at position $\vec{r}$ and time t) for turbulent flow as $\vec{v}(\vec{r}, t) = \vec{u} + \vec{\xi}$ ($\vec{u} \equiv \langle \vec{v} \rangle$) and $p(\vec{r}, t) = \langle p \rangle + p^{\dagger}$ ($p^{\dagger}(\vec{r}, t) = 0$), correspondingly. Here, $\vec{\xi}$ and $p^{\dagger}$ are *fluctuating parts* of velocity and pressure.

Therefore, the continuity equation for this flow becomes $\partial u_i(\vec{r}, t)/\partial r_i = 0$ ($i = 1, 2, 3$) and the equation of motion for the fluctuating velocity becomes

$$\frac{\partial \xi_\mu}{\partial t} + \frac{\partial u_\mu}{\partial r_\tau} \xi_\tau + \frac{\partial u_\tau}{\partial r_\tau} \xi_\mu + \frac{\partial (\xi_\mu \xi_\tau - \langle \xi_\mu \xi_\tau \rangle)}{\partial r_\tau} = -\frac{\partial p^{\dagger}}{\rho \partial r_\mu} + \nu \frac{\partial^2 \xi_\mu}{\partial r_\mu^2},$$

$$\mu, \tau = 1, 2, 3 \qquad (6.5)$$

(we used the summation rule by repeated Greek indices only).

It is clear that for a fluid in the volume V (with boundary surface ∂V) the total kinetic energy of fluid motion is $E_{\text{tot}} = \sum_\mu \int_V (\rho v_\mu^2/2) dV$. An equation for E_{tot} can be derived from the NSE;

it has the form of

$$\frac{dE_{tot}}{dt} = \int_V \rho \left(v_\mu (f_{ex})_\mu - \varepsilon \right) dV, \quad \varepsilon = \frac{\nu}{2} \sum_\mu \sum_\tau \left(\frac{\partial v_\mu}{\partial r_\tau} + \frac{\partial v_\tau}{\partial r_\mu} \right)^2.$$

(6.6)

We multiply (6.5) for ξ_μ and ξ_ρ by $\xi_\rho(\vec{r}, t)$ and ξ_μ, respectively; further, we average obtained equations, add results together and set $\mu = \rho$. Finally, we have

$$\frac{\partial \langle \xi_\mu^2 \rangle}{\partial t} + u_\tau \frac{\partial \langle \xi_\mu^2 \rangle}{\partial r_\tau} = -\frac{\partial}{\partial r_\tau} \left(\langle \xi_\mu^2 \xi_\tau \rangle + \frac{2}{\rho} \langle \xi_\tau p^\dagger \rangle \right)$$

$$- 2 \langle \xi_\mu \xi_\tau \rangle \frac{\partial u_\mu}{\partial r_\tau} + \nu \frac{\partial^2 \langle \xi_\mu^2 \rangle}{\partial r_\tau^2} - 2\nu \left\langle \left(\frac{\partial \xi_\mu}{\partial r_\tau} \right)^2 \right\rangle.$$

(6.7)

If we sum this relation over μ, we obtain the balance equation for the turbulent kinetic energy E (per unit fluid mass), which is defined by $\sum_\mu \langle \xi_\mu^2 \rangle = 2E$. The first and third terms on the right-hand side of (6.7) describe the diffusion of turbulent energy through space by nonlinear and viscous action, correspondingly; the second term there describes a flow of energy from the mean field to the fluctuating velocity field; the fourth term is connected with irreversible dissipation of kinetic energy into heat.

6.2. The Homogeneous Turbulence and Its Spectrum

6.2.1. *The properties of the correlation tensors*

If the turbulence is homogeneous, the mean velocity $\vec{u} = \langle \vec{v} \rangle$ may be assumed zero, for it follows from homogeneity and (1.35) that $\partial \langle v_i(\vec{r}, t) \rangle / \partial t = 0$ ($i = 1, 2, 3$), since ensemble averaging and space–time differentiation commute. Therefore, $\langle v_i(\vec{r}, t) \rangle$ is space and time independent and may, by suitable choice of reference frame, be chosen zero. We will use this choice in the following consideration.

Let us investigate the velocity-correlation tensors in detail. It may be shown that the following properties are true:

(I) $\varrho_{ij}(\Delta\vec{r}) = \varrho_{ij}(-\Delta\vec{r})$ (it is a consequence of homogeneity),

(II) $\partial\varrho_{\mu\tau}/\partial(\Delta r_\mu) = \partial\varrho_{\mu\tau}/\partial(\Delta r_\tau) = 0$ (it follows from incompressibility),

(III) $|\varrho_{ij}(\Delta\vec{r})|^2 \leq \varrho_{ii}(0)\varrho_{jj}(0)$ (this property follows from Schwarz's inequality).

We use the convention of summation on repeated Greek indices (only).

The vorticity correlation tensor $\langle\varpi_i(\vec{r} + \Delta\vec{r})\varpi(\vec{r})\rangle$, where $\varpi_i \equiv (\text{rot }\vec{v})_i$, can be related to ϱ_{ij} with the help of property (II):

$$\langle\varpi_i(\vec{r} + \Delta\vec{r})\varpi_j(\vec{r})\rangle = \nabla^2\varrho_{ij}(\Delta\vec{r}) - \delta_{ij}\cdot\nabla^2\varrho_{kk}(\Delta\vec{r}) + \frac{\partial^2\varrho_{kk}(\Delta\vec{r})}{\partial r_i\partial r_j}.$$
(6.8)

If the turbulence is isotropic, we can simplify ϱ_{ij}. Else if $\vec{a}_1, \vec{a}_2$ are arbitrary vectors, then

$$\daleth(\vec{a}_1, \vec{a}_2, \Delta\vec{r}) = (a_1)_\alpha(a_2)_\beta\varrho_{\alpha\beta}(\Delta\vec{r}) = \langle\vec{a}_1\cdot\vec{\xi}(\vec{r} + \Delta\vec{r})\vec{a}_2\cdot\vec{\xi}(\vec{r})\rangle$$

is a scalar under rotations if the turbulence is isotropic. Hence, $\daleth$ can depend only on the scalars that can be formed out of combinations of its arguments: a_1^2, a_2^2, $(\Delta\vec{r})^2$, $\vec{a}_1\cdot\vec{a}_2$, $\vec{a}_1\cdot\Delta\vec{r}$, $\vec{a}_2\cdot\Delta\vec{r}$, $\vec{a}_1\cdot\vec{a}_2\times\Delta\vec{r}$. Since $\daleth$ is linear in $\vec{a}_1, \vec{a}_2$, it follows that

$$\daleth = (\vec{a}_1\cdot\Delta\vec{r})(\vec{a}_2\cdot\Delta\vec{r})A_1(\Delta r) + (\vec{a}_1\cdot\vec{a}_2)A_2(\Delta r)$$
$$+ (\vec{a}_1\cdot\vec{a}_2)\times\Delta\vec{r}A_3(\Delta r),$$

so that we have

$$\varrho_{\alpha\beta}(\Delta\vec{r}) = (\Delta r)_\alpha(\Delta r)_\beta A_1(\Delta r) + \delta_{\alpha\beta}A_2(\Delta r)$$
$$+ \epsilon_{\alpha\beta\mu}(\Delta r)_\mu A_3(\Delta r),$$
(6.9)

where $\epsilon_{\alpha\beta\mu} = +1$ if α, β, μ is an even permutation of 1, 2, 3, $\epsilon_{\alpha\beta\mu} = -1$ if α, β, μ is an odd permutation of 1, 2, 3 and $\epsilon_{\alpha\beta\mu} = 0$ otherwise (it is the consequence of $(\text{rot }\vec{\xi})_\alpha = \epsilon_{\alpha\beta\mu}\partial\xi_\mu/\partial r_\beta$).

If the turbulence is invariant to reflections in space in addition to being invariant to rotations, then the quantity $\vec{a}_1\cdot\vec{a}_2\times\Delta\vec{r}$ is not invariant and the term involving $A_3(\Delta r)$ cannot appear. We assume

reflection-invariant isotropic turbulence, but in a general case, reflec-
tion non-invariance consideration is required.

The property (II) implies a relation between the scalar functions
A_1 and A_2. It is conventional to rewrite (6.9) as

$$\varrho_{\alpha\beta}(\Delta\vec{r}) = \langle\xi^2\rangle\left(\frac{f(\Delta r) - g(\Delta r)}{(\Delta r)^2}(\Delta r)_\alpha(\Delta r)_\beta + g(\Delta r)\delta_{\alpha\beta}\right),$$

$$\varrho_{kk}(\Delta r = 0) = \langle\xi_1^2\rangle = \langle\xi_2^2\rangle = \langle\xi_3^2\rangle \equiv \langle\xi^2\rangle,$$

$$\langle\xi_k\xi_m\rangle = 0 \quad \text{if } k \neq m, \tag{6.10}$$

and the incompressibility constraint becomes

$$g(\Delta r) = f(\Delta r) + \frac{\Delta r}{2}\frac{df}{d(\Delta r)}. \tag{6.11}$$

It follows from (6.10) that $f(0) = g(0) = 1$. The functions $f(\Delta r)$ and
$g(\Delta r)$ are called the *longitudinal* and *latitudinal velocity correlation
functions* correspondingly:

$$\langle\xi^2\rangle f(\Delta r) = \langle\xi_f(\vec{r} + \Delta\vec{r})\xi_f(\vec{r})\rangle, \quad \langle\xi^2\rangle g(\Delta r) = \langle\xi_g(\vec{r} + \Delta\vec{r})\xi_g(\vec{r})\rangle, \tag{6.12}$$

where ξ_f is the velocity component parallel to $\Delta\vec{r}$ and ξ_g is one
of the components perpendicular to $\Delta\vec{r}$. The property (I) implies
$df(0)/d(\Delta r) = dg(0)/d(\Delta r) = 0$, so that the expansion about $\Delta r = 0$
gives

$$f(r) = 1 - \frac{(\Delta r)^2}{2\lambda^2} + O((\Delta r)^4), \quad g(r) = 1 - \frac{(\Delta r)^2}{\lambda^2} + O((\Delta r)^4),$$

where $\lambda^{-2} = -d^2f(0)/d(\Delta r)^2 = \langle(\partial\xi_1/\partial r_1)^2\rangle/\langle\xi^2\rangle$, λ is the *Taylor
microscale*, which is related to $\langle\varpi^2\rangle$:

$$\langle\varpi^2\rangle = -\nabla^2\varrho_{ii}(\Delta\vec{r} = 0)$$

$$= -\langle\xi^2\rangle\frac{1}{\Delta r}\frac{d^2(\Delta rf + 2\Delta rg)}{d(\Delta r)^2}\Big|_{\Delta r=0} = 15\lambda^{-2}\langle\xi^2\rangle. \tag{6.13}$$

The physical importance of the Taylor microscale is indicated by
the relationship connecting it to the *Batchelor rate of dissipation of
energy* in isotropic turbulence, which is $\varepsilon = \nu\langle\varpi^2\rangle = 15\lambda^{-2}\nu\langle\xi^2\rangle$,
where ν is the kinematic viscosity coefficient.

Now, we relate f and g to the coefficients A_1 and A_2 in equation (6.9) by considering the special case, where we take $\Delta \vec{r}$ to lie along the r_1-axis. We put $\vec{r} = 0$ and $\Delta \vec{r} = (\Delta r_1, 0, 0)$. Then we obtain

$$\varrho_{11}(\Delta r) = A_1 (\Delta r)^2 + A_2 = \langle \xi^2 \rangle f, \quad \varrho_{11}(\Delta r) = A_2 = \langle \xi^2 \rangle g.$$

Then $\varrho_{\alpha\beta} = \langle \xi^2 \rangle (f - g)(\Delta r)_\alpha (\Delta r)_\beta / (\Delta r)^2 + \langle \xi^2 \rangle g \delta_{\alpha\beta}$. We can eliminate g by differentiating this relation with respect to $(\Delta r)_\alpha$ and invoking the continuity relation in the form of

$$\frac{\partial}{\partial r_\alpha} \varrho_{\alpha\beta}(\vec{r}, \vec{r}^\dagger; t, t^\dagger) = \frac{\partial}{\partial r_\beta^\dagger} \varrho_{\alpha\beta}(\vec{r}, \vec{r}^\dagger; t, t^\dagger) = 0.$$

Finally, we obtain

$$\varrho_{\alpha\beta}(\Delta \vec{r}) = \langle \xi^2 \rangle f \delta_{\alpha\beta} + \frac{\Delta r \langle \xi^2 \rangle}{2} \frac{df}{d(\Delta r)} \left(\delta_{\alpha\beta} - \frac{(\Delta r)_\alpha (\Delta r)_\beta}{(\Delta r)^2} \right).$$

For the particular case $\Delta r = 0$, this formula gives

$$\sum_\tau \varrho_{\tau\tau}(\Delta r = 0) = \langle \xi^2 \rangle f(0) \cdot Sp(\delta_{\tau\beta}) = 3 \langle \xi^2 \rangle = 2E. \qquad (6.14)$$

6.2.2. Cumulants and NSE in the Fourier representation

Now, we apply the Fourier analysis of the velocity field to investigate homogeneous turbulence. A Fourier component is not a local coordinate but is a collective coordinate which specifies the total excitation in some scale over the whole flow. In this way, it is possible to give precise meaning to concepts such as energy of small-scale motions. The disadvantage of Fourier analysis also stems from its collective nature. Local coherent flow structures are not easily recognizable except by a detailed examination of phase relations among Fourier components.

A statistically homogeneous velocity field can be represented as

$$\vec{\xi}(\vec{r}, t) = \int \vec{V}(\vec{k}, t) \exp(i \vec{k} \cdot \vec{r}) d\vec{k},$$

where $\vec{V}(\vec{k}, t) = (2\pi)^{-3} \int \vec{\xi}(\vec{r}, t) \exp(-i\vec{k}\cdot\vec{r}) d\vec{r}$ is a random generalized function. Moments of V are

$$\langle V_i(\vec{k}_1) V_j(\vec{k}_2) \rangle \equiv (2\pi)^{-6} \int d\vec{r} \int d\vec{s} \langle v_i(\vec{r}) v_j(\vec{s}) \rangle$$
$$\times \exp(-i\vec{k}_1 \cdot \vec{r} - i\vec{k}_2 \cdot \vec{s})$$
$$= S_{ij}^{(2)}(\vec{k}_1) \delta(\vec{k}_1 + \vec{k}_2),$$

$$\langle V_i(\vec{k}_1) V_j(\vec{k}_2) V_k(\vec{k}_3) \rangle = S_{ij}^{(3)}(\vec{k}_1, \vec{k}_2) \delta(\vec{k}_1 + \vec{k}_2 + \vec{k}_3), \dots,$$

$$S_{ij}^{(2)}(\vec{k}_1) = (2\pi)^{-3} \int \varrho_{ij}(\Delta\vec{r}) \exp(-i\vec{k}_1\Delta\vec{r}) d(\Delta\vec{r}),$$

$$\Delta\vec{r} = \vec{r} - \vec{s},$$

$$S_{ijk}^{(3)}(\vec{k}_1, \vec{k}_2) = (2\pi)^{-6} \int d\vec{s} \int d\vec{z} \langle v_i(\vec{r} + \vec{s}) v_j(\vec{r} + \vec{z}) v_k(\vec{r}) \rangle$$
$$\times \exp(-i\vec{k}_1\vec{s} - i\vec{k}_2\vec{z}), \dots \qquad (6.15)$$

where we use the definition of δ-function: $\delta(\vec{k}_2) = \int \exp(-i\vec{k}_2 \cdot \vec{s}) d\vec{s}$. The fourth-order moment has the following form:

$$\langle V_\alpha(\vec{k}_1) V_\beta(\vec{k}_2) V_\gamma(\vec{k}_3) V_\delta(\vec{k}_4) \rangle$$
$$= S_{\alpha\beta}(\vec{k}_1) S_{\gamma\delta}(\vec{k}_3) \delta(\vec{k}_1 + \vec{k}_2) \delta(\vec{k}_3 + \vec{k}_4)$$
$$+ S_{\alpha\gamma}(\vec{k}_1) S_{\beta\delta}(\vec{k}_2) \delta(\vec{k}_1 + \vec{k}_3) \delta(\vec{k}_2 + \vec{k}_4)$$
$$+ S_{\alpha\delta}(\vec{k}_1) S_{\beta\gamma}(\vec{k}_2) \delta(\vec{k}_1 + \vec{k}_4) \delta(\vec{k}_2 + \vec{k}_3)$$
$$+ U_{\alpha\beta\gamma\delta}(\vec{k}_1, \vec{k}_2, \vec{k}_3) \delta(\vec{k}_1 + \vec{k}_2 + \vec{k}_3 + \vec{k}_4), \qquad (6.16)$$

where $U_{\alpha\beta\gamma\delta}$ are called *fourth-order cumulants*:

$$U_{\alpha\beta\gamma\delta}(\vec{k}_1, \vec{k}_2, \vec{k}_3) = (2\pi)^{-9} \iiint d\vec{s} d\vec{z} d\vec{w}$$
$$\times [\langle \xi_\alpha(\vec{r} + \vec{s}) \xi_\beta(\vec{r} + \vec{z}) \xi_\gamma(\vec{r} + \vec{w}) \xi_\delta(\vec{r}) \rangle$$
$$- \varrho_{\alpha\beta}(\vec{s} - \vec{z}) \varrho_{\gamma\delta}(\vec{w}) - \varrho_{\alpha\gamma}(\vec{s} - \vec{w}) \varrho_{\beta\delta}(\vec{z})$$
$$- \varrho_{\alpha\delta}(\vec{s}) \varrho_{\beta\gamma}(\vec{z} - \vec{w})] \exp(-i(\vec{k}_1\vec{s} + \vec{k}_2\vec{r} + \vec{k}_3\vec{w}))$$

(cumulants of the second and third orders are the same as that of the moments of these orders).

Reality of the velocity field $\xi(\vec{r})$ implies $S^{(2)}_{\alpha\beta}(\vec{k}) = (S^{(2)}_{\alpha\beta}(-\vec{k}))^*$, $S^{(3)}_{\alpha\beta\gamma}(\vec{k}_1, \vec{k}_2) = (S^{(3)}_{\alpha\beta\gamma}(-\vec{k}_1, -\vec{k}_2))^*$, etc., while the incompressibility property $\nabla\xi(\vec{r}, t) = 0$ becomes

$$k_\alpha S^{(2)}_{\alpha\beta}(\vec{k}) = k_\beta S^{(2)}_{\alpha\beta}(\vec{k}) = 0, \tag{6.17}$$

etc. Isotropy leads to much simplification in the structure of the cumulants. The theory of isotropic invariants implies $(a_1)_\alpha (a_2)_\beta S^{(2)}_{\alpha\beta}(\vec{k})$ can be a function of $\vec{a}_1 \cdot \vec{a}_2$, $\vec{a}_1 \cdot \vec{k}$, $\vec{a}_2 \cdot \vec{k}$, $\vec{k} \cdot \vec{k}$, so that linearity in $a_{1,2}$ implies $S^{(2)}_{\alpha\beta}(\vec{k}) = A_1(k)k_\alpha k_\beta + A_2(k)\delta_{\alpha\beta}$. Furthermore, $A_2(k) = -k^2 A_1(k)$ (it follows from $k_\alpha S^{(2)}_{\alpha\beta}(\vec{k}) = 0 = A_2 k_\beta + A_1 k^2 k_\beta$) and $S^{(2)}_{\alpha\beta} = \mathcal{D}_{\alpha\beta} A_2(k)$, $\mathcal{D}_{\alpha\beta}(\vec{k}) = \delta_{\alpha\beta} - k_\alpha k_\beta / k^2$. We know that $Sp\left(\mathcal{D}_{\alpha\beta}(\vec{k})\right) = 2$. Therefore,

$$Sp\left(S^{(2)}_{\alpha\beta}(\vec{k})\right) = Sp\left(\mathcal{D}_{\alpha\beta}(\vec{k})A_2(k)\right) = 2A_2(k).$$

From (6.14), we have

$$E = \frac{3\langle\xi^2\rangle}{2} = \frac{1}{2}Sp\left(\varrho_{\alpha\beta}(0)\right) = \frac{1}{2}Sp\left(\int S^{(2)}_{\alpha\beta}(\vec{k})d\vec{k}\right)$$

$$= Sp\left(\int_0^\infty \frac{k^2}{2} A_2(k)\,dk \int \mathcal{D}_{\alpha\beta}(\vec{k})d\Omega_k\right)$$

$$= \frac{4\pi}{3}Sp\left(\delta_{\alpha\beta}\int_0^\infty A_2(k)\,k^2\,dk\right)$$

$$= \int_0^\infty 4\pi k^2\,A_2(k)\,dk = \int_0^\infty E(k)dk. \tag{6.18}$$

Consequently, we obtain from the equality $S^{(2)}_{\alpha\beta} = \mathcal{D}_{\alpha\beta}A_2(k)$ that

$$S^{(2)}_{\alpha\beta}(\vec{k}) = \frac{E(k)}{4\pi k^2}\mathcal{D}_{\alpha\beta}(\vec{k}). \tag{6.19}$$

Here, the scalar function $E(k)$ is interpreted as the kinetic energy density in modes of wavenumber k so that $3\langle\xi^2\rangle/2 = \int_0^\infty E(k, t)dk$. The quantity $E(k) = 4\pi k^2 A_2(k)$ is the *wavenumber spectrum*, i.e., the distribution of energy with wavenumber; the quantity $A_2(k)$ is the density of contributions in wavenumber space to the total energy.

Let us use the considered facts for investigation of the NSE in the Fourier representation. The Navier–Stokes equations for $\vec{V}(\vec{k}, t)$ are

$$\left(\frac{\partial}{\partial t} + \nu k_1^2\right) V_\alpha(\vec{k}_1, t) = -i \cdot (k_1)_\alpha p(\vec{k}_1, t)$$

$$- i \int (k_2)_\beta V_\beta(\vec{k}_1 - \vec{k}_2, t) V_\alpha(\vec{k}_2, t) d\vec{k}_2. \quad (6.20)$$

Application of the incompressibility property $k_\alpha V_\alpha(\vec{k}, t) = 0$ allows to rewrite this equation in the form of

$$\left(\frac{\partial}{\partial t} + \nu k_1^2\right) V_\alpha(\vec{k}_1, t) = -\frac{i}{2} \cdot (k_1)_\alpha \mathcal{D}_{\alpha\beta\mu}(\vec{k}_1)$$

$$\times \int V_\beta(\vec{k}_2, t) V_\mu(\vec{k}_1 - \vec{k}_2, t) d\vec{k}_2,$$

$$\mathcal{D}_{\alpha\beta\mu}(\vec{k}) = k_\beta \mathcal{D}_{\alpha\mu}(\vec{k}) + k_\mu \mathcal{D}_{\alpha\beta}(\vec{k}). \quad (6.21)$$

Dynamical equations for the cumulants have similar forms:

$$\left(\frac{\partial}{\partial t} + 2\nu k_1^2\right) S^{(2)}_{\alpha\beta}(\vec{k}_1, t)$$

$$= -\frac{i}{2} \mathcal{D}_{\alpha\beta\mu}(\vec{k}_1) \int S^{(3)}_{\tau\beta\mu}(-\vec{k}_1, \vec{k}_2) d\vec{k}_2$$

$$- \frac{i}{2} \mathcal{D}_{\tau\beta\mu}(-\vec{k}_1) \int S^{(3)}_{\alpha\beta\mu}(\vec{k}_1, \vec{k}_2) d\vec{k}_2, \quad (6.22)$$

$$\left(\frac{\partial}{\partial t} + \nu(k_1^2 + k_2^2 + |\vec{k}_1 + \vec{k}_2|^2)\right) S^{(3)}_{\alpha\beta\mu}(\vec{k}_1, \vec{k}_2)$$

$$= -\frac{i}{2} \mathcal{D}_{\alpha\rho\sigma}(\vec{k}_1) \int U_{\beta\mu\rho\sigma}(\vec{k}_2, -\vec{k}_1 - \vec{k}_2, \vec{k}_3) dk_3$$

$$+ \cdots - i \mathcal{D}_{\alpha\rho\sigma}(\vec{k}) S^{(2)}_{\beta\rho}(\vec{k}_2) S_{\mu\sigma}(-\vec{k}_1 - \vec{k}_2) + \cdots. \quad (6.23)$$

The equation for sth-order cumulants involves $(s + 1)$th-order cumulants, so that the system of equations is not closed; any finite subsystem of this infinite set of equations possesses more unknown functions that are determined by the subsystem. The situation is similar to BBGKY hierarchy in some sense.

If the turbulence is isotropic, equation (6.22) reduces to a very simple equation for energy $E(k, t)$. To obtain this fact, let us rewrite (6.22) as

$$\frac{\partial E(k, t)}{\partial t} + 2\nu k^2 E(k, t) = -4\pi k^2 k_\alpha \Im \left(\int S^{(3)}_{\mu\alpha\mu}(\vec{k}, \vec{k}_1) d\vec{k}_1 \right) \equiv \mathcal{B}(k, t).$$

$$(6.24)$$

Conservation of energy by nonlinear interactions implies $\int_0^\infty \mathcal{B}(k, t)\, dk = 0$ (see, for example, [12]). Then after integration (6.24) over k and from relation $3\langle \xi^2 \rangle = 2 \int_0^\infty E(k) dk$ (see above), we obtain

$$\frac{3}{2} \frac{d\langle \xi^2 \rangle}{dt} = -\varepsilon(t), \quad \varepsilon(t) = 2\nu \int_0^\infty k^2 E(k, t) dk = -\frac{dE}{dt}, \quad (6.25)$$

i.e., the quantity $\varepsilon(t)$ is the rate (per unit volume) of kinetic energy dissipation (see text after formula (6.13)).

The relation (6.25) shows that *high wavenumber Fourier components are dissipated more effectively by viscosity than low wavenumber components.*

6.2.3. *The spectrum of homogeneous isotropic turbulence*

Now, we consider the energy-spectrum tensor $S^{(2)}_{\alpha\beta}(\vec{k})$ in detail. The size of an eddy of a turbulent flow with the wavenumber k is $\sim 2\pi/k$. The spectrum of homogeneous isotropic turbulence (i.e., the set of wavenumber k) consists of three regions (by the sizes of the eddies): $k \ll L_\ell^{-1}$ (the large eddies), $k \sim L_\ell^{-1}$ (the energy-containing eddies), $k \gg L_\ell^{-1}$ (the small eddies), where $L_\ell = \int_0^\infty f(r) dr = 2 \int_0^\infty g(r) dr$ is the *longitudinal integral scale* (see (6.12)).

For investigation of the large eddies, one may consider the expansion of the cumulants about $\vec{k} = 0$; the analyticity of cumulants follows if spatial correlations decay exponentially at infinity. For example, we have

$$S^{(2)}_{\alpha\beta} = \widetilde{S}^{(2)}_{\alpha\beta} + \widetilde{S}^{(2)}_{\alpha\beta\gamma} k_\gamma + \widetilde{S}^{(2)}_{\alpha\beta\gamma\mu} k_\gamma k_\mu + O(k^3).$$

Then from (6.17), for small k we obtain $k_\alpha \widetilde{S}^{(2)}_{\alpha\beta} + O(k^2) = 0$. Consequently, since $\vec{k}/k$ is arbitrary, it follows, that $\widetilde{S}^{(2)}_{\alpha\beta} = 0$. From the inequality $q_\alpha q_\beta^* S_{\alpha\beta}^{(2)}(\vec{k}) \geq 0$, where $\vec{q}$ is any complex vector (this inequality follows from the definition (6.15)), we have

$$q_\alpha q_\beta^* k_\gamma \widetilde{S}_{\alpha\beta\gamma}^{(2)} + O(k^2) \geq 0,$$

which implies $\widetilde{S}_{\alpha\beta\gamma}^{(2)} = 0$. One may demonstrate that $d\widetilde{S}_{\alpha\beta\gamma\mu}^{(2)}/dt = 0$, if third-order cumulants are analytical. If the turbulence is isotropic, then for $k \to 0$, $E(k) = (2\pi/3)\widetilde{S}_{\alpha\alpha\beta\beta}^{(2)}k^4 + o(k^4)$ (the so-called *permanence of the large eddies*). As a consequence, we have $dJ/dt = 0$, where $J = \langle \xi^2 \rangle \int_0^\infty r^4 f(r)dr$ is the *Loitsyanskiy invariant*.

There exist other assumptions about behavior cumulants about $k = 0$, for example, $S_{\alpha\beta}^{(2)}(\vec{k}) = c_0 \mathcal{D}_{\alpha\beta} + O(k^2)$ ($c_0 = $ const); then for isotropic turbulence $E(k) = 4\pi c_0 k^2 + o(k^2)$ and cumulants decay as $(\Delta r)^{-4}$.

Let us consider the energy-containing eddies. In this case, the characteristic velocity of these eddies is $\sqrt{\langle \xi^2 \rangle}$ and their characteristic length is L_ℓ. They make the dominant contribution to $\int_0^\infty E(k)dk$. The timescale for decay of the turbulence is $L_\ell/\sqrt{\langle \xi^2 \rangle}$ which is also the circulation time of an eddy in the energy-containing region, i.e., the turbulence decays on the same timescale as eddies in this range execute their motion.

Equation (6.25) may be rewritten as

$$\frac{3}{2}\frac{d\langle \xi^2 \rangle}{dt} = -A\langle \xi^2 \rangle \cdot \frac{\sqrt{\langle \xi^2 \rangle}}{L_\ell},$$

where $A = A_t \equiv \varepsilon L_\ell(\langle \xi^2 \rangle)^{-3/2} \sim O(1)$ even for large Re, i.e., for turbulent flow (in contrast to $A = A_\ell \equiv O(1/Re)$, $Re = \sqrt{\langle \xi^2 \rangle}L_\ell)/\nu$ for laminar flow). The quantity $L_\ell/\sqrt{\langle \xi^2 \rangle}$ is the turnover time of the decay of size L_ℓ and typical velocity $\sqrt{\langle \xi^2 \rangle}$. Therefore, from the last equation, it follows that the energy-containing region is approximately critically damped.

In the range of small eddies, $k \gg 1/L_\ell$, there is no direct interaction between the turbulence and the motion of the large, energy-containing eddies. This is because the small scales have been generated by a long series of small steps, losing information at each step. The spectrum in this range of large wavenumbers is nearly isotropic, as only the large eddies are aware of the directions of mean gradients. The spectrum here does not depend on how much energy is present at large scales (where most of the energy is contained), or the scales at which most of the energy is present. The spectrum depends only on the parameters that determine the nature of the small-scale flow.

The principal dynamical effect of small eddies back on larger scales is through their contribution to the energy dissipation rate ε. In order for ε to be asymptotically independent of Re, it is necessary that the wavenumber range contributing to *enstrophy* $\mathcal{E} = (1/2)\langle|\vec{\xi}_{(3)}|^2\rangle$ ($\vec{\xi}_{(3)} = \text{curl}\,\vec{v}$ is the vorticity of flow) grows with Reynolds parameter. The excitation of small scales is accomplished by the nonlinear terms in the NSE. The qualitative character of the energy spectrum $E(k)$ may be obtained from dimensional analysis. We consider this spectrum in the following section.

6.2.4. *The energy cascade*

We may interpret equation (6.24) in the following sense: the energy in the turbulent system at small k (large scales) is transferred by the nonlinear term $\mathcal{B}(k, t)$ to large k (small scales) (this process is called the *energy cascade*), where it is dissipated (turned into heat) by the viscous term. Let us rewrite (6.24) as

$$\frac{\partial E(k, t)}{\partial t} = -2\nu k^2 E(k, t) + \int_0^\infty \widetilde{\mathcal{B}}(k, k_1, |\vec{k} - \vec{k}_1|)dk_1, \qquad (6.26)$$

where the kernel $\widetilde{\mathcal{B}}$ is defined from the explicit form $\mathcal{B}(k, t)$. The solution of (6.26) has the form of

$$E(k, t) = E(k, t_0)\,\exp\big(-2\nu k^2(t - t_0)\big), \qquad t_0 = \text{const}. \qquad (6.27)$$

Hence, each mode $\vec{k}$ decays individually with inverse time $2\nu k^2$.

It is clear that the higher the value of k, the faster the decay. When we consider (6.26), we can expect the nonlinearity to transfer energy from where it enters the system (typically at small values of k and $2\nu k^2$) up to large values of k, where the viscous damping will be very rapid. Hence, the effect of the nonlinear inertial transfer should concentrate on the dissipation process in those regions of wavenumber space where it will be most efficient.

Let us consider the range of wavenumbers involved. The largest possible eddy will be bounded by the size of the system, and so the smallest possible wavenumber $k_{min} = 2\pi/\mathcal{L}$, where $\mathcal{L}$ is the largest relevant linear dimension of the turbulent system. We can expect the upper cut-off in the wavenumber to be determined by the viscous dissipation. The only relevant physical parameters available to us are the kinematic viscosity ν and the dissipation rate ε; therefore, from the dimensional analysis, we introduce a characteristic length scale $\eta = \nu^3/\varepsilon$ (and an associated velocity scale $v_\eta = (\nu\varepsilon)^{1/4}$). Hence, $k_{max} = 1/\eta$ is an approximate measure of the maximum possible wavenumber (the corresponding local Reynolds number $Re(k_{max}) = \eta v_\eta/\nu = 1$). The smallest wavenumbers are determined by the nature and size of the particular turbulent system under consideration, whereas the largest ones are determined by the general properties ν and ε. Thus, the ratio of k_{max} to k_{min} can be made as large as we please and, in the limit of infinite Re, infinitely large.

Now, we consider the relation between turbulent kinetic energy and dissipation rates in modes. The energy is determined (see formula (6.18) by the lowest wavenumbers, while the dissipation rate is determined by the highest wavenumbers (see formula (6.25)); these two ranges do not overlap even at quite modest values of the Reynolds number. Thus, it follows that the inertial term (which is the link between the two ranges) can be made to dominate over as large a range of wavenumbers as we like simply by increasing the Re and hence the dissipation wavenumber (through formula $k_{max} = \eta^{-1} = \varepsilon^{1/4}\nu^{-3/4}$). The consequences of this fact are crucial for the physics of turbulence, for it makes us to consider the inertial transfer of energy without worrying about the details of input (at low $|\vec{k}|$) or output (at high $|\vec{k}|$). In the NSE for the Fourier components of the velocity field (6.21), the nonlinear term is interpreted in the following way. Two velocity coefficients with different frequencies $\vec{k}_2$

and $\Delta\vec{k}_{12} = \vec{k}_1 - \vec{k}_2$ are coupled together to make a contribution to the Fourier coefficient with frequency $\vec{k}_1$. The total contribution from nonlinearity is then given by the integral over all such interactions. This coupling of wavevectors in triads is usually referred to as the "triangle condition". In principle, each Fourier mode of the velocity field is coupled to every other mode, and we are faced with a very difficult physical problem.

We introduce by simplifying the assumption: the distant wavevectors are weakly coupled, and any particular mode $\vec{k}_1$ effectively interacts with modes $\vec{k}_2$ and $\Delta\vec{k}_{12}$, $\vec{k}_1$, $\vec{k}_2$, $\Delta\vec{k}_{12}$ are the same orders of magnitude ("integration over modes of the nearest neighborhood"). The physical basis for such an assumption can be seen by considering the effect of the nonlinearity in the configurational space. The interaction of two eddies can be decomposed into the convection of one by the other and the shearing of one by the other. The first of these effects results only in a phase change of the associated Fourier coefficient and is not dynamically significant. The second effect results in the internal distortion of the eddies with the transfer of energy to a smaller size of disturbance. If the interacting eddies differ greatly in size, then it is physically plausible to argue that the dynamically irrelevant phase change is the main effect. From there, it is only a step to argue that we can assume that the nonlinear coupling of modes is to some degree local in $\vec{k}$-space.

We can interpret the nonlinear term in the energy equation in terms of energy flows between the modes, with each such mode coupling being the average effect of many eddy–eddy interactions. The combinations of some degree of localness in the basic interaction with the effect of averaging lead to the important idea that energy cascade may become independent of the way in which the turbulence was created. Therefore, the energy spectrum at high wavenumbers may take a universal form.

These ideas were first formalized by A.N. Kolmogorov (see, e.g., [1, 39, 41]) in two hypotheses; these hypotheses are essentially similarity principles for the energy spectrum and can be formulated in the following way.

(I) *At sufficiently high wavenumbers, the spectrum can only depend on the fluid viscosity, the dissipation rate, and the wavenumber;* then, on dimensional grounds, the energy spectrum can be written as

$$E(k) = v_\eta^2 \eta f(k\eta) = \nu^{5/4} \varepsilon^{1/4} f(k\eta), \qquad (6.28)$$

where $f(k\eta)$ is an unknown function of the universal form, and the dissipation length scale η is given by the formula $\eta = \nu^3/\varepsilon$.

(II) $E(k)$ *should become independent of the viscosity as the Reynolds number tends to infinity*; thus, the unknown function f in (6.28) must take the form

$$f(k\eta) = \alpha(k\eta)^{-5/3} = \alpha\nu^{-5/4}\varepsilon^{5/12}k^{-5/3}, \qquad (6.29)$$

where α is a constant.

Consequently, spectrum $E(k)$ from (6.28) can be written as

$$E(k) = \alpha\varepsilon^{2/3}k^{-5/3} \qquad (6.30)$$

in the limit of infinite Reynolds number. This is the *Kolmogorov 2/3 law.*

For large, but finite, Reynolds numbers, we can postulate the existence of an interval of an inertial subrange of wavenumbers such that $2\pi/\mathcal{L} \ll k \ll k_{max}$, for which the spectrum, as given by (6.28), is independent of the viscosity. Then we can modify (6.29) to take the form $f(k\eta) = \alpha(k\eta)^{-5/3}\widehat{f}(k/k_{max})$, where $\widehat{f}$ is a new function such that $\widehat{f}(0) = 1$. Finally, we obtain for energy spectrum: $E(k) = \alpha\varepsilon^{2/3}k^{-5/3}F(k/k_{max})$ for $k \gg 2\pi/\mathcal{L}$.

The universal range of wavenumbers (in which (6.28) is supposed to be valid) is the "equilibrium range of wavenumbers". This terminology is based on the argument that the small eddies will evolve much more rapidly than the large eddies which contain most of the energy. Thus, eddies in the universal range can adjust so quickly to changes in external conditions that they can be assumed to be always in a state of "local equilibrium"; it should be pointed out that the process under consideration is, in the thermodynamic sense, far from equilibrium. In practice, many real flows are stationary; we can consider the stationary state even within the concept of isotropic turbulence by means of an artifice. Let us introduce to the energy equation an input term $w(k)$, where $\int_0^\infty w(k)dk = \varepsilon$:

$$w(k) - 2\nu k^2 E(k,t) + \int_0^\infty \widetilde{\mathcal{B}}(k, k_1, |\vec{k} - \vec{k}_1|)dk_1 = 0. \qquad (6.31)$$

In order to have a well-posed problem with well-separated input and dissipation ranges of wavenumbers, we choose a wavenumber $k_\flat$ such that

$$\int_0^{k_\flat} w(k)dk \simeq \varepsilon \simeq - \int_{k_\flat}^\infty 2\nu k^2 E(k)dk. \qquad (6.32)$$

This means that we require the input term to be peaked near $k = 0$, and that Reynolds number should not be too low. With all this in mind, we can obtain two energy-balance equations:

$$\int_0^{k_\flat} dk \int_{k_\flat}^\infty \widetilde{\mathcal{B}}(k, k_1, |\vec{k} - \vec{k}_1|)dk_1 + \int_0^{k_\flat} w(k)dk = 0,$$

$$\int_{k_\flat}^\infty dk \int_0^{k_\flat} \widetilde{\mathcal{B}}(k, k_1, |\vec{k} - \vec{k}_1|)dk_1 - \int_0^{k_\flat} 2\nu k^2 E(k)dk = 0,$$

where we use the property $\int dk \int \widetilde{\mathcal{B}}(k, k_1, |\vec{k} - \vec{k}_1|)dk_1 = 0$ for $0 \leq k, k_1 \leq k_\flat$ and $k_\flat \leq k, k_1 \leq \infty$, respectively.

The first of these equations means that the energy supplied directly by the production term to modes $k < k_\flat$ is transferred by the nonlinearity to modes $k_1 > k_\flat$. Thus, in this range of wavenumbers, $\mathcal{B}(k)$ (the nonlinear term of equation (6.24)) behaves like a dissipation and absorbs energy. From the second equation, we see that the nonlinearity transfers energy from modes $k_1 < k_\flat$ to modes $k > k_\flat$. Therefore, in this range of wavenumbers, $\mathcal{B}(k)$ behaves like a source of energy and this input is dissipated by the viscous term.

A consideration of these energy-balance equations, taken in conjunction with (6.32), allows us to put interpretation on ε as the rate at which the inertial forces transfer energy from low to high wavenumbers, assuming that the turbulence is stationary.

6.2.5. *Intermittency*

In [1], we read:

"Anyone who has carried out the classical Reynolds pipe-flow experiment (and this is often done as part of an undergraduate course in fluid dynamics) will have encountered a very strange effect. The usual way to do this experiment is to set the pump to give some particular rate of flow, which is then noted, as is the accompanying pressure drop along the pipe. The latter can be measured on a manometer connected to pressure tappings in the wall of the pipe. Then the pump speed is increased and the new flow rate and pressure drop noted, and so on. Eventually, a speed is reached where the manometer levels begin to oscillate wildly. This behavior continues over a range of speeds, but ultimately a speed is reached where the manometer reading steadies again and thereafter remains steady (however much the speed is increased). This wild oscillation of the manometer is the "strange effect" referred to above. It can be explained by the fact that the transition from laminar flow to turbulence is not an abrupt phenomenon, occurring at a particular Reynolds number, but takes the form of an alternation between the two states. That is, for a range of values of the Reynolds number (known as the "transition range"), patches of turbulence are interspersed with patches of laminar flow. Hence, if an anemometer were to be set up in the flow it would register a turbulent signal alternating with a steady signal. This is an example of what we mean by intermittency".

The correct notion of intermittency may be quantified when the random function $v(t)$ is stationary. It is then convenient to work with the high-pass signal $v_{\omega_0}^>(t)$, defined (in the temporal domain) by

$$v(t) = \int_{\mathbb{R}^3} \widehat{v}_\omega \exp(i\omega t)\, d\omega, \quad v_{\omega_0}^>(t) = \int_{|\omega|>\omega_0} \widehat{v}_\omega \exp(i\omega t)\, d\omega.$$

We shall say that the random function $v(t)$ is *intermittent* if the flatness

$$F(\omega_0) = \langle \left(v_{\omega_0}^>(t)\right)^4 \rangle \cdot \langle \left(v_{\omega_0}^>(t)\right)^2 \rangle^{-2}$$

grows without bound with the "filter frequency" ω_0. The function $F(\omega_0)$ does not depend on the time argument.

Let us consider the signals in Fig. 6.2.

The function $v_\gamma(t)$ is obtained from $v(t)$ by setting it equal to zero except during "on intervals" which are randomly selected in such a

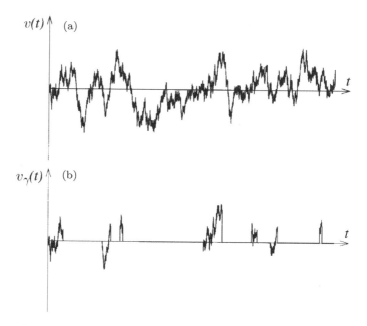

Fig. 6.2. A stationary signal (a); the same "chopped-off" a fraction $1 - \gamma$ of the time (b) [39].

way that the time the signal is on represents a fraction γ of the total time. Assuming that all relevant moments exist, it follows that $\langle v_\gamma^2 \rangle = \gamma \langle v^2 \rangle$, $\langle v_\gamma^4 \rangle = \gamma \langle v^4 \rangle$. Consequently,

$$F_\gamma = \langle v_\gamma^4 \rangle / \langle v_\gamma^2 \rangle^2 = \gamma^{-1} \langle v^4 \rangle / \langle v^2 \rangle^2.$$

True intermittency seldom achieves the "lacunar" character of Fig. 6.2 (b), but the flatness is still a useful measure of intermittency for signals having a bursty aspect. Of course, instead of the flatness, it is possible to use other non-dimensional ratios, such as the moment of order 6 divided by the cube of the moment of order 2. Odd order moments are, however, inappropriate, since they may vanish for reasons of symmetry or accidentally.

In recent years, it has been discovered that intermittencies are associated with various types of regular or quasi-deterministic behavior. The generic term is "coherent structures". We shall discuss the subject of coherent structures in some detail later on: it is nowadays regarded as a crucial aspect of shear flow turbulence. But for the moment, we shall turn our attention to the fundamental kind of intermittency in the turbulence (turbulent cascade).

In turbulence, energy is injected into eddies of size ℓ_0 and cascaded down through intermediate scales ℓ_n to the dissipation scale ℓ_d. Let us assume that this process can be represented by the discrete series of length scales $\ell_n = 2^{-n}\ell_0$, $n = 0, 1, 2, \ldots$ with the corresponding discrete wavenumbers $k_n = 1/\ell_n$. The argument is then conducted in terms of magnitudes, with numerical factors being dropped except when they would accumulate multiplicatively. Thus, if the energy (i.e., kinetic energy of turbulent fluctuations per unit mass of fluid) in scales is approximately equal to ℓ_n, we can define the root mean square velocity difference across a distance ℓ_n by v_n, where $E_n \approx v_n^2$ (we should note that, with this definition, the correlations of v_n will be closely related to the structure functions). The eddy turnover time can also be defined as $t_n \approx \ell_n/v_n$. In [39], a fundamental assumption is made that a sizeable fraction of the energy in scales ℓ_n is transferred to scales ℓ_{n+1} in time t. Hence, the energy transfer rate (for the n-order eddies) is given by $\varepsilon_n \approx E_n/t_n \approx v_n^3/\ell_n$. For stationary turbulence, conservation of energy then implies that $\varepsilon_n = \langle \varepsilon \rangle$ ($\ell_0 > \ell_n > \ell_d$), where $\langle \varepsilon \rangle$ is the mean dissipation rate. At this point, in [39], it was concluded that $\langle \varepsilon \rangle$ can be interpreted equally well as a rate of energy injection or as a rate of energy transfer, and that the latter is the dynamically relevant quantity for the inertial range.

From the above formulas, we have $v_n \approx \langle \varepsilon \rangle^{1/3} \ell_n^{1/3}$ and $E_n \approx \langle \varepsilon \rangle^{2/3} \ell_n^{2/3}$, which, after Fourier transformation, yields the Kolmogorov spectrum (6.30).

The so-called β-model (see [1, 39]) is introduced in terms of rather arbitrary assumption that the average number of offspring of any eddy is N (i.e., an eddy of scale ℓ_n is assumed to give rise to N eddies of scale ℓ_{n+1}, irrespective of the value of n). Then the fractional reduction in volume from one generation to the next is given by $\beta = N\ell_{n+1}^3/\ell_n^3 = N/2^3 \leq 1$. If it is further assumed that the largest eddies fill all the space available to them, then in the nth generation only the fraction $\beta_n = \beta^n$ of the space will be occupied by eddies of scale. Now, we repeat our previous arguments, but restrict them to the active volume of fluid for each generation n. That is, equation $v_n \approx \langle \varepsilon \rangle^{1/3} \ell_n^{1/3}$ still holds locally for v_n in an active region. But the relationship between the globally averaged energy density and the locally averaged v_n is postulated to be $E_n \approx \beta^n v_n^2$ or, after simple transformations, $E_n \approx \langle \varepsilon \rangle^{2/3} \ell_n^{2/3} (N/2^3)^{n/3}$.

With the further assumption $N = 2^D$ and the use of the definition $\ell_n = 2^{-n} \ell_0$ to eliminate n, the last equation can be written as $E_n \approx \langle \varepsilon \rangle^{2/3} \ell_n^{2/3} (\ell_n/\ell_0)^B$, where $B = (D-3)/3$ and D is identified as the (non-integer) fractal dimension [56].

So, the main idea of the β-model is as follows: at each stage of the Richardson cascade, the number of "daughters" of a given "mother-eddy" is chosen such that the fraction of volume occupied is decreased by a factor β ($\beta \in]0; 1[$) (Fig. 6.3). The factor β is an adjustable parameter of the model. It is necessary to note that the energy spectrum in the inertial range has the form $E(k) \sim k^{-(5/3 + c(D))}$, where $c(D)$ is the function of fractal dimension D; if $c(D) = 0$, our spectrum has the form of the Kolmogorov one: $E(k) \sim k^{-5/3}$.

A natural extension of the β-model is the *bifractal* one: there are now two sets of fractal dimension $\mathcal{M}_1$ and $\mathcal{M}_2$, imbedded in the physical space $\mathbb{R}^3$. Near $\mathcal{M}_1$, the velocity has scaling exponent h_1

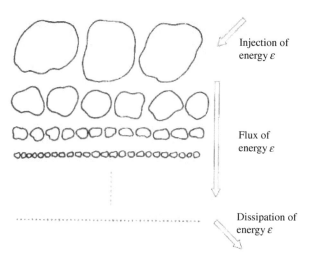

Injection of energy ε

Flux of energy ε

Dissipation of energy ε

Fig. 6.3. The (Richardson) cascade with energy transfer over scale of eddies.

and near $\mathcal{M}_2$ it has scaling exponent h_2. Specifically, we assume that

$$\frac{\delta v_\ell(\vec{r})}{v_{\ell_0}} = \left(\frac{\ell}{\ell_0}\right)^{h_i}, \quad \vec{r} \in \mathcal{M}_i, \quad \dim \mathcal{M}_i = D_i, \quad i = 1, 2,$$

where v_ℓ is the typical value of the velocity associated to scales $\sim \ell$ and $\delta v_\ell(\vec{r})$ denotes the velocity increment between point $\vec{r}$ and another point a distance ℓ away.

Besides the bifractal model, there exist the more general multi-fractal ones (see, e.g., [57, 58]).

6.3. Direct Numerical Simulation of Turbulence

In principle, the time-dependent, three-dimensional Navier–Stokes equations contain all of the physics of a given turbulent flow. However, the smallest scales of turbulence are generally many orders of magnitude smaller than the largest scales of turbulence, the latter being of the same order of magnitude as the dimension of the object about which the fluid is flowing. Furthermore, the ratio of smallest to largest scales decreases rapidly as the Reynolds number increases. To make an accurate numeric simulation of a turbulent flow, all physically relevant scales must be resolved. Even modern supercomputers cannot guarantee the effective solving of these problems. Therefore, special methods for simulation of turbulent flows were developed (for example, based on the algebraic, one- and two-equation models and the second-order closure ones, see, e.g., [59]).

The last decades have been marked by a new approach [60, 61] in the study of turbulence, namely by the direct numeric simulation of the process of hydrodynamic flow on the basis of the solution of hydrodynamic and kinetic equations. The results of such an approach demonstrate an important difference from the traditional statistical methods. In particular, the significant role of large-scale "ordered" structures on the turbulence development was demonstrated.

The basis of the new ("rational") approach may be formulated as follows:

(1) for direct numerical simulation (DNS) of the characteristics of the fully developed shear turbulence by high Re — the dynamics

of the large-ordered structures (LOS) and small-scale stochastic turbulence (ST) — we assume the validity of two hypotheses: (H_1) independence of LOS and ST and (H_2) weak influence of molecular viscosity (or, more generally, the mechanism of dissipation) on properties of LOS;

(2) at the same time, the properties of flows within the boundary layers and within the thin layers of mixing, at the viscous interval of turbulence, as well as those of flows by the moderate Reynolds numbers and in the domain of laminar-turbulent transition, are primarily determined by the molecular diffusion, and for these flows, it is necessary to consider Navier–Stokes models;

(3) the pulsational motions in turbulence are of chaotic type and have an unstable, irregular character, thus constituting a stochastic process. In this respect, one can speak only about obtaining the mean characteristics of that type motion by means of the statistical processing of results using Monte–Carlo approaches.

6.3.1. *Turbulence and ordered structures*

At the dawn of the study of turbulence, the phenomenon was treated as a stochastic process (determined by random distributions of fluctuating quantities). However in papers [60, 61, 114], Belotserkovskii demonstrated that turbulent flows incorporate ordered motions of almost coherent structures and these motions have a defining role in the development of the turbulence.

Different experimental and theoretical studies [62–67] have shown that a wide class of turbulent flows with transverse shear is characterized by the unsteady ordered motion of large-scale formations ("large vortices") which oscillate weakly (*ordered motion of stochastic structures*) and possess a stable and typical space-time structure. In the case of jet problems, the inner flow zone is turbulent (stochastic) in nature and is formed by disordered small-scale fairly intense fluctuations, although its structure is approximately uniform.

Thus, in free shear flows (such as wakes and jets), a double ("intermittent") turbulence structure is observed [68]. The major zone occupied by the turbulent fluid is comparatively small and uniform.

However, an external system of steady "slow" (non-turbulent) large vortices is superposed onto it. This transfers the turbulent fluid from one part of the flow to another. Thus, ordered structures are typical and form the basis of shear turbulent motion. Following [69], turbulent structures are subdivided into the following types:

(1) *dynamic structures of bifurcation origin*, which exist near the laminar-turbulent transition point (structural stochastically: chaos born of order);

(2) *quasi-equilibrium structures* formed within the regions of developed turbulence where chaotic motion is so highly developed that a system is close to the thermodynamic equilibrium (the ordered structure of such a stochastic motion: order born of chaos);

(3) *flow structures* occupying a position halfway between the two enumerated above.

In free shear turbulent flows (e.g., jets, wakes, mixing layers), ordered large-scale formations are observed for arbitrarily large Reynolds numbers. The strength, scale and shape of this low-frequency ordered motion are quasi-deterministic for a given flow and should be described by the hydrodynamic equations (not statistically). The dimensions of large vortices are comparable with the characteristic flow dimension and are much larger than the scales of the vortices forming the turbulent motion. The turbulence within the central portion of a wake is characterized by a high degree of local isotropy (with respect to the strengths of velocity pulsations in various directions, and hence with respect to the energy characteristics [2]).

In numerical analysis, it is important to simulate transfer processes correctly. There are three types of energy-related phenomena characteristic of real turbulent motion [70]:

(T_1) generation of large-scale vortices, which depend on the specific properties of the flow as a whole;

(T_2) fragmentation of the vortices into smaller-scale ones caused by nonlinearity, and the transmission of energy down the spectrum without considerable loss (the cascade process);

(T_3) viscous energy dissipation for the smallest scales.

Until recently, only types (T_2) and (T_3) have been subjected to analysis by statistical methods and the theory was developed ignoring the vortex-generation mechanism (the necessary fluctuation energy was added from outside by introducing a random external force). However, the vortex-generation aspect of large-scale turbulence (which is mostly deterministic) is very important since it incorporates the source of the turbulence as well as the mechanism for sustaining it.

Evidently, large-scale transfer of a turbulent fluid is primarily implemented by an ordered motion of groups of large vortices that distort the boundaries of a turbulent flow field and transfer the turbulent fluid in the transverse direction (see [62, 68]). Hence, the motion of ordered and large-scale turbulent structures is a major factor in determining the kinematic, dynamic and energy characteristics of a flow as a whole; also, it determines the properties of the depth turbulence responsible for energy dissipation.

However, for free developed shear turbulent flows with high Reynolds numbers, the backward influence of small-scale turbulence (as well as of the molecular diffusion) on the major characteristics of the large-scale flow is obviously insignificant since these are effects of different orders of magnitude: small-scale structures are "almost" universal, and the process of energy transfer itself is one-sided.

Now, consider the basic results of the investigation of coherent structures [71]. A fluid with a shear and/or temperature gradient is a thermodynamically non-equilibrium system. Although the density ρ, the internal energy ϵ and the velocity $\vec{v}$ are then defined in the usual manner, the remaining thermodynamic quantities, such as entropy and pressure, are defined as the same functions ρ and ϵ, as they are at the thermal equilibrium, but retain their exact physical meaning only up to the terms that are small compared with the shear and/or temperature gradient. The developed locally isotropic turbulence in the inertial range, which does not exhibit a uniform distribution of kinetic energy over the wavenumber space, but rather the strongly non-uniform Kolmogorov spectrum $\sim k^{-11/3}$, is not a system in thermodynamic equilibrium. In order words, turbulent disorder is quite different from the complete chaos that is characteristic of thermodynamic equilibrium in that it exhibits a certain degree of order.

That this order exists also implies that at any given time, some spatial structure or structures will exist in the flow. While each type of laminar flow is characterized by a specific spatial structure, in turbulent flows, ordered structures may be generated at random points and random times. These structures are randomly oriented and exist for random periods of time. At present in the synergetics concerning self-organization processes, which include the creation of transitions "order ⇆ chaos", these macrostructural elements are defined "coherent structures".

To identify coherent structures and analyze their individual properties, one has to use specific techniques for a conditional averaging of ensembles of visualized flows, say, by relating them to singular points of the topology of streamlines as well as by selecting the elements of an ensemble from the maximum of the histogram of the time intervals between them. In [69], coherent structures classify in *residual flows* (the "successors" of ordered subcritical stationary or quasi-periodic forms encountered in turbulent flows), *non-equilibrium flows* (which in the case of a developed turbulence is far from thermodynamic equilibrium, such as Kolmogorov turbulence, may possess universal self-similarity properties) and the theoretically imaginable, *quasi-equilibrium flows* (synergetically born from chaos close to the thermodynamically equilibrium; for example, the formation of zonal flows out of a two-dimensional turbulence in an ideal fluid at the surface of a rotating sphere).

Thermodynamic non-equilibrium in, say, incompressible fluid is generated by the velocity shear. For the sake of simplicity, we start by considering plane-parallel flows and the boundary-layer flows on a flat plate in a uniform oncoming flow; or, as another example, we consider flows with constant velocity shear. In these cases, the most natural coherent structures are rolls whose axes are perpendicular to the plane of shear.

Owing to secondary instabilities, vortex tubes may meander and suffer varicose expansions and contractions (in an ideal compressible fluid, this results in variations in density, and, in the presence of gravity, in the portions of the tubes with lower density floating up due to buoyancy forces). Similar structures (toroidal rolls) may appear in the flow regions with a more complex geometry. In a viscous fluid, neighboring locally antiparallel vortex tubes may reconnect, forming pairs of oppositely oriented vortices with residual "bridges" between them.

Sometimes, these vortex structures may form quasi-periodic one-dimensional chains or 2D- and 3D-lattices. Defects may arise within the latter, i.e., defects "frozen" into the lattice or involved into "Brownian motion" on it; if the number of defects is large, then a "turbulence of defects" capable of evolving into fully developed turbulence may merge.

Taking into account that the topology of coherent structures is in the best manner represented by the vorticity vector lines, we propose to define coherent structure as preferable nonlinear superposition of large-scale spatial components of the turbulent vorticity field which are the least unstable with respect to perturbations existing in the turbulent flow. Turbulent-flow coherent structures should be distinguished from the quite different laminar-flow spatial structures (e.g., axisymmetrical toroidal vortices in three-dimensional space, two-dimensional dipole vortices in a plane), described by some general self-similar solutions of the nonlinear vorticity equation with initial singularities. Identification of real coherent structures in the nature, e.g., in the ocean surface layer (as well as the reflections in it of coherent structures existing in the atmosphere boundary layer) is very important. The structures may appear as successors to both primary and secondary instabilities of the Ekman boundary layers in the ocean and atmosphere.

The reproduction of coherent structures within the turbulent flows by the numerical solution of hydrodynamic equations is difficult, especially for high Reynolds numbers, when the spectrum of turbulence scales is very wide and cannot be fully resolved from the maximum to the minimum scales even using the fastest modern computers. Therefore, the *large eddies simulation* (LES) must be implemented separately from a numerical description of a small-scale turbulence with full account of the molecular viscosity, the latter being at present feasible only for $Re \leq 1000$.

As we mentioned above, the original approach to the LES problem has been developed by Belotserkovskii [44, 61, 72]. The approach is based on a direct numerical solution (DNS) of the ideal-fluid hydrodynamic equations combined with the approximate account of energy dissipation due to small-scale turbulence. The turbulent flow under study is represented as relatively ordered slow motion of coherent structures — slightly unstable large eddies transporting portions of a fluid with developed turbulence from place to place. In particular, this approach may be used to describe the "residual" coherent structures of bifurcational origin ("dynamic structures").

6.3.2. *Simulation of coherent structures in turbulent flow*

The basic problems of organizing the computational process for the DNS of large-scale flows with developed turbulence for very high Reynolds numbers are: Problem (P_I) — calculation of the unsteady motion of ordered and large-scale turbulent structures; Problem (P_{II}) — numerical simulation of the stochastic component of a turbulent flow (small-scale turbulence). Let us consider these problems briefly.

Problem (P_I). Since the motion is large scale and ordered, it can be described by numerical schemes based on the non-stationary hydrodynamic equations (integral laws of conservation) for the ideal-medium model (Euler's difference equations) and possessing an approximate dissipation mechanism. The major objective of the dissipation mechanism is to extinguish small-scale oscillations and to ensure both the stability of the solution and the required resolution of the flow microstructures. Both experimental and numerical studies have shown that "oriented" compact asymmetric differences in the integral laws of conservation used without semiempirical models of turbulence satisfy these conditions because the properties of large-scale motions are mainly determined by volume convection (they have a wavy or dynamic nature) and depend on the solution as a whole. Hence, the calculations should be carried out throughout the flow field on real difference grids and should be followed by the determination of the required averaged characteristics of turbulent flow by an appropriate statistical reduction of the results.

It should be noted that the possibility of determining the statistical characteristics of an unsteady large-scale turbulent flow by smoothed equations is far from evident and the correctness of the problem formulation should be carefully analyzed.

For the present, the problem solution allows us to simulate the motion of ordered structures, large (energy-transferring) turbulent vortices, the process of turbulence generation and other processes based on large-scale transfer. Also, by using an appropriate representation of the dissipation mechanism, we may study the cascade process of energy transfer, i.e., the descent to small scale, locally isotropic turbulence. If the dissipation rate corresponding to large structures is determined correctly, then the energy characteristics of turbulence during the descent will evidently be simulated correctly since the process is rather conservative.

The deterministic approach allows us to single out the ordered and large-scale formation characteristics of structural turbulence; however, one has to specify the method for averaging oscillatory fields.

Problem (P_{II}). This problem deals with the simulation of the local resolution, i.e., the process of dissipation, the way in which energy propagates along the turbulent core of jets, etc. We believe such flows should be calculated using statistical methods or by introducing appropriate turbulent-viscosity coefficients phenomenologically. We may also use algorithms based on oscillatory equations [74], for which the averaged-flow parameters are determined by solving Problem P_I.

It should be noted that at present, the above flows should only be calculated within limited subregions by cutting off large-gradient zones out of the general flow pattern. This approach is based on the assumption that if the turbulence scale is small as compared with the averaged-flow scale, then the local structure of turbulence is sufficiently universal for various flows and is determined by the local conditions only. Thus, the flows may be calculated by using clearly defined models and sufficiently fine numerical grids, and hence the required computer capacity may be lowered.

Usually, the Navier–Stokes equations are averaged with respect to the Reynolds number over all scales of turbulence simultaneously and over large time intervals (the regular averaged motion is singled out), and this requires all the structures to be simulated at once. Hence, it is impossible to construct a model of turbulence simulating different classes of motion. Unlike this approach, the method of Belotserkovskii is based on splitting the general motion into large- and small-scale structures (the non-random ordered motion of macrostructures is singled out). The motion of ordered and large-scale turbulent vortices (with dimensions $\lambda \geq h_g$, where h_g is the spacing of a difference grid) is determined by the direct integration of the full hydrodynamic equations, while simulation (smoothing) is only used for the subgrid small-scale fluctuations which cannot be resolved explicitly by numerical integration and possess rather universal properties.

For the construction of a numerical algorithm to calculate large-scale structures, as the initial point, we use a system of equations approximating the laws of conservation in the integral form:

$$\frac{\partial F}{\partial t} = - \oint_{\mathcal{S}_\Omega} \vec{Q}_F d\vec{\mathcal{S}}, \quad F = \{M, \vec{P}, E\}, \tag{6.33}$$

where $\vec{Q}_F$ is a vector of the densities of flows of value F (additive characteristics of a medium), M is a mass, $\vec{P}$ is a momentum, E is a total energy and $\mathcal{S}_\Omega = \partial\Omega$ is a side surface of the volume Ω. These equations with account of boundary conditions are solved for every elementary cell with using quadrature formula $(F_{n+1} - F_n)/\tau = -L_n(\oint_{\mathcal{S}_\Omega} \vec{Q}_F d\vec{\mathcal{S}})$, where L_n is an operator of smoothing of characteristics of medium in the elementary cell. To improve the characteristics of the monotony of difference schemes within the regions of large gradients, or those of the complicated structure of flow, the hybrid asymmetric schemes are used (these schemes remain the positiveness of an operator). Constructed using such a method, the schemes are divergent conservative and dissipative stable.

6.3.3. *Euler's difference equations*

Euler's difference equations (derived using differential approximations for the simplest splitting schemes in the method of large particles) have the form [75]:

$$\frac{\partial \rho}{\partial t} + \nabla(\rho \vec{V}) = 0, \tag{6.34}$$

$$\frac{\partial \rho u}{\partial t} + \nabla(\rho u \vec{V}) + \frac{\partial p}{\partial x} = \frac{\partial}{\partial x}\left(\rho \varepsilon_x \frac{\partial u}{\partial x}\right) + \frac{\partial}{\partial y}\left(\rho \varepsilon_y \frac{\partial u}{\partial y}\right), \tag{6.35}$$

$$\frac{\partial \rho v}{\partial t} + \nabla(\rho v \vec{V}) + \frac{\partial p}{\partial y} = \frac{\partial}{\partial x}\left(\rho \varepsilon_x \frac{\partial v}{\partial x}\right) + \frac{\partial}{\partial y}\left(\rho \varepsilon_y \frac{\partial v}{\partial y}\right), \tag{6.36}$$

$$\frac{\partial \rho E}{\partial t} + \nabla(\rho E \vec{V}) + \frac{\partial p}{\partial y} = \frac{\partial}{\partial x}\left(\rho \varepsilon_x \frac{\partial E}{\partial x}\right) + \frac{\partial}{\partial y}\left(\rho \varepsilon_y \frac{\partial E}{\partial y}\right), \tag{6.37}$$

where $\varepsilon_x = |u|\Delta x/2$ and $\varepsilon_y = |v|\Delta y/2$ are the coefficients of the toughness of approximation.

These series expansions are accurate to $O(\Delta t, h^2)$. Note that the structure of the coefficients of the toughness of approximation, $\varepsilon \sim |v|h$, in (6.35)–(6.37) is similar to that of the turbulent-viscosity coefficients for the scale λ ($\nu_{t,\lambda} \sim v_\lambda \lambda$). If a turbulent eddy of the scale λ moving with velocity v_λ is replaced by a large particle moving with velocity v_h, then the expressions for v_λ and $\nu_{t,\lambda}$ are identical. If $\varepsilon_x \sim \varepsilon_y \sim \nu_{t,\lambda}$, then equations (6.34)—(6.37) for $\rho = $ const (an incompressible fluid) transform into the exact Navier–Stokes equations, where the molecular viscosity ν_m is replaced by the effective turbulent viscosity ν_t.

The left-hand sides of (6.34)–(6.37) correspond to the exact Euler's differential equations and the right-hand sides to the dissipation terms, which describe the perturbation background, appear due to the smoothing of the subgrid fluctuations and the approximation of original system of differential equations by finite-difference ones and depend on the internal structure of the representation. The dynamics of this background is the source of the fluctuations. It follows that equations (6.34)–(6.37) in concrete calculations are dissipative (though they were derived from the ideal fluid model). The specific form of the dissipation mechanism depends on the nature of the approximation, and generally, its structure may be controlled.

Let us compare Euler's difference equations with the Navier–Stokes difference ones. Using the splitting scheme from the method of large particles for the Navier–Stokes model, the momentum equation derived in [75] for a compressible medium has the form of

$$\frac{\partial \rho u}{\partial t} + \frac{\partial (p + \rho u^2)}{\partial x} = \left(\mu \left(\frac{\lambda}{\mu} - 2 \right) + \rho |u| \frac{\Delta x}{2} \right) + O(\Delta t, \Delta x^2),$$

where $\mu = \rho \nu$, λ is the second viscosity coefficient, and ν is the molecular viscosity. It follows from this that the scheme effects also have an influence on the approximation of the viscous terms. In order to ensure the calculation stability for flows with large Re, the effective viscosity must be high and usually exceeds the real viscosity. Thus, we see that for high Re, the numerical solution of boundary-value problems based on the NSE encounters both technical and theoretical difficulties. Thus, if the ideal gas equations are supplemented with dissipation terms, then for rather arbitrary assumptions about the nature of the dissipation, the generalized solution for the flow macrostructures corresponding to the limiting regime $\nu \to 0$ may be obtained, with a certain accuracy, from the equations incorporating an approximate dissipation mechanism instead of the Navier–Stokes equations.

Our approach to vortex-motion macrostructures does not employ semi-empirical models of turbulence. In this case, the dissipation mechanism incorporated in the averaging operator itself (which arose due to the use of oriented finite-difference representations) acts as a filter that cuts off the subgrid fluctuations. It is this dissipation mechanism in the averaged form that accounts for the contribution of small-scale subgrid vortices. Direct calculations on the grid of fully averaged dynamic macroscopic equations must ensure a stable solution, the build-up of the energy integral and the stabilization of the unsteady flow parameters. This yields the effective dissipation mechanism and satisfies the requirement that the splitting into "large" and "small" scales is invariant. Other important characteristics of turbulent flow, such as the energy dissipation rate and the integral scale, may also be determined within the domain of large-scale vortices. Thus, the above large-scale approach allows us to study the major properties of free fully developed turbulence.

6.3.4. *Large-scale turbulence in ocean*

As an example of using the above-analyzed approach to modeling of turbulence, we consider the problem of the origin and evolution of the flow in the barotropic ocean having constant depth, under the action of suddenly arising large-scale field of wind. The basic equations describing two-dimensional oceanic circulation acquire the following dimensionless form:

$$\frac{\partial u}{\partial t} + \varepsilon^2 \left(\frac{\partial u^2}{\partial x} + \frac{\partial uv}{\partial y} \right) - fv = -\frac{\partial p}{\partial x} + \gamma^3 \cdot \triangle u - \delta u + \tau_x,$$

$$\frac{\partial v}{\partial t} + \varepsilon^2 \left(\frac{\partial v^2}{\partial y} + \frac{\partial uv}{\partial x} \right) + fv = -\frac{\partial p}{\partial y} + \gamma^3 \cdot \triangle v - \delta v + \tau_y,$$

$$\frac{\partial u}{\partial x} + \frac{\partial v}{\partial y} = 0,$$

where axis x is directed to the east, axis y to the north, u, v are components of the averaged over depth velocity (along x- and y-axes), p is the averaged over depth pressure, (τ_x, τ_y) are the components of the wind's tangential stress, $f = (a_0/L)\mathrm{tg}(\theta_0) + \beta y$ is the Coriolis' parameter in the β-plane approximation, a_0 is Earth's radius, θ_0 is a latitude, L is the characteristic size of the basin; the problem's parameters:

$$\varepsilon = \sqrt{\nu/\beta}/L, \quad \gamma = \sqrt{A_L/\beta}/L, \quad \delta = r/\beta L, \quad U = \tau_0/\beta H L,$$

where parameter ε characterizes the effect of velocity's convection; the parameter γ reflects the presence of a horizontal turbulent friction with coefficient A_L; the parameter δ is responsible for the bottom-side friction with coefficient r.

According to the experimental data, the spectrum of values Re and ε is sufficiently wide. In [73], the region of increasing of Reynolds number $Re = 0.01 \ldots 3$, and the region of decreasing of convection parameter $\varepsilon = 0.035 \ldots 0.0126$ (these values are close to real one) were considered. With the increase of Re for fixed $\varepsilon = 0.02$ from calculations (based on the analyzed above approach), the following regimes were obtained:

(1) steady $(Re \sim 0.25\text{--}0.5)$;
(2) periodic $(Re \sim 0.5\text{--}1)$;
(3) stable irregular $(Re \sim 2.5)$;
(4) unstable irregular (*the chaos*) $(Re \sim 3)$.

In the steady case, the formation of the vortex is just beginning; in the periodic regime, the vortex goes out of the boundary layer in northern direction (upwards) and merges with the corner vortex (here, the loss of stability in boundary layer is already initiated, but the northward motion is strictly periodical); in the stable irregular regime, which arises by the intensification of β-effect, the vortices are split up and move chaotically within the basin (the periodicity is not observed, and the system's energy oscillates randomly about some fixed value). And at last in the unstable irregular regime, due to the increase of Re, the sizes of vortices are essentially decreasing, and they move in a random way.

6.4. Summary

In this chapter, the basic properties of small-scale turbulence motion in fluid and scenarios of the turbulence origin were examined. The method of the description of turbulent flows with the help of correlation functions was considered and the equation for evolution of kinetic energy density in modes of wavenumber was obtained. Cascade process of transfer of energy on a scale was analyzed and the description of methods of calculation of turbulent flows was investigated. In Section 6.1, the specific properties of turbulent flows were considered. The mechanism of origin of turbulence via doubling of the periods was described. Homogeneous isotropic turbulence was investigated and the hydrodynamical equations for fluctuating velocity were obtained. Section 6.2 is devoted to the consideration of properties of correlation functions, cumulants and Navier–Stokes equations in the Fourier representation. The energy cascade and the equation for energy on wavenumber were obtained. The properties of intermittency were considered. In Section 6.3, the direct numerical simulation of turbulence was considered. The basic idea in the considered approach of modeling turbulence is based on the Belotserkovskii ideas, that large and small scales in turbulent flow can be examined separately and we can use the Euler equation for simulation of large structures.

During the work on the material of this chapter, authors used the following books and papers: [1, 3, 12, 15, 18, 54], etc. We especially recommend for readers Refs. [20, 39 and 41] for a detailed study of the problems mentioned in this chapter. For the study of numerical methods, the reader should refer to the works of O.M. Belotserkovskii and V.M. Chechetkin (see the following References Section).

Part III
Coherent Hydrodynamic Structures

Chapter 7

Vortex Dynamics on the Plane

In the previous part, we considered the sophisticated mathematical formalism of fluid instabilities which reduce to postlaminar regimes of flows. We need to explain the inner structure of such flows. The nature of turbulence is connected closely with vortices in fluid and multivortex systems which can form structures of essentially varied scales.

7.1. Joyce–Montgomery Equation for Vorticity

We rewrite the two-dimensional (2D) Euler–Helmholtz equation as $\partial \omega / \partial t = [\psi, \omega]$, where $\omega(x, y) = \partial v_y / \partial x - \partial v_x / \partial x_y$ is a local vorticity of flow, $\psi(x, y)$ is a streamline, $v_x = \partial \psi / \partial y$, $v_y = -\partial \psi / \partial x$ are components of 2D flow velocity $\mathbf{v} = (v_x, v_y)$, $[\cdot, \cdot]$ is a canonical Poisson bracket.

The simplest vortex solutions of EHE are *point vortex* $\omega_{\delta,1}(x, y; t) = \gamma_1 \delta(x - x(t)) \delta(y - y(t))$ and system of N point vortices $\omega_{\delta,\sum}(x, y; t) = \sum_{k=1}^{N} \gamma_k \delta(x - x_k(t)) \delta(y - y_k(t))$, where $\gamma_{k=1,\dots,N}$ is an intensity of kth point vortex, which is equal to the *local circulation* of the flow: $\int_{S_k} \omega_{\delta,k} \, dS$, $S_k = \operatorname{supp} \omega_{\delta,k}(x, y; t)$, $S_k \cap S_\ell = \varnothing$ if $k \neq \ell$.

We consider the 2D motion of an incompressible fluid characterized by the velocity field $\mathbf{v} = \{v_x, v_y\}$. The stream function ψ is defined by the relation $\mathbf{v} = \nabla \psi \times \mathbf{e}_z$, where $\mathbf{e}_z$ is the basis vector of the z-axis normal to the flow plane xOy (the existence of ψ is a consequence of the incompressibility assumption $\operatorname{div} \mathbf{v} = 0$ and is independent of whether the fluid is viscous or inviscid). The stream function is related to the local (at the point $\mathbf{r} = \{x, y\}$) vorticity $\omega(\mathbf{r})\mathbf{e}_z = \operatorname{rot} \mathbf{v}(\mathbf{r})$ via Poisson's equation $\omega(\mathbf{r}) = -\triangle_2 \psi(\mathbf{r})$. Assume that the vorticity can be represented as a collection of point vortices with the same circulation γ; i.e., $\omega_N(\mathbf{r}) = \sum_{i=1}^{N} \gamma \delta(\mathbf{r} - \mathbf{r}_i(t))$, where $\mathbf{r}_i(t)$ is the point in the plane xOy, where the vortex i is located at the time t. Then the velocities of the vortices are $\mathbf{v}_i = d\mathbf{r}_i/dt = \nabla \psi_N(\mathbf{r}_i, t) \times \mathbf{e}_z$, where $\psi_N(\mathbf{r})$ is the solution of Poisson's equation for the system of N vortices. For an unbounded fluid, the solution is given by $\psi_N(\mathbf{r}) = -(2\pi)^{-1}\gamma \sum_{i=1}^{N} \ln |\mathbf{r} - \mathbf{r}_i|$. As a result, in the absence of self-action, i.e., the action of the given vortex on itself, the velocity of the point vortex i can be represented as

$$\mathbf{v}_i = \sum_{j \neq i}^{N} \mathbf{v}_{(i)}^{(j)}, \quad \mathbf{v}_{(i)}^{(j)} \equiv \frac{\gamma}{2\pi} \frac{\mathcal{B} \cdot (\mathbf{r}_i - \mathbf{r}_j)}{|\mathbf{r}_j - \mathbf{r}_i|^2}, \quad \mathcal{B} = \begin{pmatrix} 0 & I \\ -I & 0 \end{pmatrix},$$

where $\mathcal{B}$ is a symplectic operator over the $2N$-dimensional configuration space. The dynamics of point vortices in a plane can be described by the Euler equations $d(\partial L/\partial \dot{\xi}_k)/dt = \partial L/\partial \xi_k$ ($\xi_k = \{\eta_k, \bar{\eta}_k\}$, $\eta_k = x_k + iy_k$), which are derived by varying the action $J_L \equiv \int_0^t L(\xi, \dot{\xi})dt$ on the Lagrangian–Chapman functions

$$L(\eta_j, \bar{\eta}_j, \dot{\eta}_j, \dot{\bar{\eta}}_j) = \frac{1}{2i} \sum_{k=1}^{N} \gamma_k (\bar{\eta}_k \dot{\eta}_k - \eta_k \dot{\bar{\eta}}_k)$$

$$-\frac{1}{2} \sum_{k=1}^{N} \sum_{j=1}^{N}{}' \gamma_k \gamma_j \ln \left((\eta_k - \eta_j)(\bar{\eta}_k - \bar{\eta}_j) \right).$$

By using the Legendre transformation, the equations of motion, $\dot{\eta}_k = (2\pi i)^{-1} \sum_j{}' \gamma_j/(\bar{\eta}_k - \bar{\eta}_j)$, $k, j = 1, \ldots, N$, in the vortex system can be represented in the Hamilton form (γ): $\gamma \dot{x}_j = \partial H/\partial y_j$, $\gamma \dot{y}_j = -\partial H/\partial x_j$, where $H(\mathbf{r}_j|\gamma) = \gamma^2 \sum_{j<k} U_{jk}$ is the Hamilton–Kirchhoff function of the system and $U_{jk}(\mathbf{r}_j, \mathbf{r}_k)$ is the potential for the interaction between vortices. In the case under study (for unit intensity of the vortices), $U_{jk} = U_{jk}^{(0)} \equiv -(2\pi)^{-1}\ln |\mathbf{r}_j - \mathbf{r}_k|$. This function corresponds to potential inviscid flows described by the Euler fluid dynamics equation $\partial \mathbf{v}/\partial t + \mathbf{v}\nabla \mathbf{v} = -\operatorname{grad} \chi$ and the continuity equation $\operatorname{div} \mathbf{v} = 0$, where $\chi = p/\rho + G$ (here, p is the pressure, ρ is the fluid density, and G is the potential energy density

of the external force field). Indeed, the solutions of these equations can be represented as $v_x \propto \partial U^{(0)}/\partial y$, $v_y \propto -\partial U^{(0)}/\partial x$, where $U^{(0)} = (-2\pi)^{-1}\ln\sqrt{x^2 + y^2}$, $\chi(x,y) = \kappa_1\sin(\kappa_0 t) + \kappa_2\cos(\kappa_0 t) + \kappa_3$, $\kappa_j(x,y)$ are functions of $\mathbf{r} = \{x,y\}$ ($j = 1,2,3$ and $\kappa_0 = \mathrm{const}$) determined in the course of solving the Euler equations by substituting the above expressions for v_x, v_y, $\chi(x,y)$ into the equations and equating the coefficients of $\sin(\kappa_0 t)$ and $\cos(\kappa_0 t)$.

The formation of large structures that are similar to those in the spectral energy range can be explained and described by invoking near-equilibrium statistical mechanics of vortices. For simplicity, the presentation is again given in terms of 2D flows. Consider a bounded system $\mathcal{D}$ of area G occupied by vortices (the circulations γ_i are assumed to have identical signs). Since the coordinates x and y are canonically conjugate, the measure of the corresponding phase space of the N-vortex system is finite and can be represented as

$$G^N = \int \prod_j^N dx_j dy_j.$$

The energy state density is given by

$$\mu(E) = \int \delta(E - H(x_1, y_1, \ldots, x_N, y_N)) \prod_{j=1}^N dx_j dy_j.$$

The relation to an equilibrium N-vortex function has the form

$$F_N(\mathbf{r}_1, \ldots, \mathbf{r}_N) = (\mu(E))^{-1}\delta(E - H(x_1, y_1, \ldots, x_N, y_N)),$$

$$\int F_N \prod_{j=1}^N dx_j dy_j = 1.$$

The volume of the phase spaces corresponding to energies H not exceeding a given value E is

$$\varphi(E) = \int_{E_m}^E \mu(E)dE.$$

The function $\varphi(E)$ grows monotonically from zero to $\varphi_{\max} = G^N$ as the argument increases from $E = E_m$ to $E = +\infty$. Thus, $\mu(E) = d\varphi(E)/dE$ has a local maximum at $E = E_0$. For a microcanonical ensemble, the entropy S and the temperature T have the form $S = \ln \mu(E)$ and $T = dE/dS$. Obviously, for $E > E_0$, the function $S(E)$ is decreasing. Therefore, the temperature of the system is negative on this interval ($T < 0$). Thus, the qualitative behavior of vortices depending on E in the domain is as follows: vortex clusters are formed with a concentration depending substantially on the temperature field, and the vortex density in the clusters increases with the state energy starting with E_0. If $E < E_0$, then $T > 0$ and the vortices accumulate near the walls of the domain.

Indeed, consider, for example, a system of point vortices in a bounded radially symmetric domain with the measure $V = R^2$. The energy state density has the form

$$\mu(E, V) = \int_0^R \cdots \int_0^R \delta \left(E + (2\pi)^{-1} \sum_{i<j} \gamma_i \gamma_j \ln |\mathbf{r}_i - \mathbf{r}_j| \right) \prod_{i=1}^N d\mathbf{r}_i.$$

The entropy of the system is then given by

$$S_1(E, V) = N \ln V + S_0(E),$$

$$S_0(E) = E + \Gamma_0 (4\pi)^{-1} \ln V \left(\Gamma_0 \equiv \sum_{i<j} \gamma_i \gamma_j \right).$$

Consequently, the pressure in the system is then given by $P = T(\partial S_1/\partial V)_E = NTV^{-1}(1 + \Gamma_0/(4\pi NT))$. The critical temperature $T_0 = -\Gamma_0/(4\pi N) < 0$ corresponds to the transition to vortex gas statistics with $P = 0$.

An equilibrium vortex distribution (somehow similar to the Boltzmann distribution in classical statistical mechanics) can be obtained in the mean-field approximation. The entropy of the system can then be written as $S = -N \int F_1(\mathbf{r}) \ln F_1(\mathbf{r}) d\mathbf{r}$. Varying the entropy $\delta S - \beta \delta E - \alpha \delta \Gamma = 0$ (where α, $\beta \equiv 1/T$ are Lagrange multipliers) at fixed values of the mean energy $E = (1/2)N(N-1)$ $\gamma^2 \int U(|\mathbf{r} - \mathbf{r}'|) F_2(\mathbf{r}, \mathbf{r}') d\mathbf{r} d\mathbf{r}'$ (i.e., with the mean–energy conservation laws satisfied) and of the complete circulation $\Gamma = N\gamma$ gives *Joyce–Montgomery equation* (JME) $\triangle_2 \psi = -\varkappa \exp(-\beta \gamma \psi)$. Here, $\triangle_2 \psi = -\langle \omega \rangle = -N\gamma F_1(\mathbf{r})$ and $\varkappa$ is an arbitrary constant. If we take into account that the angular momentum $L = \int \langle \omega \rangle r^2 d\mathbf{r}$ is fixed while the entropy is varied and if we use the simplifying assumption $\varkappa = -2\Omega_L$ (which does not change the situation qualitatively), then the Joyce–Montgomery equation becomes $\triangle_2 \widetilde{\psi} + \lambda(\exp(\widetilde{\psi}) - 1) = 0$, where $\lambda = -\beta\gamma\varkappa$, $\widetilde{\psi}(r, \vartheta) = -\beta\gamma(\psi - \Omega_L r^2/2)$ and Ω_L is an additional Lagrange multiplier (all the quantities used are assumed to be suitably non-dimensionalized).

7.2. Vortex Systems with Negative Temperature

If coherent hydrodynamic structures are understood as coupled large-scale turbulent fluid masses with vorticity correlated in phase (coherent) in the entire domain occupied by a structure, then, according to the above formalism of a microcanonical ensemble, we conclude that these structures are characteristic of the inertial range and correspond to systems' negative temperatures, at which the point the vortices merge into multivortex groups. The local maximum of $\mu(E)$ near the critical point E_0 corresponds to the bifurcating solution of the JME at $\lambda = \lambda_0$ with a branch corresponding (in the linear isotropic approximation) to the quasi-periodic behavior of the stream function (the characteristic length size of an individual structure satisfies $\triangle z_n \propto z_{n+2} - z_n$, where z_n are the roots of the equation $J_0(z) = 0$). Obviously, a decrease in the temperature $T(< 0) \to T_0$ of the structure corresponds to $\lambda \searrow \lambda_0$, while as $T(< 0) \to 0$, $\lambda(> \lambda_0)$ increases indefinitely.

Assume that a system of vortices (with the standard pair potential $U_{12}(\mathbf{r}_1, \mathbf{r}_2) \propto \ln |\mathbf{r}_1 - \mathbf{r}_2|^{-1}$ for the interaction between vortices) has a spatially inhomogeneous temperature field with several local extrema. Then under certain conditions, the system can have several states, any of which is close in a certain sense to a local thermodynamic equilibrium and, hence, to the existence of a system of $K(> 1)$ multivortex structures, each of which corresponds to its own stream function $\widetilde{\psi}^{[k]}$ (additive with respect to the point vortices) that solves its own JME with allowance for the surroundings. The complete function $\widetilde{\psi}$ of the system can be represented as the sum of solutions of the corresponding collection of K equations, taking into account the influence of the structures on each other (specifically, via the mean pressure field).

Thus, the quasi-equilibrium states of the turbulent system described by (generally, a set of) Joyce–Montgomery equations can be identified with coherent structures observed in actual postlaminar flows. Note that the expediency of simulating free developed turbulence on the basis of the scale separation hypotheses as applied to near and far wakes, ocean currents, and flows between coaxial cylinders was examined by Belotsekovskii and Chechetkin in many papers. Specifically, it was indicated that the distribution of large vortices does not reflect the random structure of perturbations, but rather corresponds to the fluid dynamics laws in the case when inertial forces dominate viscous stresses. Then the integral pair of forces arising from the pressure field and velocity-related dynamic forces leads to the formation of a large vortex. In principle, the non-local structure of actual vortices can be described by applying the above formalism of point vortices and their unions under certain assumptions.

For large bounded systems, the conditions for the formation of large periodic structures are easy to analyze. For this purpose, consider the Joyce–Montgomery equation in the Hammerstein integral form

$$\psi(\mathbf{r}) = \lambda \int_{\mathcal{D}} U^*(\mathbf{r}, \mathbf{r}') \exp\big(\psi(\mathbf{r}')\big) d\mathbf{r}',$$

where the pair potential $U^*(\mathbf{r}, \mathbf{r}') \sim \ln(1/|\mathbf{r} - \mathbf{r}'|)$ is defined in a domain $\mathcal{D}$ (except for the singular sets). For simplicity, we consider the case of $\Omega_L \equiv 0$. The following basic assumptions are made: (i) the zero term in the Frechet–Taylor expansion of the integral operator on the right-hand side is bounded; i.e., $\int U^*(s) s \, ds = \sigma_R < \infty$; and (ii) the equation has a constant solution $\psi(\mathbf{r}) = \zeta_0 (= \text{const})$ (corresponding to the homogeneous system). For ζ_0, we obtain the transcendental equation $\zeta_0 = \lambda \exp(\zeta_0) \int U^*(\mathbf{r}, \mathbf{r}') d\mathbf{r}'$. According to assumption (i), $\zeta_0 = \lambda^* \exp(\zeta_0)$, where $\lambda^* \equiv 2\lambda\pi\sigma_R$. For $\lambda^* > 0$, this equation has two real solutions for ζ_0 if $\ln \lambda^* < -1$, and no real solution if $\ln \lambda^* > -1$. Thus, the boundary of solutions with a constant stream function is determined by the condition $\ln \lambda^* = -1$. Linearizing the Joyce–Montgomery equation about $\psi = \zeta_0$ (setting

$\psi = \zeta_0 + \delta\psi$, $|\delta\psi| \ll \zeta_0$), we obtain the second-kind Fredholm linear equation

$$\widehat{\Xi}(\delta\psi) \equiv \delta\psi(\mathbf{r}) - \widetilde{\lambda}_0 \exp(\zeta_0) \int_{\mathcal{D}} U^*(|\mathbf{r} - \mathbf{r}'|)\delta\psi(\mathbf{r}')\, d\mathbf{r}' = 0,$$

$$\widetilde{\lambda}_0 = \frac{\zeta_0 \exp(-\zeta_0)}{2\pi\sigma_R}. \tag{7.1}$$

Its solution is sought in the form $\delta\psi(\mathbf{r}) \propto \exp(i\mathbf{k}\mathbf{r})$ ($\mathbf{k}$ is the wave vector). Based on the Sommerfeld integral representation of Bessel functions, a criterion for the existence of such periodic solutions is the approximate equality $2\pi\lambda \exp(\zeta_0) \int xU^*(x)J_0(kx)\, dx \approx 1$. This "structural criterion" makes it possible to determine the period of the solution (in fact, the size of the coherent structure) depending on the temperature $1/\beta$ and vorticity at the homogeneous stage of the system evolution (and also on the pair potential). If the criterion is positive, the solution can be obtained (improved) by substituting, as before, the Puiseux series expansion about $\widetilde{\lambda}_0$, namely, $\delta\psi(\mathbf{r}) = (\lambda - \widetilde{\lambda}_0)^{1/2}(\delta\psi)_1(\mathbf{r}) + (\lambda - \widetilde{\lambda}_0)^{2/2}(\delta\psi)_2(\mathbf{r}) + \cdots$ into the Joyce–Montgomery integral equation. The first-approximation equation (for the coefficient $(\delta\psi)_1(\mathbf{r})$) coincides in the form with equation (7.1). Its solution $(\delta\psi)_1(\mathbf{r}) = C \sin(\mathbf{k}\mathbf{r})$, where C is a constant and the wavenumber $|\mathbf{k}|$ is determined by the structural criterion. The coefficients $(\delta\psi)_j$ for $j = 2, 3, \ldots$ are determined by solving the following chain of recurrence linear inhomogeneous Fredholm integral equations:

$$\widehat{\Xi}((\delta\psi)_2) = \int U^*(|\mathbf{r} - \mathbf{r}'|)$$

$$\times \left(\Xi^{[0]}(\mathbf{r}') + \widetilde{\lambda}_0 \exp(\zeta_0)\, \Xi^{[2]}(\mathbf{r}')\big((\delta\psi)_1(\mathbf{r}')\big)^2 \right) d\mathbf{r}', \ldots,$$

$$\widehat{\Xi}((\delta\psi)_n) = \int U^*(|\mathbf{r} - \mathbf{r}'|)\left(2\widetilde{\lambda}_0 \exp(\zeta_0)\, \Xi^{[2]}(\mathbf{r}')(\delta\psi)_1(\mathbf{r}')(\delta\psi)_{n-1}(\mathbf{r}') \right.$$

$$\left. + P_n\big((\delta\psi)_1(\mathbf{r}'), \ldots, (\delta\psi)_{n-2}(\mathbf{r}')\big) \right) d\mathbf{r}', \ldots; \quad \Xi^{[k]}:$$

$$\exp(\psi(x)) = \sum_{k=0}^{\infty} \Xi^{[k]}\big(\psi(x)\big)^k,$$

where $P_n(x_1, \ldots, x_{n-2})$ is a polynomial of degree n.

The effect of, for example, second-order terms in the Puiseux series can be interpreted as an increase in the local density of vortices arising from the homogeneous distribution of periodic structures.

7.3. Kinetic Origin of Multivortex Systems

Vortex-like hydrodynamic flows of various scales have attracted much interest for a fairly long time, which is motivated by the fact that fluid motions of this type can be strongly related to both turbulence and laminar flows of various origins, as can be inferred from physical observations and field experiments. Therefore, it can be expected that vortex dynamics methods in fluid media can be used to explain the mechanism of the transition to postlaminar flow regimes.

The description of vortex motions is based on the Navier–Stokes approximation in the form of Helmholtz (or Gromeka–Lamb) equations involving the vorticity. In a sense, vorticity is a natural variable, especially for viscous flows, since the latter manifest their vorticity within boundary layers and at distances comparable with the size of the body. However, the numerical solution of the Helmholtz equations faces a fundamental difficulty related to the specification of boundary conditions for vorticity, since they require complete information on skin friction, which is not known in advance. On the other hand, a mathematically correct formulation of the boundary problem does not involve a boundary condition at all: its presence in the problem makes it overdetermined. As a result, the vorticity on the boundary of the body is usually specified approximately, frequently with invoking rather specific correction techniques (based on empirical data); eventually, the shear stress on the boundary of the body is also determined approximately. Thus, the formulation of the boundary problem for the Helmholtz equations involves an element of artificiality (its detailed analysis shows that the exact boundary conditions can be represented in terms of surface delta functions). As a result, the system of fluid dynamics equations has to be supplemented with additional complicating assumptions, for example, with the presence of unsteady Oseen–Lamb vortices with a non-trivial internal structure, whose evolution is described using contour dynamics methods.

The standard Navier–Stokes equations are not very suitable for describing turbulent flows. Consequently, even in terms of vorticity, they can hardly be used to explain the transition to a postlaminar flow regime. An argument supporting this conclusion is provided by the structure of constructive models designed for numerical algorithms. Specifically, they involve Reynolds averaging, in which case the Navier–Stokes equations or their solutions are averaged with respect to some (unknown) parameters with the help of some (also unknown) function. Moreover, since the resulting averaged system is not closed, a certain closure model is required. The presence and character of these auxiliary operations are so substantial that the original formulations of fluid dynamics laws for laminar flows become subsidiary.

For the fluid dynamic modeling of postlaminar regimes, direct numerical simulation based on the Euler equations seems to be a much more attractive and consistent approach, at least for large-scale dynamic processes. Large vortices are responsible for the basic energy characteristics of turbulent flows and determine their structure. The distribution of such vortices does not reflect the random structure of perturbations but rather corresponds to fluid dynamics laws, with the inertial terms in the Navier–Stokes equations dominating the viscous stresses. Then the vortex structure is generated by two forces originating from the pressure and velocity fields. The creation of large vortices and the subsequent dynamics of the flow must be described by the Euler equations. At the same time, the simulation of small-scale vortex motions based on this technique seems to require the introduction of Rankine, Kirchhoff, and other discrete vortices and the construction of a theory of their evolution (for example, based on JME-type equations with the concept of indefinite temperature for the corresponding Onsager statistical system).

Therefore, it seems reasonable to explore the possibility of describing vortex structures in a hydrodynamic flow from the point of view of the kinetic approach based on Boltzmann-type and more general equations derived by simplifying the correlation functions from the BBGKY hierarchy. As a result, the basic dynamic aspects of the formation of both coherent structures and high-frequency vortex regimes at high (subcritical and postcritical) Reynolds numbers can be revealed and the limits of the applicability of present-day methods used for approximate flow computation (based on the Euler, Navier–Stokes, Burnett, etc., equations) can be determined. The constructive design of algorithms and the numerical simulation of vortex dynamics based on the direct solution (or in combination with other available techniques) of the Boltzmann kinetic equations or similar equations with a modified collision term (in order to form a suitable spectrum structure of the associated linear operators) have long been addressed by numerous authors (description of turbulence with new terms added to the BBGKY hierarchy, the study of vortex structures and turbulence in postcritical regimes by directly solving the Boltzmann equation, the investigation of the Taylor–Couette problem by applying the formalism of kinetic equations).

One may examine the properties of a stationary modified Boltzmann kinetic equation with allowance for long-range interparticle correlations in order to analyze the existence of its special solutions (associated with closed cycles on the state manifold of the system of bifurcation equations), which can be associated with vortex structures in the fluid. For this purpose, it is necessary to use the Lyapunov–Schmidt projection formalism and consequences of the bifurcation theory for systems of nonlinear differential equations. The results of this analysis can be used to establish conditions for the transition to a postlaminar flow regime and to refine the properties of a developed regime of small-scale turbulence generation.

The possibility of the formation of vortex structures in the steady Couette flow was explored in [76] (at the kinetic level). It was shown that such structures can arise in a fluid flow under certain conditions that can be interpreted in terms of a Hopf bifurcation occurring in the solution of the stationary Boltzmann kinetic equation with allowance for correlations of interacting particles on near-hydrodynamic intervals. Note that the hard sphere approximation was used as a model for the interaction of particles in the collision integral. Strictly speaking, the collision integral then takes the Enskog form, and, in the general case, the representation of the kernel of the given integral operator with a pair distribution function in the coarse-grained approximation differs from the standard (multiplicative) representation. However, this model was chosen only to simplify the presentation. In the case of physically more realistic interparticle potentials (e.g., inversely power-like ones), the generalized eigenvalues of the discrete spectrum of have to treated as pseudoeigenvalues (in the sense of Friedrichs spectral concentration). Apparently, this does not lead to substantial changes in the technique for analyzing the formation of a Hopf bifurcation of the solution to the modified Boltzmann equation.

7.4. Summary

In this chapter, we demonstrated that there exists an important task of current importance associated with the issue under consideration, which is to study problems of various types concerning the evolution of vortex solutions of the EHE that have a non-local spatial structure on the basis of the statistical approaches. However, to construct a continual-discrete generalization of the theory (when vorticity supports are finite structures), we need to restate existing methods and create new ones for analyzing vortex dynamics. It seems reasonable to study them on the basis of available material concerning the behavior of clusters of point vortices. Such vortex clusters can serve as a basis for developing efficient numerical algorithms for solving fluid dynamics problems, such as the simulation of turbulent regimes of shear flows and the evolution of large-scale coherent structures. The quasi-equilibrium dynamics of vortex structures is described by the Joyce–Montgomery equation (JME), which can be treated as an analog of the Poisson equation for systems of electromagnetic or gravitating particles. Thus, a justified approach is to develop a corresponding methodology for analyzing hydrodynamic flows. For this purpose, we need to begin with determining the form of solutions (namely, a functionally invariant general solution and a maximally large set of particular solutions) for the system of the EHE and JME. In Section 7.1, we considered the microcanonical ensemble of Onsager vortices and introduced a Joyce–Montgomery equation for vorticity in fluid. In Section 7.2, we investigated the bifurcation of solutions to the Joyce–Montgomery equations, and mathematically substantiated the existence of coherent quasi-equilibrium structures in flows and even provided research techniques for their evolution. In Section 7.3, we considered the possibility of origin of multivortex systems in the fluid on the basis of the Boltzmann-like kinetic equation theory.

During the work on the material of this chapter, authors used the following the books and papers: [76–80] etc. We especially recommend for readers Refs. [81–84] for a detailed study of problems mentioned in this chapter.

Chapter 8

Poisson Systems in Hydrodynamics

To investigate coherent multivortex systems, we must introduce a rigorous definition of coherency in terms of thermodynamics. With this aim, we generalize Hamiltonian dynamical systems.

8.1. Poisson Systems and States of Relative Equilibrium

Now, we introduce a rigorous definition of coherent structure on the Poisson manifolds and on its basis define the coherent structure of hydrodynamic type. With this aim, we consider an abstract evolutionary equation $\partial\mathfrak{f}/\partial t = \widehat{SM}(\mathfrak{f})\delta H(\mathfrak{f})$, where $H(\mathfrak{f})$ is an autonomous (Hamilton) functional in state space $\mathcal{F} \ni \mathfrak{f}$, $\delta H(\mathfrak{f})$ is a variational derivative of H, and also $\delta H(f) \in T_\mathfrak{f}^*\mathcal{F}$ (here, $T_\mathfrak{f}^*\mathcal{F}$ is a cotangent to $\mathcal{F}$ space); an operator $\widehat{SM}(\mathfrak{f}) : T_\mathfrak{f}^*\mathcal{F} \to T_\mathfrak{f}\mathcal{F}$ is a *structure mapping*, connected with skew-symmetric (and satisfied Jacobi condition) generalized Poisson brackets

$$[K_1, K_2](\mathfrak{f}) \equiv \langle \delta K_1(\mathfrak{f}), \widehat{SM}(\mathfrak{f})\delta K_2(\mathfrak{f})\rangle,$$

where $\langle\cdot,\cdot\rangle$ is an inner product in the state space. If an operator $\widehat{SM}(\mathfrak{f})$ is an irreversible one, then its kernel is non–empty, and $\widehat{SM}(\mathfrak{f})\delta C(\mathfrak{f}) \equiv 0$, where $C(\mathfrak{f})$ are (generalized) Casimir functionals, which satisfied an equation $[H, C] = 0$. The triple $(\mathcal{F}, \widehat{SM}(\mathfrak{f}), [K_1, K_2](\mathfrak{f}))$ defines a *Poisson system* (it may be considered as a generalization of the classical Hamilton system).

The Euler–Helmholtz equation in the two-dimensional case is $\partial\omega/\partial t = \widehat{SM}(\omega)\delta H(\omega)$, or, in simplified form:

$$\partial\omega/\partial t + \widehat{\Lambda}_2\nabla_2\psi \cdot \nabla_2\omega = 0,$$

where $\widehat{\Lambda}_2$ is an operator of symplectic structure (2×2-matrix). This equation is an example of the Poisson system with the Hamilton functional $H = \frac{1}{2}\int \psi\omega d\mathbf{x}$ (*de facto*, it is the Lamb energy of flow in a bounded region with Dirichlet conditions on the boundary), and the Poisson bracket:

$$[K_1, K_2](\omega) = \langle\delta K_1(\omega), \widehat{SM}(\delta K_2(\omega))\rangle, \quad \widehat{SM} \equiv -\nabla_2\omega \cdot \widehat{\Lambda}_2\nabla_2,$$

$$\widehat{\Lambda}_2 = \begin{pmatrix} 0 & 1 \\ -1 & 0 \end{pmatrix}, \quad \mathbf{v} = \widehat{\Lambda}_2\nabla_2\psi.$$

Variation derivative $\delta H(\omega)$ may be calculated on the basis of the above form of the Lamb energy $H = H(\mathbf{v}) = \frac{1}{2}\int \mathbf{v}^2 d\mathbf{x}$. By definition, the first variation $\delta H[\omega; \delta\omega] = \langle\delta H/\delta\omega, \delta\omega\rangle \equiv \int (\delta H/\delta\omega) \cdot \delta\omega\, d\mathbf{x}$, but then

$$\delta H[\omega; \delta\omega] = \int \mathbf{v} \cdot \delta\mathbf{v}\, d\mathbf{x} = \int (\nabla_2 \times \psi) \cdot \delta\mathbf{v}\, d\mathbf{x}$$

$$= \int \psi \cdot (\nabla_2 \times \delta\mathbf{v})\, d\mathbf{x} = \int \psi \cdot \delta\omega\, d\mathbf{x}.$$

Thus, $\delta H(\omega) = \psi$. Consequently, the second term in the left-hand side of EHE transforms to the well-known form: $(\partial\psi/\partial x_1)(\partial\omega/\partial x_2) - (\partial\psi/\partial x_2)(\partial\omega/\partial x_1)$.

If we consider the problem of determination of energy extremum with additional conditions $\text{extr}_\omega\{H(\omega) \mid I_1(\omega) = I_1; \cdots ; I_s(\omega) = I_s\}$, where $I_{k=\overline{1,\ldots,s}}$ are additional integrals of motion, obeying in involution ($[I_m, I_n] = 0$, $m, n = 1, \ldots, s$), then solutions of this problem are *states of relative equilibrium* (SRE) $\Omega(I_1, \ldots, I_s)$ (its define invariant manifolds of the flow). For investigation of properties of vortex flows, we can use integrals of energy, circulation, enstrophy, angular momentum, linear momentum, etc.:

$$H(\omega) = \frac{1}{2}\int \psi\omega d\mathbf{x}, \quad \Gamma(\omega) = \int \omega d\mathbf{x}, \quad W(\omega) = \frac{1}{2}\int \omega^2 d\mathbf{x},$$

$$P(\omega) = \frac{1}{2}\int |\mathbf{x}|^2\omega d\mathbf{x}, \quad L(\omega) = (L_1(\omega), L_2(\omega))$$

$$= \left(\int x_1\omega d\mathbf{x}, \int x_2\omega d\mathbf{x}\right).$$

In particular, for determination of a Rankine vortex, we must solve the problem on minimization with restraints $\min_\omega\{H(\omega) \mid P(\omega) = I_1^{(0)}, \Gamma(\omega) = I_2^{(0)}\}$. Then we obtain $\Omega_{\text{Rank.}}(i_1^{(0)}) = \left(I_2^{(0)}\right)^2/(4\pi I_1^{(0)}) \cdot \chi(R = \sqrt{4\pi I_1^{(0)}/I_2^{(0)}})$, where $\chi(R)$ is an indicator-function of the circle region $\{\mathbf{x} \mid |\mathbf{x}| \leq R\}$ (in this case energy function is $H(I_1^{(0)}, I_2^{(0)}) = (I_2^{(0)})^2/(16\pi) - (I_2^{(0)})^2/(8\pi)\ln(4I_1^{(0)}/I_2^{(0)})$, and the obtained vortex rotates steady as a rigid body with angular velocity $\omega_{\text{Rank.}} = -I_2^{(0)}/(8\pi I_1^{(0)})$).

The investigation of states of relative equilibrium for Navier–Stokes equations for many physically interesting case may be reduced to the analysis of stability of inviscid families of SRE. For example, we consider the NSE $\dot{\omega} = (-\nabla_2\omega \cdot \widehat{\Lambda}_2\nabla_2)\delta H(\omega) + \nu\Delta_2\omega$ on the plane bounded domain D, with boundary condition $\omega = 0\big|_{\partial D}$. Supposing first $\nu = 0$, and we then investigate the variational problem with additional condition on the enstrophy (as example): $\text{extr}_\omega\{H(\omega) \mid W(\omega) = I^{(0)}\}$. This problem has infinite family of SRE $\mathcal{R}_k$ $(k = 0, 1, \ldots)$, and state $\omega^\dagger \in \mathcal{R}_k$ if and only if $-\Delta_2\omega^\dagger = \mu_k\omega^\dagger$, $\{-\mu_k\}_{k=0,1,\ldots}$ are eigenvalues of the Laplace operator in domain D ($\mu_0 > 0$, $\mu_m < \mu_n$ if $m < n$ $(m, n \in \mathbb{N})$. All families $\mathcal{R}_k$ are invariant for vortex NSE, but the amplitude of SRE is non-steady: for any $\omega^\dagger \in \mathcal{R}_{k=0,1,\ldots}$, we can write an evolutionary solution: $\omega(t) = \omega^\dagger\exp(-\nu\mu_k t)$. For dissipation of energy, we obtain $dH/dt \leq -2\nu\mu_0 H$ and $dW/dt \leq -2\nu\mu_0 W$. The family $\mathcal{R}_0$ is the attractor for NSE, but all families with $k \geq 1$ are non-stable [85, 86].

Let us introduce the basic definition of coherency in a hydrodynamical system based on the theory of variation of functionals in Poisson manifolds. *The coherent structure* (CS) for (generalized) hydrodynamical system is the state of relative equilibrium (*elementary CS*) or ordered system of elementary coherent structures (maybe, several types) (*complex CS*) of the problem, connected with variation of Casimir functionals or integral of motion, with additional conditions on other several Casimirs (we suppose any CS is the quasi-stationary solution of the Euler equation).

8.2. Young Measures in Hydrodynamics

We consider Hamiltonian system $\{H(\mathbf{x}, \mathbf{p}) \mid \mathcal{M}_N\}$ with $2N$-dimensional phase space $T^*\mathcal{M}_N = \{(x_k, p_k)\}_{k=1}^{k=N}$. An example is a system of N Onsager vortices in the plane, since its Kirchhoff–Routh function is reduced to the Hamiltonian form by renormalization. The fact that the coordinates of the kth microstate (x_k, p_k) are canonical implies the validity of the Liouville theorem $\partial \dot{x}_k / \partial x_k + \partial \dot{p}_k / \partial p_k = 0$ ($\forall k = 1, \ldots, N$). With the use of this theorem, we can construct equilibrium statistical mechanics for the given Hamiltonian system, introducing $\mu^N = \prod_{k=1,\ldots,N} dx_k dp_k$ as an invariant measure. For the collection of quantities $\{\mathcal{C}_j(\mathbf{x}, \mathbf{p})\}_{j=1}^{j=n}$ conserved in the dynamics of the Hamiltonian system, the quantities $\mu_n^N = (\Omega_n)^{-1} \prod_{k=1}^N \Xi(\mathcal{C}_1, \ldots, \mathcal{C}_n) dx_k dp_k$ are also invariant measures (here, $\Xi(\ldots)$ is an arbitrary function and Ω_n is a normalizing constant). For an isolated system, a microcanonical distribution $\mu_{n,\text{micro}}^N$ that takes into account the given number of invariants of motion can be introduced by setting $\Omega_n \to \Omega_{n,\text{micro}}(\mathcal{C}_{1'}, \ldots, \mathcal{C}_{n'})$ and $\Xi(\mathcal{C}_1, \ldots, \mathcal{C}_n) = \prod_{j=1}^n \delta(\mathcal{C}_j(\mathbf{x}, \mathbf{p}) - \mathcal{C}_{j'})$ in the above general formula. Here, $\Omega_{n,\text{micro}}(\mathcal{C}_{1'}, \ldots, \mathcal{C}_{n'}) \prod_{j=1}^n \Delta\mathcal{C}_j$ is the volume of a portion of the phase space such that $\mathcal{C}_{j'} \leq \mathcal{C}_j \leq \mathcal{C}_{j'} + \Delta\mathcal{C}_j$ for sufficiently small $\{\Delta\mathcal{C}_j\}$). The use of the apparatus of microcanonical distributions leads to the sequential construction of equilibrium statistical mechanics for an N-particle Hamiltonian system, specifically, for Onsager vortices. To describe the nonstationary evolution of this vortex system, we need to use the kinetic theory based on the Liouville equation for the $(2N+1)$-dimensional distribution function with possible reduction to an equation for a single-vortex distribution function by applying the BBGKY method. However, since the kinetic approach is highly complicated, it is reasonable to put it aside for special problems, for example, ones requiring an especially detailed description of the evolution for part of a system, and to invoke the formalism of a canonical assemble for hydrodynamic systems.

In this context, questions arise associated with the fact that the Euler equation is an infinite-dimensional dynamical system and, accordingly, on it we cannot construct a canonical structure (similar to that based on the pair of adjoint coordinates $(\mathbf{x}, \mathbf{p})$ in the finite-dimensional case). However, for this system, we can introduce the non-canonical Poisson brackets $\{F_1, F_2\} \equiv \int \omega(\mathbf{x})[\delta F_1/\delta\omega, \delta F_2/\delta\omega]\, d\mathbf{x}$ (here, $F_{1,2}[\omega]$ are functionals, $\delta/\delta\omega$ is the functional derivative, $[G_1(\mathbf{x}), G_2(\mathbf{x})] \equiv (G_1)_{x_1}(G_2)_{x_2} - (G_1)_{x_2}(G_2)_{x_1}$, and $\mathbf{x} = (x_1, x_2)$). With the help of these brackets, the dynamics of the system can be described by an equation of the form $\partial\omega/\partial t = \{\omega, H[\omega]\}$ (where ω is local vorticity and H is the Hamiltonian of the system). Moreover, there exists an infinite-dimensional set of nuclear (Casimir) functionals $\mathfrak{C}$ of the bracket operator $\{F, \mathfrak{C}\} = 0$ (for an arbitrary functional $F[\omega]$).

The macroscopic state in the considered vortex system in the plane ("reducible to a canonical one") can be defined as follows. From the strictly deterministic vorticity $\omega(\mathbf{x})$ at a given point of the plane, we pass to the probability density $\varrho(\mathbf{x}, \lambda)$ of the existence of a continuous (pseudodiscrete) "vorticity level" $\lambda \in \mathbb{R}^1$ at the point $\mathbf{x} \in D \subseteq \mathbb{R}^2$ (with the normalization condition $\int \varrho(\mathbf{x}, \lambda)\, d\lambda = 1$). The locally averaged coarse-grained vorticity is expressed in probability terms as

$$\omega^\circ(\mathbf{x}) = \int \varrho(\mathbf{x}, \lambda)\lambda\, d\lambda, \quad \omega^\circ(\mathbf{x}) = \omega(\mathbf{x}) \quad \varrho(\mathbf{x}, \lambda) = \delta(\lambda - \omega(\mathbf{x})).$$

In fact, we introduce a Young measure of levels, i.e., a family of probability measures $\lambda = \{\lambda_{\mathbf{x}}\}_{\mathbf{x}\in D}$ associated with a sequence of measurable functions $F_k : \mathbb{R}^2 \to \mathbb{R}^1$ satisfying the condition

$$\lim_{k\to\infty} \int_D \varrho(\mathbf{x}, F_k(\mathbf{x}))\, d\mathbf{x} = \int_\Omega \int_{\mathbb{R}^1} \varrho(\mathbf{x}, \tau)\, d\lambda_{\mathbf{x}}(\tau)d\mathbf{x}.$$

The function $\omega = \omega°(\mathbf{x})$ is a stationary solution of the two-dimensional Euler–Helmholtz equation

$$\frac{\partial \omega}{\partial t} + \mathbf{v}\nabla\omega = 0, \quad \mathbf{v} = \nabla\psi \times \mathbf{e}_z, \quad \psi = -\Delta_2^{-1}\omega; \qquad (8.1)$$

moreover, the stream function must satisfy the condition $\psi|_{\partial D} = 0$ ($\psi, \mathbf{v}$ and $\mathbf{v}$ are consistent with $\omega = \omega°(\mathbf{x})$). Invariants (corresponding to $\mathcal{C}_j$ for the microcanonical description) in the evolution of the hydrodynamic system described by (8.1) are the kinetic energy of the Lamb flow (Hamiltonian function) $E[\omega] = -\frac{1}{2}\int \omega\Delta_2^{-1}\omega\,d\mathbf{x} = \text{const}$ and a set of Casimir invariants $\mathfrak{C}_{s[\omega]} = \int_D s[\omega(\mathbf{x})]d\mathbf{x}$, where $s[\omega]$ is an arbitrary admissible function (usually, $s[\omega] = \omega^m$, where $m = 1$ and $m = 2$ correspond to physically obvious values of the circulation Γ and the enstrophy W. For $m \geq 3$, the corresponding moments of the vorticity can be called *generalized enstrophies* of the mth order, which characterize the fine local structure of the hydrodynamic system). The conservation of all $\mathfrak{C}_{m\geq 1}$ is equivalent to the conservation of the area of the sublevels $A(\lambda) = \int_D \chi(\omega(\mathbf{x}) \leq \lambda)\,d\mathbf{x}$, where $\chi(D_\lambda)$ is the indicator function of the set $D_\lambda \subset D$.

The μ-phase space of the hydrodynamic system is partitioned into cells of two types: "macroscopic" $(x_1, x_1+dx_1)_i \times (x_2, x_2+dx_2)_i$, $i = 1, \ldots, N$, and "microscopic" — a single macroscopic cell contains $\ell(\geq 2^2)$ (identical) microcells $D_\ell \times D_\ell$. Let n_{ij} denote the number of microcells occupied by the vorticity level λ_j in the ith macrocell. Then the number of microstates associated with the macrostate $\{n_{ij}\}$ is

$$w(\{n_{ij}\}) = \prod_j N_j! \prod_i \frac{\ell!}{n_{ij}!},$$

where $N_j = \sum_i n_{ij}$ is the total number of microcells occupied in the level λ_j and they satisfy the cell normalization condition $\Sigma_j n_{ij} = \ell$. This condition specifies the Lynden–Bell statistic of type IV according to the classification in [87] (distinguishable particles with an exclusion principle) and, in vortex statistics, in fact, plays the role of an analog of the well-known Pauli exclusion principle (taking into account that the μ-phase and configuration spaces of the components

of the vortex system coincide in the plane). The entropy of the state
of the system with such a structure is computed by taking the log-
arithm of the function $w(\ell, n)$ and passing to the continual limit in
view of the Stirling formula

$$S[\varrho] = -\int \varrho(\mathbf{x}, \lambda) \ln \varrho(\mathbf{x}, \lambda) \, dx d\lambda. \tag{8.2}$$

8.3. Equilibrium States in Statistical Mechanics of the Euler Equation and Variation of Functionals in Poisson Systems

In the two-dimensional case, under some *a priori* constraints imposed
on the energy and a certain collection of Casimir functionals,
coherent hydrodynamic structures can be associated with solutions
of extremum problems for the functionals in the Poisson system
corresponding to the Euler equation in the plane.

It seems that the most detailed and clear description of the
mathematical apparatus for examining the properties of "quasi-
stationary states" of the Euler equation (of medium motions corresponding to a constrained maximum of the system's entropy) can
be given within the framework of Miller–Robert–Sommeria (MRS)
theory. Here, the basic assumptions are as follows: (i) the evolution of the two-dimensional hydrodynamic flow is described by the
Euler equation (with no external forces or dissipation); (ii) the values of all the Casimir invariants are known; and (iii) the system
reaches its statistical equilibrium at the most probable state. Under
these assumptions, a statistically equilibrium state for the Euler
equation in the plane can be obtained by constrained maximization of entropy (8.2) with additional Lagrange conditions that the
Lamb total energy, the total vorticity distribution in the system,
and the normalizing condition on the probability density $\varrho(\mathbf{r}, \lambda)$ are
preserved.

It should be emphasized that, in the given approach, small-
scale structures have no effect on the dynamics of large vortices
(since the latter are described using the coarse-grained vorticity
and the allowance for the fluctuation influence of the small-scale

dissipative environment goes beyond the model). Obviously, it is the large vortices that determine the energy of the flow (the Lamb energy of the coarse-grained vorticity is a constant). Thus, the postulates of the MRS model, in fact, coincide with the basic assumptions used in the development of Belotserkovskii turbulence models, which are based on the assumption that quasi-deterministic structures exist on large flow scales.

Consider the problem of finding a constrained extremum of the entropy $S[\varrho]$:

$$\text{extr}_\varrho \left\{ S[\varrho] \mid \int \varrho(\mathbf{x}, \lambda)\, d\mathbf{x} = A(\lambda),\ \frac{1}{2} \int w^\circ \psi\, d\mathbf{x} = -E,\ \int \varrho\, d\lambda = 1 \right\},$$

where $\psi = -\Delta_2^{-1} w^\circ$ is the stream function. In variations, we have

$$\delta S = \beta \delta E + \int \alpha_1(\lambda)\delta A(\lambda)\, d\lambda + \int \alpha_2(\mathbf{x})\delta \left(\int \varrho(\mathbf{x}, \lambda)\, d\lambda \right) d\mathbf{x},$$

where $\beta(= 1/T)$, $\alpha_1(\lambda)$, $\alpha_2(\mathbf{x})$ are functional Lagrange multipliers. The corresponding probability density has the form

$$\varrho(\mathbf{x}, \lambda) = (Z(\mathbf{x}))^{-1} \exp(-\alpha_1(\lambda) - \beta\lambda\psi),$$

$$Z(\mathbf{x}) = \int \exp(-\alpha_1(\lambda) - \beta\lambda\psi)\, d\lambda.$$

Thus, the coarse-grained vorticity is a function of the variable ψ (therefore, equation (8.1) has the solution w°):

$$w^\circ(\mathbf{r}) = (Z(\mathbf{r}))^{-1} \int \lambda \exp(-\alpha_1(\lambda) - \beta\lambda\psi(\mathbf{r}))\, d\lambda \equiv f[\psi(\mathbf{r})].$$

Consider the simplest non-trivial case of a discrete distribution of the coarse-grained vorticity in the system $w^\circ = P\lambda_1 + (1 - P)\lambda_2$ (where P and $1 - P$ are the probabilities that the vorticity is in a state at levels λ_1 and λ_2, respectively). Then the entropy expression becomes $S^{(2)}[\lambda_1, \lambda_2] = -\int (P \ln P - (1 - P) \ln(1 - P))\, d\mathbf{r}$. For $\lambda_2 \equiv 0$ (vorticity of level $\lambda_1 > 0$ against the background of a vortex-free medium), we have

$$S^{(2)}[\lambda_1, 0] = -\int((\omega^\circ/\lambda_1)\ln(\omega^\circ/\lambda_1) + (1 - \omega^\circ/\lambda_1)\ln(1 - \omega^\circ/\lambda_1))\, d\mathbf{r}.$$

Moreover, the extremum of the entropy is reached on a two-parameter family of equilibrium functions (having a form formally similar to the Fermi distribution in quantum statistics) of the coarse-grained vorticity

$$\omega_{LB}^\circ(\mathbf{r}) = \lambda_1(1 + \zeta\exp(\beta\lambda_1\psi(\mathbf{r})))^{-1},$$

where $\zeta = \exp(\lambda_1\alpha_1)$. Obviously, in the limit $\omega^\circ \ll \lambda_1$ (i.e., as $\ln(1 - \omega^\circ/\lambda_1) \to 0$), we then obtain an analog of the Boltzmann distribution $\omega_B^\circ \equiv \omega_{\beta,\lambda_1}^\circ(\mathbf{r}) = C_0\exp(-\beta\lambda_1\psi(\mathbf{r}))$ (and, accordingly, $S^{(2)}[\lambda_1, 0] \to S_B = -\int(\omega^\circ/\lambda_1)\ln(\omega^\circ/\lambda_1)\, d\mathbf{r}$).

On the basis of what was said above, we can conclude that, in the coarse-grained representation with the effective Lynden–Bell–Pauli principle (which means that the average occupation numbers of a phase microcell are bounded), the Joyce–Montgomery equation for the mean field, which describes the stationary dynamics of the vortex system (assuming that it has a unique non-trivial vorticity level), takes the following form in terms of the stream function:

$$-\Delta_2\psi(\mathbf{r}) = \omega_{LB}^\circ = \lambda_1\varrho(\mathbf{r}, \lambda_1) = \lambda_1(1 + \zeta\exp(\beta\lambda_1\psi(\mathbf{r})))^{-1}.$$

In addition to the MRS statistical theory, a rather large number of approaches to the variation of functionals for a Poisson Euler system that take into account various aspects of the problem of existence of statistically equilibrium states can be found in the literature. As major examples, we note the following ones:

(i) the Ellis–Haven–Turkington variational principle [88]:

$$\max_\varrho\left\{S_\chi[\varrho] \mid E, \Gamma, \Gamma, \int \rho d\lambda = 1\right\};$$

(ii) the Naso–Chavanis–Dubrulle variational principle [89]:

$$\max_\varrho\left\{S[\varrho] \mid E, \Gamma, \int \rho\lambda^2 d\mathbf{x}d\lambda, \int \rho d\lambda = 1\right\};$$

(iii) the minimization principle for the coarse-grained enstrophy:

$$\min_{\omega^\circ} \left\{ \int (\omega^\circ)^2 d\mathbf{x} \,\bigg|\, E, \Gamma \right\} \bigg|_{\omega^\circ(\mathbf{x}) = \int \varrho \lambda d\lambda} ;$$

(iv) the maximization principle for the coarse-grained pseudoentropy:

$$\max_{\omega^\circ} \{ S_g[\omega^\circ] \mid E, \Gamma \}.$$

Here, we used the following notation:

$$\Gamma = \int \omega^\circ d\mathbf{x}, \quad S[\varrho] = \int \varrho(\mathbf{x}, \lambda) \ln \varrho(\mathbf{x}, \lambda) d\mathbf{x} d\lambda,$$

$$S_g[\omega^\circ] = - \int \mathfrak{C}(\omega^\circ) d\mathbf{x},$$

$$S_\chi[\varrho] = - \int \varrho(\mathbf{x}, \lambda) \ln[\varrho(\mathbf{x}, \lambda)/\chi(\lambda)] \, d\mathbf{x} d\lambda,$$

$$\chi(\lambda) = \exp\left(-\sum_{k=1} \alpha_k \lambda^k \right).$$

In all the variational principles, except for MRS theory, only a distinguished set of quantities associated with the Casimir functionals are assumed to be conserved (in addition to the energy); accordingly, dissipation on small scales, the possibility of a phase transition in the system, etc., can be taken into account.

The structure and evolution of large-scale coherent vortices arising in turbulent hydrodynamic flows have to be studied by applying an approach combining numerical experiments and a detailed analysis of the fine structure of these objects (for example, the details of the behavior of the corresponding statistical ensemble). Moreover, it seems necessary to select a collection of physical processes that influence the dynamics of the structures under study (specifically, to separate different scales and to distinguish basic energy-containing structures). Allowance for small-scale turbulence in hydrodynamic computations is a task that can be addressed by applying various approaches (including the ones based on kinetic theory), and it has to be treated separately as an important, but not a crucial component of the general problem of simulating turbulent flows. In this context, it should be noted that the approach described in this chapter can help describe the inertial range of the turbulence spectrum, i.e., in the transition from large to small scales.

8.4. Summary

In this chapter, we considered near-equilibrium statistical mechanics of the Onsager-type vortices and analyzed variational problems connected with the origin of multivortex structures. In Section 8.1, we introduced the states of relative equilibrium for Poisson systems. In Section 8.2, we considered the applications of Young's measures in vortex hydrodynamics. In Section 8.3, we analyzed the variational problems of some types concerning coherent hydrodynamic flows.

During the work on the material of this chapter, authors used the following the books and papers: [90–94] etc. We especially recommend for readers Refs. [95–98] for a detailed study of problems mentioned in this chapter.

Chapter 9

Introduction of Metric Structure in Hydrodynamic Systems

The possibility of introducing a metric on manifolds associated with hydrodynamic flows various types has been repeatedly discussed in the literature, however, the results to date have either an excessive generality (which makes the development of techniques suitable for describing flows of specific types), or require the introduction of significant restrictions and problem-oriented problems with an *a priori* rigidly defined additional setting conditions, the physical content of which requires special analysis. If we consider the hydrodynamic flow as a statistical system in a state close to equilibrium, then its geometric properties can be investigated using the general Amari–Weinhold technique, using the possibility of introducing of a Riemannian topology on Gibbs manifolds (these manifolds are determined with the help of the relations between the thermodynamic potentials $p = p(\rho, T)$ and $s = s(u, \rho)$). In particular, if we turn to the model of a real flow in the form of a set of point vortices of the Onsager type, one can obtain meaningful differential-functional relations that are expressions for the specific heats of the vortex population and the equations of geodesic, binding states equilibrium levels of the system.

In this chapter, we consider the fundamentals of the geometrodynamic Hamiltonian formalism in relation to the stability of hydrodynamic structures.

9.1. Lagrangian and Hamiltonian Formalism of the Description of the Hydrodynamic System in Terms of Monge Potentials

Locally, the state of the hydrodynamic system may be described by density, velocity and density of entropy (in the general case of a compressible medium) at the given point $(\mathbf{x}, t) \in K_{N+1} \subset \mathbb{R}^{N+1}$ ($\mathbf{x} \in \mathbb{R}^{N \leq 3}$, $t \in \mathbb{R}^1_+$), that is by the set of quantities $\{\rho(\mathbf{x}, t); \mathbf{v}(\mathbf{x}, t); s(\mathbf{x}, t)\}$. The hydrodynamic equations that describe their variation in time have the form:

$$\rho_t + (\rho \mathbf{v})_\mathbf{x} = 0, \quad (\rho \mathbf{v})_t + (\rho \mathbf{v}^2)_\mathbf{x} = -p_\mathbf{x},$$

$$(\rho s)_t + (\rho \mathbf{v} s)_\mathbf{x} = 0, \tag{9.1}$$

where $p = p(\mathbf{x}, t)$ is a scalar pressure field. We use the representation of velocity fields with the help of the Monge potentials $\{M_\alpha\}|_{\alpha=1,\dots,m} \in Y_m$, separating in the expansion of the velocity field gradient term and a set of quasi-solenoidal ones:

$$\mathbf{v} = -(M^4) \cdot (M^1)_\mathbf{x} - s \cdot (M^2)_\mathbf{x} - (M^3)_\mathbf{x} \quad \text{if } m = 4,$$

where $M^\alpha(\mathbf{x}, t)$ are some scalar variables. For isentropic flows with $s = \text{const}$, we must change $M^3 + s \cdot M^2 \to M^2$, $M^4 \to M^3$ (in this case, $m = 3$). For the uniqueness of the choice of these fields, it is necessary to introduce the additional conditions: following [99], we select the following conditions as the first three ones: $\widehat{D}(M^4) = \widehat{D}(M^1) \equiv 0$, $\widehat{D}(M^2) \equiv T$, where $\widehat{D}(\dots) \equiv (\dots)_t + \mathbf{v} \cdot (\dots)_\mathbf{x}$ is the substantial derivative operator, $T = T(\mathbf{x}, t)$ is the field of thermodynamic temperature. We take for the last additional condition the relation defining specific enthalpy of flow: $w(\mathbf{x}, t) = (M^4) \cdot (M^1)_t + s \cdot (M^2)_t + (M^3)_t - \mathbf{v}^2/2$. Hydrodynamic equations (9.1) with the above additional conditions are equivalent to the corollaries from the Seliger–Whitham variational principle $\delta \int \int \pi d\mathbf{x} dt = 0$, where the density of the kinetic potential (pressure in the Monge representation) can be represented in the following form (the basic thermodynamic relation

$u = (w - p/\rho)|_{p=\pi}$ is used):

$$\pi(\{M^\alpha\}; \{(M^\alpha)_\mathbf{x}\}, \{(M^\alpha)_t\}) = \rho(w - u)$$

$$= \rho(((M^4) \cdot (M^1))_t + s \cdot (M^2)_t$$

$$+ (M^3)_t - \frac{1}{2}(-M^4 \cdot (M^1)_\mathbf{x} - s \cdot (M^2)_\mathbf{x}$$

$$- (M^3)_\mathbf{x})^2 - u(\rho, s))(m = 4),$$

$$\pi(\{M^\alpha\}; \{(M^\alpha)_\mathbf{x}\}, \{(M^\alpha)_t\}) = \rho(((M^3) \cdot (M^1))_t + (M^2)_t$$

$$- \frac{1}{2}(-M^3 \cdot (M^1)_\mathbf{x} - (M^2)_\mathbf{x})^2$$

$$- u(\rho, s))(m = 3). \tag{9.2}$$

Here, $u(\rho, s)$ is the specific internal energy of the flow (the caloric equation of the state of the medium is assumed to be known).

We consider the Hamiltonian representation, for which we define the conjugated to quasi-velocity variables $\{(M^\alpha)_t\}$ the "Monge representation impulses" $\{P_\alpha\}$:

$$P_1 = \frac{\partial \pi}{\partial((M^1)_t)} = \rho M^4, \quad P_2 = \frac{\partial \pi}{\partial((M^2)_t)} = \rho s,$$

$$P_3 = \frac{\partial \pi}{\partial((M^3)_t)} = \rho \ (m = 4);$$

$$P_1 = \frac{\partial \pi}{\partial((M^1)_t)} = \rho M^3, \quad P_2 = \frac{\partial \pi}{\partial((M^2)_t)} = \rho \ (m = 3).$$

We introduce the Hamilton function $H_m(\{M^\alpha\}, \{P_\alpha\})$ with the help of the Legendre transformation of Lagrangian ($H_m \equiv \sum_{\alpha=1}^{m-1} P_\alpha \cdot (M^\alpha)_t - \pi(\{M^\alpha\}; \{(M^\alpha)_\mathbf{x}\})$):

$$H_{m=4} = \frac{1}{2P_3}\left(\sum_{\alpha=1}^{3} P_\alpha \cdot (M^\alpha)_\mathbf{x}\right)^2 + P_3 \cdot u(P_2, P_3),$$

$$H_{m=3} = \frac{1}{2P_2}\left(\sum_{\alpha=1}^{2} P_\alpha \cdot (M^\alpha)_\mathbf{x}\right)^2 + P_2 \cdot u(P_2).$$

We consider the Hamiltonian space $W^m = (Y_m, H_m)$, $H_m :$ $T^*Y_m \to \mathbb{R}^1$ with the fundamental tensor $g_m^{\alpha\beta}(\{M^\mu\}, \{P_\mu\}) = \frac{1}{2}\partial^2 H_m/\partial P_\alpha \partial P_\beta$; the corresponding Riemannian element of the interval is $d\sigma_{W^m}^2 = \sum_{\alpha,\beta}(g_{\alpha\beta})_m dM^\alpha \otimes dM^\beta$. As an example, we give an explicit form of the contravariant metric coefficients for the simplest case of $m = 3$, $N = 2$ (index m omits):

$$g^{11} = \frac{(M^1)_{x_1}^2 + (M^1)_{x_2}^2}{P_2}, \quad g^{12} = g^{21} = -P_1\frac{(M^1)_{x_1}^2 + (M^1)_{x_2}^2}{P_2^2},$$

$$g^{22} = (P_1)^2\frac{(M^1)_{x_1}^2 + (M^1)_{x_2}^2}{(P_2)^3} + 2\frac{du(P_2)}{dP_2} + P_2\frac{d^2u(P_2)}{dP_2^2},$$

and the determinant $\det(g^{\alpha\beta}) = |(M^1)_{\mathbf{x}}|^2(P_2)^{-1}(2u' + P_2u'') \not\equiv 0$. It should be noted that the coefficients $g^{\alpha\beta}$ (as well as their analogs for $m = 4$) do not depend directly on the Monge potentials (new "configuration variables"). This greatly simplifies the further consideration of the geometrodynamic properties of the vortex motion of the hydrodynamic medium.

9.2. The Canonical Connections of the Hamiltonian Space

The N-linear connection on T^*Y_m is characterized by a pair d-tensor fields $D\Gamma(N) = (H^\alpha_{\beta\gamma}, C_\alpha^{\beta\gamma})$, that is, the system of generalized Christoffel coefficients (H, C), which in the general case, are functions of Monge potentials $\{M^\mu\}$ and canonically conjugate to them as pseudo-impulses $\{P_\mu\}$. However, one should pay attention to the specific structure of the dependence only on the impulses of the components of the fundamental tensor $g^{\alpha\beta} = g^{\alpha\beta}(\{P_\mu\})$, obtained above. In this case, there is a nullification of coefficients

$$H^\alpha_{\beta\gamma} \equiv \frac{1}{2}g^{\alpha\eta}(\delta_\zeta g_{\eta\gamma} + \delta_\gamma g_{\beta\zeta} - \delta_\zeta g_{\beta\gamma}),$$

where

$$\delta_\mu \equiv \partial/\partial M^\mu + N_{\mu\nu}\partial/\partial P_\nu$$

is an element of the adapted basis for direct decomposition $T_u T^* Y_m = N_u \oplus V_u \ (\forall u \in T^* Y_m)$. Here,

$$N_{\mu\nu} = \frac{1}{4}\{g_{\mu\nu}, H\} - \frac{1}{4}\left(g_{\mu\xi}\partial^2 H / \partial M^\nu \partial P_\xi + g_{\nu\eta}\partial^2 H / \partial M^\mu \partial P_\eta\right)$$

are coefficients of the nonlinear connection of the Hamiltonian space W^m (notation $\{g_{\mu\nu}, H\}$ is used for Poisson brackets on the T^*Y_m). Thus, the horizontal paths of the N-linear connection D are described by a system of differential equations

$$(M^\alpha)_{tt} + H^\alpha_{\beta\gamma}(M^\beta)_t(M^\gamma)_t = 0, \quad (P_\alpha)_t - N_{\mu\alpha}(M^\mu)_t = 0.$$

However, these equations obtain the confluent forms because of the absence of direct functional relationship of variables M^α in Hamiltonian and components of metric tensor (solutions of horizontal paths equations are trivial: $M^\alpha = c_1 t + c_2$, $P_\beta = c_3$).

Accordingly, more interestingly, we will analyze the vertical paths (v-paths) from the physical point of view (in a fixed point $(M_\alpha)_0 \in Y_m$) with respect to the N-linear connection $D\Gamma$. These v-paths are characterized by a system of differential equations that are analogs of the (non-confluent) Euler–Lagrange equations:

$$\frac{d^2 P_\alpha}{dt^2} - C^{\beta\gamma}_\alpha(\{M^\eta\}, \{P_\eta\})\Big|_{M_\alpha = (M_\alpha)_0} \frac{dP_\beta}{dt}\frac{dP_\gamma}{dt} = 0,$$

$$C^{\beta\gamma}_\alpha = -\frac{1}{2}g_{\alpha\zeta}\left(\frac{\partial g^{\zeta\gamma}}{\partial P_\beta} + \frac{\partial g^{\beta\zeta}}{\partial P_\gamma} - \frac{\partial g^{\beta\gamma}}{\partial P_\zeta}\right). \tag{9.3}$$

Values of some generalized Christoffel coefficients $C^{\beta\gamma}_\alpha$ are as follows:

$$C^{11}_1 = -\frac{P_1|(M^1)_{\mathbf{x}}|^2}{(P_2)^4} - \frac{(P_1)^2|(M^1)_{\mathbf{x}}|^2}{(P_2)^5},$$

$$C_1^{21} = \frac{|(M^1)_{\mathbf{x}}|^2}{(P_2)^3} + \frac{(P_1)^2|(M^1)_{\mathbf{x}}|^2}{(P_2)^5},$$

$$C_1^{22} = -\frac{2P_1|(M^1)_{\mathbf{x}}|^2}{(P_2)^4} + \frac{P_1(M^1)_{\mathbf{x}}}{2(P_2)^2}$$

$$\times \left(\frac{3(M^2)_{\mathbf{x}}}{(P_2)^2} - \frac{3(P_2)^2(M^2)_{\mathbf{x}} + 3(P_1)^2(M^1)_{\mathbf{x}}}{(P_2)^4} \right.$$

$$\left. + 3u''(P_2) + P_2 u'''(P_2) \right) - \frac{(M^1)_{\mathbf{x}}^2}{2(P_2)^3},$$

$$C_1^{12} = \frac{(M^1)_{\mathbf{x}}^2}{2(P_2)^3} + \frac{2(P_1)^2(M^1)_{\mathbf{x}}^2}{(P_2)^5} - \frac{P_1(M^1)_{\mathbf{x}}}{(P_2)^2} \left(\frac{3(M^2)_{\mathbf{x}}}{(P_2)^2} \right.$$

$$\left. - \frac{3(P_2)^2(M^2)_{\mathbf{x}} + 3(P_1)^2(M^1)_{\mathbf{x}}}{(P_2)^4} + 3u''(P_2) + P_2 u'''(P_2) \right).$$

Equation (9.3) can be considered as the basic equations in the study of the stability of the dynamics of the vortex flow of a fluid, described initially by the system of Euler equations. The basic information contained in their solutions relates to the form of the geodetic trajectory in a "impulse space" (really, on a manifold of densities of scalar flow characteristics). Of special interest are closed trajectories that are naturally associated with periodic hydrodynamic structures of different scales.

9.3. The Deviation of Geodesics on the Monge Manifolds

The question naturally arises of the stability of periodic orbits in an impulse space. For research deviation from the geodesic motion described by equation (9.3), we represent $P_\alpha = (P_\alpha)_0 + \epsilon \Pi_\alpha + O(\epsilon^2)$, where $(P_\alpha)_0$ is the solution of (9.3), ϵ is a small parameter, Π_α is the deviation from the exact solution (depending on from the geodetic parameter or time). Substituting the above equation in (9.3), we get

$$
0 = \frac{d^2(P_\alpha)_0}{dt^2} - C_\alpha^{\beta\gamma}(\{M^\eta\}, \{(P_\eta)_0\})\Big|_{M^\alpha=(M^\alpha)_0} \frac{d(P_\beta)_0}{dt} \frac{d(P_\gamma)_0}{dt}
$$
$$
+ \epsilon \left(\frac{d^2\Pi_\alpha}{dt^2} - 2C_\alpha^{\beta\gamma}(\{(M^\alpha)_0\}, \{(P_\alpha)_0\}) \right) \frac{d(P_\beta)_0}{dt} \frac{d\Pi_\gamma}{dt}
$$
$$
- C_{\alpha,\eta}^{\beta\gamma} \left(\{(M^\varsigma)_0\}, \{(P_\varsigma)_0\} \right) \frac{d(P_\beta)_0}{dt} \frac{d(P_\gamma)_0}{dt} \Pi_\eta(t) \right) + O(\epsilon^2).
$$

Carrying out the transformation of the expression in parentheses with the factor ϵ, we obtain an analog of the Jacobi equation (it can be called the Jacobi–Cartan equation) for the deviation vector with the components Π_α:

$$
\frac{D_p^2(\Pi_\alpha)}{dt^2} + \left(\frac{d(P_\beta)_0}{dt} \right) \left(\frac{d(P_\gamma)_0}{dt} \right) S_\alpha^{\beta\gamma\eta}(\Pi_\eta) = 0, \qquad (9.4)
$$

where $D_p\Pi_\alpha/dt \equiv d\Pi_\alpha/dt - C_\alpha^{\beta\gamma}(\{(M^\xi)_0\}, \{P_\xi\})(\Pi_\beta)(d(P_\gamma)_0/dt)$, $S_\alpha^{\beta\gamma\eta}$ is the ("third") curvature d-tensor of path:

$$
S_\alpha^{\beta\gamma\eta} = \frac{\partial C_\alpha^{\beta\gamma}}{\partial P_\eta} - \frac{\partial C_\alpha^{\beta\eta}}{\partial P_\gamma} + C_\alpha^{\mu\beta}C_\mu^{\beta\eta} - C_\alpha^{\mu\eta}C_\mu^{\beta\gamma}.
$$

Equation (9.4) describes the evolution of the deviation vector from the geodesic motion and when considering the (λ, ϵ)-congruence (λ is the affine parameter along the streamline proportional to the time t) closed trajectories, it is possible to trace the change in the density characteristics $(\rho, \rho M^4, \rho s)|_{m=4}$ or $(\rho, \rho M^3)|_{m=3}$ of the hydrodynamic structure, containing this congruence. At the same time, it is not assumed that the system has strict limitations in the spatial sense, that is, this system has non-local correlation properties (which is characteristic of coherent structures of different genesis). If the solution of the Jacobi–Cartan system has stable limit cycles, then this allows us to state that the system has a set of certain pseudo-stationary states (associated with these cycles). It seems very interesting to investigate the properties of these pseudo-stationary states and to reveal the relationship with quasi-stationary states, associated with coherent structures in multi-vortices dynamics. If the norm of the solution of the Jacobi–Cartan equation grows with time, this obviously indicates the decay of the coherent hydrodynamic system. If, at some point, the equation of geodesics has a singularity, and the deviation vector tends to zero in the norm, then we have a collapse of the vortex system.

We consider the simple special cases of Jacobi–Cartan equation, which demonstrate the main features of the physical processes described with the help of deviation of geodesic lines. We assume that the values of $(M^{1,2})_{\mathbf{x}} = \widetilde{M}^{1,2}$ are constants, then the coefficients of the connection $C_\gamma^{\alpha\beta}$ and components of $S_\alpha^{\beta\gamma\eta}$ depend only on $(P_0^{1,2})$:

$$c_1^{11} = -(P_1)_0 \widetilde{M}^1/(P_2)_0^4 - (P_1)_0^2 \widetilde{M}^2/(P_2)_0^3 \quad (m = 3),$$

etc. Respectively, in the case under study, equation (9.4) takes the form of a system of ordinary differential equations of the second order:

$$\frac{d^2\Pi_\alpha}{dt^2} + K_1\big((P_1)_0, (P_2)_0\big)\frac{d\Pi_\alpha}{dt} + K_2\big((P_1)_0, (P_2)_0\big)\Pi_\alpha$$
$$+ \frac{d(\Pi_\beta)_0}{dt}\frac{d(\Pi_\gamma)_0}{dt} S_\alpha^{\beta\gamma\eta}\big((P_1)_0, (P_2)_0\big)\Pi_\eta = 0$$

with coefficients depending on variables $(P_1)_0$ and $\big((P_2)_0\big)_t$ explicitly and being implicit functions of variable t. In fact, these coefficients may be considered constant for a fixed point in time. Thus, considering this equation as an equation for the vector variable P_α with "frozen" coefficients, one can see its analogy with the system of equations (with dissipative terms) describing the dynamics of coupled oscillators (in principle, it is possible to create conditions for the occurrence of self-oscillations in the system described by Eq. (9.4)). If we set $P_2 = \text{const}$ additionally, then for the simplest case $m = 3$ we will get one dissipative oscillator ordinary differential equation for the variable $P_1(\equiv \rho M^3)$ (without singularities in the coefficients); its solutions, according to general criteria, are oscillations decreasing in magnitude.

We have considered the possibility of formulating the dynamics of large-scale hydrodynamic structures in terms of geometric objects associated with the Hamiltonian system derived from Euler equations of hydrodynamics. This line of research seems extremely promising, since all the existing coherence criteria are either descriptive (they allow us to determine only the qualitative level of the situation) or they are purely specific and cannot be extended to flows with similar properties. The geometrodynamic approach developed in this chapter is universal, it subtracts the deep essence of coherence as a congruence of geodesic streamlines, and also allows predicting the behavior of the system (and, possibly, influencing the creation and destruction of coherent structures by local impact methods).

9.4. Summary

In this chapter, we considered the geometrization of the description of vortex hydrodynamic systems on the basis of the introduction of the Monge–Clebsch potentials, which led to the Hamiltonian form of the original Euler equations. In Section 9.1, we constructed the kinetic Lagrange potential with the help of the flow velocity field, which has been preliminarily determined through a set of scalar Monge potentials and thermodynamic relations. The next step transformed the resulting Lagrangian by means of the Legendre transformation to the Hamiltonian function and correctly introduced the generalized impulses canonically conjugate to the configuration variables in the new phase space of the dynamical system. In Section 9.2, we defined the Hamiltonian space on the cotangent bundle over the Monge potential manifold. Calculating the Hessian of the Hamiltonian, we obtained the coefficients of the fundamental tensor of the Hamiltonian space defining its metric. Next, we determined analogs of the Christoffel coefficients for the N-linear connection. Considering the Euler-Lagrange equations with the connectivity coefficients obtained, we arrived at the geodesic equations in the form of horizontal and vertical paths in the Hamiltonian space. In the case under study, non-trivial solutions can have only differential equations for vertical paths. In Section 9.3, we analyzed the resulting system of equations of geodesic motion from the point of view of the stability of solutions, and obtained important physical conclusions regarding the initial hydrodynamic system. To do this, we investigated a possible increase or decrease in the infinitesimal distance between the geodesic vertical paths (solutions of the corresponding system of Jacobi–Cartan equations). As a result, we can formulate very general criteria for the decay and collapse of a vortex continual system.

During the work on the material of this chapter authors used the following the books and papers: [100–104], etc. We especially recommend for readers Refs. [105–107] for a detailed study of problems mentioned in this chapter.

Appendices

Appendix A

A.1. Chaplygin's Gas

Chaplygin, in [108], considered the example of an ideal gas with adiabat $pV^{-1} = \text{const}$, or $p = p_0\rho_0/\rho$ (i.e., gas with the adiabatic exponent $\gamma = -1$), where p_0 and ρ_0 are the unperturbed pressure and density, respectively. The continuity equation and Euler's momentum equation for Chaplygin's gas are

$$\frac{\partial \widehat{\rho}}{\partial t} + (\vec{v} \cdot \nabla)\widehat{\rho} + \widehat{\rho} \operatorname{div} \vec{v} = 0, \quad \frac{\partial \vec{v}}{\partial t} + (\vec{v} \cdot \nabla)\vec{v} - mc_0^2 \operatorname{div} \widehat{\rho}^{1/m} = 0,$$
$$(\text{A.1})$$

where $\widehat{\rho} = \rho/\rho_0$, $c_0^2 = p_0/\rho_0 > 0$, $m = -1/2$ is the *azimuthal number*. For the one-dimensional medium ($\rho = \rho(t, x)$, $\vec{v} = \{v(t, x), 0, 0\}$), the system taking the form of

$$\frac{\partial \rho}{\partial t} + \frac{\partial(\rho v)}{\partial x} = 0, \quad \frac{\partial v}{\partial t} + v\frac{\partial v}{\partial x} = c_0^2 m \frac{\partial \rho^{1/m}}{\partial x}; \qquad (\text{A.2})$$

for the case of two-dimensional but time-independent flow ($\rho = \rho(x, y)$, $\vec{v} = \{v_1(x, y), v_2(x, y), 0\}$), we have

$$(\vec{v} \cdot \nabla)\rho + \rho \operatorname{div} \vec{v} = 0, \quad (\vec{v} \cdot \nabla)\vec{v} = c_0^2 m \nabla \rho^{1/m}. \qquad (\text{A.3})$$

Under the condition $rot\, \vec{v} = 0$, we can from the second equation, obtain the Bernoulli integral and the following relationship between derivatives:

$$v_1^2 + v_2^2 - 2c_0^2 m \rho^{1/m} = \text{const}, \quad \frac{\partial v_2}{\partial x} = \frac{\partial v_1}{\partial y}. \qquad (\text{A.4})$$

Equations (A.2) and (A.3) belong to the elliptic type and hence describe no running waves. In system (A.2), we examine the evolution of perturbations against an initially steady background, whereas, for system (A.3), we consider perturbations against the background of a uniform flow with velocity $v_1 = v_1^{(0)}$ along the x-axis. For the case $v_1^{(0)} \gg c_0$ (the flow consists of streams extended far along the x-axis in the down-stream direction), we can neglect the terms of order of $(c_0/v_1^{(0)})^2$ and the system (A.3) takes the approximate form:

$$v_1^{(0)} \frac{\partial \rho}{\partial x} + \frac{\partial (\rho v_2)}{\partial y} = 0, \quad v_1^{(0)} \frac{\partial v_2}{\partial x} + v_2 \frac{\partial v_2}{\partial y} \approx c_0^2 m \frac{\partial \rho^{1/m}}{\partial y}. \quad (A.5)$$

Although Chaplygin's gas does not exist in nature, equations similar to (A.1)–(A.5) (with other values of m and c_0^2) describe in a long-wave approximation a great number of unstable media which are called *quasi-Chaplygin media* [109]. These media differ from ordinary ideal gases only in their having a *negative compressibility*, $\partial p/\partial \rho < 0$.

We shall consider some examples of quasi-Chaplygin media in detail.

A.1.1. *Chaplygin's gas*

Chaplygin's gas is a medium with $m = -1/2$. The unsteady-state flow of this gas is defined by the following equations:

$$\frac{\partial \rho}{\partial t} + \frac{\partial (\rho v)}{\partial x} = 0, \quad \frac{\partial v}{\partial t} + v \frac{\partial v}{\partial x} = \frac{c_0^2}{\rho^2} \frac{\partial \rho}{\partial x}, \quad c_0^2 = \frac{p_0}{\rho_0}, \quad (A.6)$$

where p_0 and ρ_0 indicate the equilibrium state of the medium. With the help of the hodograph transformation

$$t = t(\rho, v), \quad x = x(\rho, v),$$

$$\frac{\partial x}{\partial v} = v \frac{\partial t}{\partial v} - \rho \frac{\partial t}{\partial \rho}, \quad \frac{\partial x}{\partial \rho} = v \frac{\partial t}{\partial \rho} + c_0^2 \rho^{1/m-1} \frac{\partial t}{\partial v},$$

$$\frac{1}{\rho^{1/m}} \frac{\partial (\rho^2 \cdot \partial t/\partial \rho)}{\partial \rho} + c_0^2 \frac{\partial^2 t}{\partial v^2} = 0,$$

we may rewrite the system (A.6) in the form

$$\frac{1}{c_0}\frac{\partial x}{\partial q_1} = q_2\frac{\partial t}{\partial q_1} - q_1\frac{\partial t}{\partial q_2}, \quad \frac{1}{c_0}\frac{\partial x}{\partial q_2} = q_1\frac{\partial t}{\partial q_1} + q_2\frac{\partial t}{\partial q_2},$$

$$\frac{\partial^2 t}{\partial q_1^2} + \frac{\partial^2 t}{\partial q_2^2} \equiv \Delta_{q_1 q_2} t = 0,$$

(A.7)

where $q_1 = p/p_0 = 1/\rho$ and $q_2 = v/c_0$ are new independent variables (dimensionless pressure and velocity, respectively). The solution of two-dimensional Laplace equation should be represented by a function of the form $t(q_1, q_2) = \Re(T(Q))$, where $T(Q)$ is an analytical function of the complex argument $Q = q_1 + iq_2$. It can be easily shown that the coordinate $x = x(q_1, q_2)$ also satisfies the Laplace equation $\Delta_{q_1 q_2} x = 0$, and thus $x = c_0 \Re(X(Q))$, where $X(Q)$ is an analytical function of variable Q. From (A.7), we obtain the relation $dX/dQ = -iQdT/dQ$.

The solutions for Chaplygin's gas can be found by analogy with electrostatics since the time t is similar to the "plane" electrostatic potential $\phi(x, y)$. We assume that as $t \to -\infty$, perturbations are absent, as if they were concentrated at the point $Q = 1$ of a complex plane. Then we obtain $T(Q) = t_* \ln((Q - 1)/Q + 1))$ and $X(Q) = -it_* \ln(Q^2 - 1)$, where $t_* > 0$ is arbitrary constant factor; consequently,

$$t = \Re(T(Q)) = \frac{t_*}{2}\ln\frac{(q_1 - 1)^2 + q_2^2}{(q_1 + 1)^2 + q_2^2},$$

$$x = c_0\Re(X(Q)) = -c_0 t_* \text{arctg}\frac{2q_1 q_2}{1 + q_2^2 - q_1^2}.$$

By inverting these formulae, we get

$$q_1 = \frac{p(x, t)}{p_0} = \frac{\text{ch}\,\tau - \cos\chi}{\text{sh}\,|\tau|}, \quad q_2 = \frac{v}{c_0} = -\frac{\sin\chi}{\text{sh}\,|\tau|},$$

$$\tau = \frac{t}{t_*} < 0, \quad \chi = \frac{x}{c_0 t_*}.$$

In the limit as $t \to -\infty$, we obtain $p/p_0 \approx 1 - 2\exp(\tau)\cos\chi$, $v/c_0 \approx -2\exp(\tau)\sin\chi$; but when $t \to -0$ the maximum pressure $(p/p_0)_{\max} = \text{cth}(|\tau|/2)$ at the points $\chi = \pi + 2\pi n$, and the minimum pressure $(p/p_0)_{\min} = th(|\tau|/2)$ at the points $x = 2\pi n$ $(n \in \mathbb{Z})$. So, this solution corresponds to perturbations which are periodic in x.

Solitary perturbations similar to a standing growing "soliton" can be obtained by taking such an expression for the function $T(Q)$ that would not allow going round the point $Q = 1$ without changing the sign of the time t. The simplest case is the "dipole" of the form $T(Q) = 1/(Q-1)$, for which from $dX/dQ = -iQ\,dT/dQ$ we find $X(Q) = -i\big(1/(Q-1) - \ln(Q-1)\big) + \text{const}$. To describe the solutions corresponding to these functions, let us introduce the parametric representation: $q_1 = 1 + \widetilde{q}_1 \cos \widetilde{q}_2$, $q_2 = \widetilde{q}_1 \sin \widetilde{q}_2$. Now, taking the real part of $T(Q)$ and $X(Q)$ and adding the dipole value of a proper sign, we get the following result for the time and coordinate:

$$t = -\frac{t_*}{\widetilde{q}_1} \cos \widetilde{q}_2, \quad x = c_0 t_* \widetilde{q}_2 + \frac{c_0 t_*}{\widetilde{q}_1} \sin \widetilde{q}_2, \quad \widetilde{q}_2 \in \left(-\frac{\pi}{2}, \frac{\pi}{2}\right),$$

$$t = \frac{t_*}{\widetilde{q}_1} \cos \widetilde{q}_2, \quad x = c_0 t_* \pi - c_0 t_* \widetilde{q}_2 - \frac{c_0 t_*}{\widetilde{q}_1} \sin \widetilde{q}_2, \quad \widetilde{q}_2 \in \left(\frac{\pi}{2}, \frac{3\pi}{2}\right).$$

In the asymptotic limit as $t \to -\infty$, these solutions yield the profiles for the "solitary hill" and "solitary well". Pressure is apparently extreme at the point $x = 0$ where it varies with time: $(p/p_0)_{\text{hill}} = 1 + |t_*/t|$, $(p/p_0)_{\text{well}} = 1 - |t_*/t|$.

The above-considered solutions are the simplest cases of a great variety of possible solutions of system for Chaplygin's gas.

A.1.2. *Overturned shallow water*

The best known classic example of absolute aperiodic instabilities is the Rayleigh–Taylor instability for boundary between two different liquids in a gravitational field, the heavier liquid overlying the lighter one. Let us consider a specific case of one-dimensional drops — "overturned solitons" — similar to the ridges that rise in a shallow water layer clinging to the ceiling. The equilibrium here is maintained by the upward gas pressure. The mathematical description of the "overturned" shallow water can be apparently obtained from equations of shallow water, the only change being made in the sign of the gravitational acceleration.

In view of the incompressibility of water, we get the equations

$$\operatorname{div} \vec{v} = 0, \quad \frac{\partial \vec{v}}{\partial t} + (\vec{v} \cdot \nabla)\vec{v} = -\frac{1}{\rho_0}\nabla p + \vec{g}, \quad v = \{v_1, 0, v_2\}, \quad (A.8)$$

in which it can be assumed that $v = \nabla \psi(x, z, t)$ (ψ is a velocity potential, $\nabla^2 \psi = 0$), the x-axis being horizontal and the z-axis running along g vertically downwards. The boundary condition of "clinging" is $v_2 = 0$. The solution of the Laplace equation will be sought through expansion of z:

$$\psi(x, z, t) = \psi_0(x, t) + z^2 \psi_2(x, t) + \cdots .$$

Velocity components are $x_1 = v - (z^2/2)(\partial^2 v/\partial t^2) + \cdots$, $v_2 = -z(\partial v/\partial x) + (z^3/6)(\partial^3 v/\partial x^3) + \cdots$. For the liquid boundary $z = h(x, t)$, we get the equation

$$\frac{\partial h}{\partial t} = (v_2)_{z=h} - (v_1)_{z=h}\frac{\partial h}{\partial x} = -\frac{\partial(hv)}{\partial x} + \frac{1}{6}\frac{\partial(h^3 \cdot \partial^2 v/\partial x^2)}{\partial x} + \cdots ,$$

$$(A.9)$$

where the thickness in the second small term on the right-hand side can be regarded as equal to the unperturbed layer thickness $h = h_0$. If we now introduce $\rho = h/h_0$ and define $\varepsilon = h_0^2/8$, the last equation will finally take the form of

$$\frac{\partial \rho}{\partial t} + \frac{\partial(\rho v)}{\partial x} = \varepsilon\frac{\partial^3 v}{\partial x^3}. \quad (A.10)$$

It can be shown that in the same approximation, the equation of motion (A.8) will yield the relationship

$$\frac{\partial v}{\partial t} + v\frac{\partial v}{\partial x} - c_0^2\frac{\partial \rho}{\partial x} = 3\varepsilon\frac{\partial^3 v}{\partial t \partial^2 x}, \quad c_0^2 = |gh_0|. \quad (A.11)$$

Comparison with (A.5) shows that in the case of drops on the ceiling, the role of effective density is played by the layer thickness and that the azimuthal index $m = 1$. The only difference is the presence of the small dispersion term $\sim\varepsilon$; hereafter, however, the effect of dispersion may be ignored. The system (A.10)–(A.11) is written in the same approximation as the Boussinesq equation for shallow water [110]. We note that it can be used to obtain the Korteweg–de Vries equation by formally replacing $c_0^2 \to -|c_0|^2$, $\rho = 1 + \mu$ and assuming that

$\mu \approx v/|c_0|$ and $\partial/\partial t \to -|c_0|\partial/\partial x$ in the small terms, which would yield

$$\frac{\partial v}{\partial t} + \frac{v}{2}\frac{\partial v}{\partial x} + \frac{2\varepsilon}{|c_0|}\frac{\partial^3 v}{\partial x^3} = 0.$$

The exact solutions of the system (A.10)–(A.11) (the local perturbations of the medium) may be written in terms of elliptic integrals.

A.1.3. *Cylinder of a liquid with surface tension*

Let us consider the problem of the instability of a balanced gravity-free cylinder of a liquid with surface tension. By the Laplace formula, we have $p = \sigma(1/R_1 + 1/R_2)$, where σ is the coefficient of surface tension and $R_{1,2}$ represent the main radii of curvature of the surface. Under the conditions of axial symmetry, the Laplace formula can be written as follows, $a(x,t)$ standing for the boundary radius,

$$p = \frac{\sigma}{a}\left(1 + \left(\frac{\partial a}{\partial x}\right)^2 - a\frac{\partial^2 a}{\partial x^2}\right)\left(1 + \left(\frac{\partial a}{\partial x}\right)^2\right)^{-3/2}$$

and in the long-wave limit, we have $p = \sigma/a$. By analogy with (A.9), we can obtain the equation for boundary $a(x,t)$. Skipping the intermediate calculations, we put down the final result:

$$\frac{\partial \rho}{\partial t} + \frac{\partial(\rho v)}{\partial x} = 0, \quad \frac{\partial v}{\partial t} + v\frac{\partial v}{\partial x} = \frac{c_0^2}{\rho^{3/2}}\frac{\partial \rho}{\partial x}, \tag{A.12}$$

where $\rho = (a/a_0)^2$, $c_0^2 = \sigma/(2\rho_0 a_0)$ and small dispersion corrections are omitted. Thus, the instability of thin liquid cylinder is characterized by the "quasi-Chaplygin" azimuthal number $m = -2$. Solutions of the system (A.12) (periodic and localized drops) may be written in terms of elliptic integrals of the first and the second kinds also.

With the help of quasi-Chaplygin media theory, we can also describe the following phenomena in the liquids and plasma: sausage instability in an (in)compressible current-carrying skin pinches, self-focusing of light in a nonlinear medium, modulational instability of Langmuir waves in plasma, instability of the tangential velocity discontinuity, Buneman plasma instability, etc.

Appendix B

B.1. Richtmyer–Meshkov Instability

(1) The Richtmyer–Meshkov instability or RMI [111, 112] results when a shock wave passes over a density discontinuity in a fluid (contact boundary); in some sense, it is similar to the Rayleigh–Taylor instability, except that the discontinuity is accelerated impulsively by a shock, rather than gradually by a gravitational force. The RM instability results in a growth of the interface perturbations and produces with time the turbulent mixing of the fluids. The RM mixing plays an important role in many astrophysical phenomena. In particular, in supernova type II, the observations indicate an extensive mixing of the inner and outer layers of the progenitor star, and suggest RMI followed by the Rayleigh–Taylor instability as a plausible mechanism. This astrophysical system can be replicated with proper scaling in high energy density laboratory experiments and modeled numerically. Laboratory observations report the following evolution of the RMI: initially, the light fluid accelerates the heavy fluid "impulsively" and the acceleration value is determined by the shock–interface interaction; with time, a coherent structure of bubbles and spikes appears, the light (heavy) fluid penetrates the heavy (light) fluid in bubbles (spikes), and eventually a mixing zone develops. The dynamics of RMI is far from being completely understood. Only recently, an adequate description of the linear regime of compressible RMI was found, while the nonlinear motion yet remains a puzzle. Singular aspects of the interface evolution (such as secondary

instabilities, vorticity generation, direct and inverse cascades of the fluid energy) cause theoretical and numerical difficulties and preclude elementary methods of solution.

(2) When a plane shock wave is incident on a plane interface between two media of different densities, the discontinuity degenerates. Depending on the values of the thermodynamic parameters and velocities in the media on two sides of the interface, the result will be either of two cases, namely, shock waves will propagate through both the media or a shock wave will propagate in the less dense medium, while an expansion wave will travel through the denser medium. If the incident shock-wave front and the plane interface extend sufficiently far in the direction perpendicular to the shock, then the process can be treated as one dimensional. Such a process is well understood and can be described analytically. If, however, the perturbed interface is not plane but undulating, then the process becomes much more complicated. Richtmyer investigated this problem with the theoretical viewpoint in [111], and Meshkov confirmed his results experimentally [112].

The evolution of arising instability can be nominally subdivided into four stages: the linear, nonlinear, transitional and turbulent ones. The first (linear) stage is characterized by a linear increase in the perturbation amplitude with time, as was predicted in [111] within the framework of the linear approximation $a \ll \lambda$ (where a is the amplitude, and λ is the wavelength of the interface perturbation): $da/dt = ka^+ \text{At} \cdot V$, where a^+ is the amplitude of the interface perturbation after the passage of a shock wave, $k = 2\pi/\lambda$ is a wavenumber, $\text{At} = (\rho_1 - \rho_0)/(\rho_1 + \rho_0)$ is Atwood number, ρ_1 and ρ_0 are densities of the gases on the left and the right of the interface, and V is the velocity of the interface between the gases after the interaction. The rates of penetration of the light gas into the heavy one (in the form of bubbles) and of the heavy gas into the light one (in the form of jets) are identical.

The second (nonlinear) stage is distinguished by smaller bubbles and larger jets, accompanied by a slower rate of penetration of one medium into the other. The process ends in the formation of "mushroom structures" at the tips of the jets. The third (transitional) stage is marked by the development of vortices at the tips of the jets penetrating into the light gas. The rate of increase of the perturbation amplitude is smaller than that in the nonlinear stage. The gases begin to mix, and the interface becomes diffuse. At some stage, the growing "mushroom structures" progressively merge. This is the end of the transitional stage, after which the last, or turbulent, stage begins. This stage is characterized by the mixing layer which slowly increases in thickness.

(3) Of fundamental importance for the considered problem is the interaction between a shock wave and a perturbed interface. The evolution of a Richtmyer–Meshkov instability in time depends in many respects on the degradation of the resulting discontinuity, as well as on the generation of waves, their shape, and direction.

The simplest case in which a shock wave is refracted by an interface occurs when a plane shock wave is incident on a plane interface between two originally quiescent media inclined at an angle α to the direction of the shock-wave front. At the limit as $\alpha \to 0$, we obtain the one-dimensional case (characterized by the classical discontinuity degradation); when two shock waves are generated due to interface degradation, if the angle α is close to zero, then refracted and reflected shock waves, as well as the contact surface, are rotated through small angles relative to each other while still remaining planar. As a consequence of this situation, a point forms at the interface at which the incident, the refracted, and the reflected waves converge. The point moves with time over the original interface in the direction of the incident shock wave [113]. As the angle α increases, the shock waves are gradually attenuated and then transformed into compression waves, or, under certain conditions, into expansion waves. It follows that the evolution of a Richtmyer–Meshkov instability and the attendant phenomena depend on the initial experimental setup.

An undulating perturbation of the interface over small regions (for short arc lengths) can be treated as a broken surface composed of a number of planes, each inclined at a certain angle to the direction of the incident shock wave. This means that the results obtained for a constant angle α hold for a sinusoidal interface perturbation, but only for a small region of the interface, in which its curvature can be neglected. Consequently, during the initial stage of the refraction of a shock wave by an undulating surface, one can expect a gradual attenuation of the generated shock waves followed by their degeneracy.

Now, let us consider (following [114]) in more detail the case of a sinusoidal interface perturbation when the maximum angle between the tangential plane and the direction of the incident shock wave is not large enough for the irregular interaction to be observed. Then the vector of the mass flow velocity behind the front of the incident shock wave will, while traveling through the surface of the reflected shock wave, be deflected through some angle from its original direction towards the non-interacting region of the discontinuity. This is due to the fact that the normal component of the momentum flux vector relative to the shock-wave front undergoes a discontinuity, while the tangential component is conserved. This gives rise to the velocity vector component normal to the velocity behind the incident shock-wave front. A similar deflection, but in the opposite direction, occurs in the adjacent symmetrical part. Further, the process of the generation of a secondary shock wave propagating in the space between the refracted and the reflected shock wave begins. The velocities behind the secondary shock wave acquire their initial direction, but are smaller than those behind the incident shock wave. The generation and propagation of secondary shock wave result in lighter medium penetrating into heavier medium in the form of a bubble and in leveling the reflected shock-wave front when it merges with the secondary shock-wave front; the secondary shock waves start interacting with one another, and the linear stage of the Richtmyer–Meshkov instability evolution ends.

(4) The continuum mechanics problem of development of unsteady chaotic flows from ordered laminar flows arises in many problems of science and technology ranging from fluid motion in channels to controlled fusion. Despite its widespread occurrence in nature and exposure to laboratory studies, a common understanding of the causes and mechanisms of this phenomenon is currently lacking. This is due to both the diversity of physical processes involving a significant role

played by turbulence and the insufficient elaborateness of mathematical tools for the quantitative treatment of the appropriate equations describing the phenomena observed.

It is known from the experimental study of RMI development [115] that the turbulence stage begins earlier when the ratio of the perturbation amplitude to the wavelength is more than a unit. So, let us consider a theoretical analysis of the physical mechanisms of transition to turbulence in RMI based on results of numerical simulations and their comparison with the experimental data [116]. In what follows, the passage of a plane shock wave at $M = 2.5$ from helium to xenon at the pressure of 0.5 bar across a sinusoidally perturbed interface, with the perturbation wavelength 0.8 cm and amplitude 1 cm, serves as an example of the mechanism of transition to turbulence to be studied. The problem is solved in a two-dimensional Cartesian formulation for a model of an ideal nonheat-conducting inviscid gas described by the system of Euler equations. The interaction of the plane shock wave with the perturbed interface is referred to as the refraction process of the shock wave at the interface. In the case under study, at any instant during the refraction, one can mark a point along the interfacial line where the incident shock wave intersects the interface. We will call it the refraction point. Apart from the incident shock wave and the tangential discontinuity, it issues a refracted shock wave propagating in xenon and reflected shock wave traveling in helium counter to the incoming flow behind the incident shock wave. Note the importance of the refraction process for the subsequent course of events. In this case, the process has an irregular character; the reflected shock wave has a highly curved shape, and, near the refraction point, part of its front merges with the incident shock wave, enhancing it.

In the direction perpendicular to the motion of the primary shock wave, the most curved part of the reflected shock wave advances beyond the refraction point, which leads to the interaction between parts of the waves reflected from the two adjoining half-wavelengths well before the end of the refraction process. This implies that the linear stage of evolution of the instability doesn't occur in the case under consideration.

At the very beginning of refraction, the first high-pressure region appears near the crest value of the perturbation of xenon in helium, due to the normal incidence of the shock wave onto the interface. Since the curved reflected shock front makes an obtuse angle with the incoming flow almost everywhere, the flow deflects towards the minimum value of the perturbation of xenon in helium after passing the front. For similar reasons, the flow behind the tangential discontinuity in xenon turns toward the perturbation maximum. The explanation is that the xenon ahead of the refracted shock wave is quiescent in the laboratory frame of reference, and the tangential components of a velocity vector are conserved across the shock front. A collision of flows deflected symmetrically toward the perturbation maximum gives rise to the second local high-pressure region in xenon.

In the vicinity of the tangential interface, a low-pressure region appears, due to the deflection of one part of the flow towards the perturbation maximum and the other part towards its minimum. By the end of the refraction process, the primary incident shock wave enhanced by the reflected shock wave, after interacting with the interface near the perturbation minimum, generates another reflected shock wave, which starts moving towards the perturbation maximum and counter to the flow behind the reflected shock wave. At the same time, the curved part of the shock wave reflected from the adjoining half-wavelength of initial perturbation reaches the perturbation maximum of the xenon and, after its interaction with local regions of excessive and reduced pressure, initiates the formation of a mushroom-shaped structure at the leading edge of the xenon jet. Thus, by the end of the refraction process, the formation of the mushroom-shaped structure of xenon in helium has already started, which is characteristic of the transition stage. Like the linear stage, the nonlinear stage does not occur in this case.

Due to the unboundedness of the initial interface in the direction perpendicular to the motion of the primary shock wave and its periodicity, the number of secondary waves emerging after the end of refraction is unlimited as well. They propagate within the space between the diverging fronts of the refracted and reflected shock waves. Their periodic interaction with one another inside the mushroom-shaped structures of xenon in helium promotes the formation of regions of excessive and reduced pressure and leads to the expansion, merging, and mixing of these structures. Note that the interaction of the secondary shock waves with the refracted shock waves, involving displacement of the tangential discontinuity, leads to the formation of a train of alternating local regions of positive and negative vorticity, making up a vortex street located between the refracted shock front and the turbulized interface.

This appears to be the physical mechanism of transition to turbulence in the variant of Richtmyer–Meshkov instability that has been considered.

(5) The passage from the one-dimensional case, characterized by the classical discontinuity degradation, to the two-dimensional case, has led to the discovery of a fundamentally new phenomenon of RMI. Will an analysis of three-dimensional flows contribute significantly to our understanding of the physical mechanism of instability?

When studying complicated physical phenomena that take place either at the galactic scale, for example, in astrophysical investigations of the substance-mixing processes in supernovas, or at the microscale, inherent in nuclear physics, for example, in realizing the inertial thermonuclear fusion, there arises the necessity of the analysis of physical mechanisms and their adequate description for various hydrodynamical instabilities. These are the Rayleigh–Taylor, Kelvin–Helmholtz and Richtmyer–Meshkov instabilities. As numerous experimental and theoretical investigations (see, e.g., [117, 121]) show, the change of the dimensional characteristics of phenomena, namely, the transition from two-dimensional to three-dimensional flows, is accompanied by the appearance of new physical effects. In lower dimensional problems, these effects are either absent or manifest themselves in a degree inaccessible for observation. Among the hydrodynamical instabilities, the three-dimensional flows formed, when developing the Richtmyer–Meshkov instability, are the least studied because, for their numerical simulation, it is necessary to have not only a large volume difference grid but also higher quality algorithms for considering violent discontinuities, in particular, shock waves and their interactions. It is this fact that substantially complicates experimental diagnostic investigations. In numerical simulation of the Rayleigh–Taylor and Kelvin–Helmholtz instabilities [122, 123], it was established that, for identical initial amplitudes of perturbations and wavelengths, the growth rate for the perturbations is higher in the three-dimensional case as compared to the two-dimensional one, while the process of formation of mushroom-shaped structures proceeds more slowly. For the Richtmyer–Meshkov instability, similar results are fair [124]. The analysis of dynamics of development of RMI for corresponding two- and three-dimensional flows was carried out in [114].

Appendix C

C.1. Instabilities and Turbulence in Astrophysics

C.1.1. *Convective instability in astrophysics*

(1) In astrophysical conditions, the problem of mixing plays the controlling role in many objects and processes. For example, the problem of mixing, as a result of the development of convection, is important in explosions of supernovas, in evolution of stars, in thick accretion discs and other objects and processes.

The development of convection is associated with Archimedean forces and energy generation processes. Convection occurs where the diffusion processes cannot ensure efficient heat conductivity. This results in the formation of a classic situation with a heavy fluid positioned above a light fluid in the field of gravitational forces, resulting in the development of Rayleigh–Taylor instability. A similar situation forms in the deceleration of shock waves in mass-losing shells of supernovas or in the scraped up interstellar gas [125, 126].

Initially, we pay attention to the development of convection during the evolution of stars. The widely used Schwarzschild criterion of convective instability says that the temperature gradient at every point must not exceed the adiabatic gradient. Consequently, the stability equation has the following form: $-(1-1/\gamma)\cdot(T/p)\cdot dp/dr > -dT/dr$, where T is temperature, p is pressure, r is the actual radius in the star, and $\gamma = c_p/c_V$ is the isentropic exponent. The fulfillment of the criterion is usually verified using the following procedure. Attention is given to a small isolated element of the volume of matter and whether it is subjected to small perturbations and whether the element intersects with the surrounding matter. The justification of the criterion is usually regarded as convincing but, in reality, it is not suitable for the accurate evaluation of the stability of the system in terms of the normal modes and the initial conditions because the system reacts to the loss of stability and to a perturbation as an integral unit. From this viewpoint, the qualitative arguments, used for the justification of the Schwarzschild criterion, are no more than an assumption that for the spherical configuration, the instability can be described by the modes of a relatively high order (i.e., the modes belonging to the spherical harmonics of high order ℓ and radial functions with many zeroes) which appear if the criterion is violated in any small vicinity of the isolated region. It may be asserted that the local approach does not satisfy the conditions of development of convection having the limiting values of the scale for characteristic unstable modes. The main assumption, representing the basis of the given approach, is the ignorance of the variation of the gravitational potential resulting from the perturbation. Although this assumption can be efficient for examination of high-order modes, it is hardly suitable for the modes belonging to $\ell = 1$ resulting from the perturbation [52]. These modes are most important for the recognition of convective instability in many important cases.

The accurate examination of non-radial adiabatic pulsations shows that the instability, formed in a homogeneous compressed sphere, has the modes of all moments, including $\ell = 1$. The main reason for the instability is that the homogeneous density is superadiabatic for any finite range of the isentropic exponent γ. Consequently, it is fully justified to expect the occurrence of large-scale convection with the modes $\ell = 1$, for example, for the case of a sphere with a viscous liquid, heated from the inside. The considerations that are valid for the homogeneous sphere must also be valid for the sphere with the equation of the state of matter described by a polytrope in the sense that in the present case, convective instabilities with the modes $\ell = 1$ also develop. This conclusion becomes evident if it is noted that the polytropic gas, for which the pressure and density are linked by the equation $p \sim \rho^{1+1/n}$, is super-adiabatic everywhere, where $\gamma < 1 + 1/n$]. Consequently, if $1 + 1/n$ exceeds γ, the Schwarzschild criterion will be violated simultaneously in the entire matter and the instability will be manifested in the form of the circulation of the largest permissible scale.

However, even in cases where the effective exponent of the polytrope decreases from a relatively high value in the external shell of the star to a sub-critical value in its central part and the conditions, finally, are different everywhere, it is clear that the most important modes are associated with convective instabilities, again those which belong to the harmonics of a lower order, including $\ell = 1$, and are described by radial functions having one or several zeroes. If these assumptions are accurate, the instabilities with the modes of actually high orders do not play such a significant role in the theory of convective instability, as assumed previously. In this sense, the *approximation of the length of mixing* for describing convection may be insufficient [52].

(2) In the last 30 years, several groups of investigators have examined the reasons why the collapse of the iron core of a star with mass $M > M_{Sun}$ results in the formation of a supernova of type II and the formation of a neutron star. To explain the supernova phenomena, two main mechanisms have been proposed: the first is associated with the rapid collapse of the shell of the star during the passage of a shock wave through the shell formed as a result of the arrest of the collapse on reaching the density higher than the nuclear density [127], the second is associated with the heating of matter behind the front of the shockwave due to the neutrino leaving the neutrino-sphere, surrounding the proton–neutron star. This results in further moment of the shock wave and the collapse of the shell [128].

However, numerical calculations show that these mechanisms are unsatisfactory because of several reasons. In the first of these mechanisms, the so-called rapid mechanism, the shock wave loses a large part of its energy in the process of splitting of the iron core into free nucleons [129]. When the shock wave reaches the neutrino-sphere, the electron neutrino carries the thermal energy and the lepton number from the front of the shockwave thus decreasing the energy of the wave and pressure in it. This results in the weakening of the shockwave and, in subsequent stages, in its arrest [128]. The second mechanism is determined by the conditions in the region between the neutrino-sphere and the shock wave and depends, in a critical manner, on the value of the neutrino luminosity and the mean spectral energy of the neutrino. In this case, the level of luminosity, required for sustaining the divergent shock wave, can be obtained only taking into account the convection both below the neutrino-sphere and above it. In turn, convection may develop relatively rapidly only on the condition that heating (and the construction of unstable stratification of matter in respect of entropy) is more rapid than the movement of the matter from the shock wave to the surface of proton–neutron star [130].

In the period after the arrest of matter in the core and the propagation of the resultant shock wave on the shell, the core of the star may also be characterized by the development of hydrodynamic instabilities associated with convection. Examination of the rapid increase of emission from SN1987 X-ray and γ-radiation, mixing of the external layers with the products of thermonuclear explosion, containing Ni^{56}, and the observed structure in the form of two jets in the center show that convective motions may take place in a certain stage of explosion. The development of several convective modes is possible in the period of restoration of the shock wave.

One of the types of convection forms in the passage of the shock wave through the external part of the core is associated with the resultant negative gradient of entropy. This convection is short-lived because the negative gradient of entropy is not maintained by the heating of the central areas of the star and is rapidly reduced during convective mixing. In addition to this, at the moment of the maximum of development of convection, the radius of photosphere for ν_e (electron neutrino) is very close to the radius of the shock wave because the region of heating behind the shockwave is very narrow. A large part of the sources ν_e and $\bar{\nu}_e$, distributed on the appropriate neutrino-spheres, whose thermodynamic condition is established by the transfer of heat and the number of leptons by the matter, accreted on the proton–neutron star. Therefore, this type of convection may enhance the divergent shockwave [131].

The numerical modeling of convection in the vicinity and below the neutrino-sphere [132], taking neutrino transfer into account, shows that the convective speed is too low in comparison with the volume speed of inflow of matter in order to cause any significant transfer of entropy and leptons. Consideration of the transfer of energy and leptons due to the neutrino from the rising element of the liquid to the background reduces the rate of growth of entropy convection 3–50 times, lepton convection 250–1000 times for the region between the neutrino-sphere and the matter with a density of $\rho = 10^{12}$ g/cm^3. However, the calculations of the evolution of the initial isothermal or adiabatic perturbation from the equilibrium position [131] show that the time to establish the equilibrium state with respect to the entropy and leptons of the liquid element with the background is slightly longer than in the hydrodynamic calculations. Nevertheless, owing to the fact that convection at a negative gradient of entropy rapidly dissipates, and in the case of lepton convection, which exists for a considerably longer period of time, the time to establish the equilibrium state is considerably shorter than the time of growth of the convective motion, the conclusion of the suppression of convection with neutrino transfer taken into account does not change in principle. As a result, the contribution to neutrino luminosity from the region in the vicinity and below the neutrino-sphere is insignificantly small.

The second type of convection, the so-called neutron fingers [128], forms in the region in which the negative gradient of the lepton number is stabilized by the positive gradient of entropy, on the condition that the element of the liquid comes into thermal equilibrium with the surrounding matter faster than to the equilibrium in respect of leptons. The first requirement is explained by the fact that as a result of rapid convection, the gradient of entropy becomes positive, whereas the negative gradient of the lepton number, which initially forms in the vicinity of the neutrino-sphere, expands into the nucleus as a result of the diffusion of leptons outwards. Since all six types of neutrino transfer energy and only two leptons (ν_e, $\bar{\nu}_e$) transfer the lepton number as a result of the large difference in the path to interaction with the matter, the second condition is also fulfilled. Numerical calculations of the development of this instability show, however, that it does not develop because of the following reasons [131]:

(i)   the condition of instability changes because the typical values of the lepton number in the large part of the nucleus are half the critical number;

(ii) the transfer of the large part of energy from the liquid element to the background takes place by means of ν_e and $\overline{\nu}_e$-neutrino and, in this case, the contribution to this process from the neutrino is small;

(iii) in the typical conditions, the number of acts of neutrino transfer of energy, required in the nucleus, is larger than for the neutrino transfer of the lepton number;

(iv) the fluxes of the ν_e and $\overline{\nu}_e$-neutrino during the large part of the evolution period of the initial perturbation are directed to opposite sides and, in this case, they are subtracted in the transfer of energy and added up during the transfer of the leptons.

Until recently, the role played by the convection in the mechanism of explosion of the supernova star has been to a large extent unclear and contradicting. To a large degree, this indeterminacy is associated with the development of methods of calculating neutrino transfer, selection of the equations of the state of matter, the dimensionality of the problem being solved, numerical methods, shortcomings of adequate models of the nuclear equation of the state of matter and, possibly, the absence of accurate understanding of the interaction between the neutrino and matter at nuclear densities.

In [133], it was shown that the area in the vicinity of the neutrino-sphere is characterized by the development of the lepton convection and that entropy convection occurs above the neutrino-sphere. In turn, in [128], it was reported that the convection inside a proton–neutron star is weak and, evidently, not important.

The results obtained in the above-mentioned papers and also in [134] show that lepton convection continues for a long time and affects the entire nucleus after 1 s. This is accompanied by the doubling of neutrino luminosity and the mean energy of the neutrino increases by 10–20%. On the other hand, it was shown in [131] that when using the multigroup diffusion approximation for calculating neutrino transfer in the approximation of the length of the mixing path for convection, the latter is characterized by low efficiency inside and around the neutrino-sphere for 30 ms, regardless of the fact that modeling lasted 0.5 s from the moment of recoil of the nucleus.

In the process of collapse of the iron core of the star, approximately 99% of the entire gravitational energy is taken away in the form of neutrino. To cause collapse of the external shell of the star, it is sufficient that only a part of this energy is transformed to the outer layers of the star by means of an effective and fast mechanism. Convection — both inside and from the outside of the neutrino-sphere — may intensify the process of transfer of energy to the front of the shock wave but, nevertheless, requires certain conditions. For example, the characteristic time of formation of convection should be shorter than the characteristic time of accretion of the matter of the outer layers on the core. In addition to this, convection requires constant supply, as in the case of entropy convection inside a proton–neutron star. Evidently, a more realistic mechanism may be associated with the effect of hydrodynamic large-scale instabilities acting at short characteristic times of the order of 0.001–0.1 s and supporting the powerful flux of high-energy neutrino radiation.

The authors of [135] investigated the possibilities of powerful emission of neutron radiation associated with hydrodynamic motion in a proton–neutron star. In fact, this study represents a development of the concepts proposed in [136]; in this preprint, the development of convective instability in a gravitating homogeneous gas sphere (star) was examined. The main idea was based on the similar time behavior of the integral over the cross-section for the entropy and the same integral for the strength of the magnetic field. In the experiments, these instabilities were detected in a Tokamak.

(3) Let us consider an example of numerical modeling of convection in stars. In [135], three-dimensional calculations of the development of hydrodynamic instability in a proton–neutron star with

an excess of entropy in the center were considered. The numerical solution of Euler equations was used to examine the time development of the scale of the heterogeneity of entropy distribution. In the formulation of the problem, the entropy shows the evolution of the rise of hotter, lighter matter. The calculations showed the development, during 4 ms, of large-scale heterogeneities of entropy with the shape resembling the "mushroom" of a nuclear explosion in the atmosphere of the earth with a distance from the center of the star of 20 km. The characteristic time of movement of the single resultant bubble to the surface of the proton–neutron star was 1 ms, which corresponds to a mean speed of $c/150$ (c is the speed of light). A total of six bubbles with a mass of $0.01M_{\text{Sun}}$ rose. The intensity of neutron radiation was $5 \cdot 10^{52}$ erg/s for the given process. The energy absorbed by the matter per 1 g in the shock wave from neutron radiation was $2.3 \cdot 10^{24}$ erg/g·s, which was comparable with the neutrino losses from the front of the shock wave. Consequently, it was concluded that the mechanism of explosive convection could sustain the divergent shock wave and, therefore, result in the burst of the supernova shell.

In the calculation of the distribution of density and temperature inside a proton–neutron star, the central density was assumed to be $\rho_{\text{center}} = 2 \cdot 10^{14}$ g/cm^3 and the central temperature $T = 10^{11}$ K. The equation of state, used in the calculations, was used in the form $p = p(\rho, s)$ (in accordance with [137]). At the initial moment of time in the vicinity of the center of the star ($r_0 = 0$), the excess of entropy was defined in accordance with the Gaussian law, determined by the non-equilibrium process of neutronization of matter after the arrest of the collapse.

The Euler equations for calculating the development of large-scale convection, used for modeling the processes taking place inside a proton–neutron star in the three-dimensional case, have the following form:

$$\rho\frac{\partial \vec{v}}{\partial t} = -\nabla p - \frac{\rho GM}{r^3}\vec{r}, \quad \frac{\partial \rho}{\partial t} + \rho \nabla \vec{v} = 0,$$

$$\frac{\partial E}{\partial t} = T\frac{\partial S}{\partial t} + \frac{p}{\rho^3}\frac{\partial \rho}{\partial t}, \quad \frac{\partial S}{\partial t} = 0,$$

where ρ is the density of matter, $\vec{v}$ is the velocity of matter, p is pressure, E is energy, and S is entropy. All these quantities are functions of three coordinates and time.

Calculations were carried for the case of slow solid-state rotation, where the ratio of the kinetic energy of rotation E_k to gravitational energy $|W|$ was selected as $E_k/|W| = 0.01$; the period of rotation of the star was 14 ms. The coordinate system was selected in such a manner that the plane of rotation of the star coincided with the plane OXY. This means that the vector of the angular velocity of rotation of the star has only one component $\omega_z = \omega = $ const. The entropy is represented in two sections of the star. In the first section, it is presented along the axis of rotation, i.e., the vector of the angular velocity of rotation is in the image plane. In other words, the OX-axis is plotted on the horizontal axis, and the OZ-axis of our frame is on the vertical axis. The initial configuration was selected for the moment $t = 0.075$ ms, and the final configuration is shown for the moment of time $t \approx 6.31$ ms. In contrast to the model examined in [138], in this case, two bubbles form initially after 3 ms and are extended along the axis of rotation in opposite directions. Four additional bubbles form slightly later, after 5 ms, and are distributed in the equatorial plane of rotation of the proton–neutron star. The bubbles, distributed along the axis of rotation, separate from hot nucleus and rise to the surface. The cold matter descends to the center in the gap between these bubbles.

The bubbles, distributed in the plane of rotation, separate from the hot nucleus later and also rise to the surface. Calculations show that the formation of the first bubbles is followed by the formation of additional bubbles whose volume is considerably smaller. They also start to rise to the surface.

The first stage of development of perturbations is the increase of asymmetry along the axis of rotation of the star which continues for 3 ms. In the plane of rotation, the asymmetry is still very weak, and the evolution is smooth. During this stage, the asymmetry of the distribution of matter is negligible and bubbles of hot matter have not yet formed. This stage is characterized by the start of the formation of bubbles created in the plane of rotation. The following stages correspond to the final formation of the bubbles in the plane of the equator and to the start of rising with bubbles to the surface. The bubbles along the axis of rotation have already left the calculation region, i.e., they have traveled the distance from the center greater than 40 km.

The numerical results lead to the following conclusion about the onset of turbulence in the shear layer with a uniform velocity gradient [139, 140]. The development of turbulence begins with the formation of large vortices. Well-developed turbulence should be modeled in the framework of the Euler equations, which correctly describe the distribution of basic length scales. Multiple small eddies merge under the action of the Zhoukovskij force.

To investigate the physical scenario of the onset of turbulence, we performed numerical simulations of free shear flows of an ideal compressible gas. For simulations, we used three-dimensional Euler gas dynamics equations in Cartesian coordinates with the ideal gas law. Let us consider the flow of matter in the integration domain $(0 \leq x \leq L_x, 0 \leq y \leq L_y, -L_z/2 \leq z \leq L_z/2)$. The initial velocity v_1 along the x-direction is used in the following form:

$$v_1 = v_0, \ \ H/2 \leq z \leq L_z/2; \ \ \ v_1 = -v_0, \ \ -L_z/2 \leq z \leq -H/2,$$

$$v_1 = v_0(2z)/H, \ \ -H/2 \leq z \leq H/2 \ \ \text{(constant gradient of } v_1).$$

The initial velocity along the y-direction is equal to zero. The initial velocity along the z-direction has a small disturbance (1% of v_1) inside the shear layer. The boundary conditions are the following: periodic conditions for the x- and y-directions, and impermeability conditions for the z-direction. Let us analyze the influence of the length of the shear layer on the evolution of the turbulence. Three variants with different shear layer lengths $L_y = 2\pi$, $\pi/2$ and $\pi/8$ and with $L_x = 2\pi$, $L_z = 2\pi$ and $H = 1$ were modeled.

Note that the value of the specific concentration for the initial time moment is equal to 1 inside the shear layer and to 0 outside it. It is shown that the evolution of the flow at the beginning has a quasi-two-dimensional nature for all calculations. Further, we can see that the nature of the evolution continues to be quasi-two-dimensional longer for lower values of L_y.

Results of calculations show that the turbulent cascade is connected to the dimensionality of the vortices. The formation of non-homogeneities in the velocity field leads to the formation of large vortices in the flow. The instability of surfaces of large vortices leads to the occurrence of secondary instability. This instability cut the large vortices up to the center. In a three-dimensional case, this process leads to their chaotic deformation (the general evolution of vortices for adiabatic flows described by the Lamb–Euler equation).

Thus, the occurrence of small scales in the vortex evolution is connected to presence of the third coordinate (length). If the length of the vortex is small, then the secondary instability develops very slowly. We can use the two-dimensional approach for special models of vortex dynamics only; the condition of physical adequacy of such approach is $R_{\text{vortex}} \gg L_{\text{vortex}}$.

C.1.2. *Turbulence in astrophysics: Accretion discs*

(1) The turbulent flow differs by the fact that thermodynamic and hydrodynamic characteristics (vector of the speed, pressure, temperature, concentration of impurities, density, speed of sound, the refractive index, etc.) are subject to random fluctuations, generated by the presence of a large number of vortices of different scale and, consequently, they change very chaotically and irregularly with time and space. In addition to this, it is important to examine the problem of the intermitting structure of turbulence. All these factors make it necessary to carry out numerical modeling of turbulence in the astrophysical conditions for every object.

An important special feature of the astrophysical conditions is the presence of gravitational forces. It is possible to derive a criterion characterizing the local stratification of the flow $N^2(z) = g\rho_*^{-1} \cdot \partial \rho_* / \partial z$, where ρ_* is the potential density, g is the acceleration due to gravity. If $N^2 > 0$, the Archimedean force is backmoving, and the turbulence should use energy for work against the effect of the Archimedean force, and, consequently, it develops only slightly and often concentrates only in the individual thin layer.

In subsequent stages, the concept of α-disc [51] has been introduced with special reference to accretion discs. The main assumption of the model of the α-disc is the assumption according to which the sink of the kinetic energy of turbulence and the thermal energy, generated in the transition of kinetic energy to thermal energy as a result of viscosity, are equal. This is one aspect.

On the other hand, the increase of the angular moment with the radius stabilizes turbulence (Rayleigh–Zel'dovich problem). The Taylor criterion was derived from the local examination of flow stability. If the rotation laws are known, it is possible to derive the Taylor criterion from the entire flow. It is assumed that high values of Taylor number Ta correspond to more rigid conditions of turbulence of the flow. For example, for the exponential law of the dependence of angular speed on radius $\omega \sim r^n$, the Taylor criterion has the form: $Ta = (8n + 16)/n^2$. The stabilization of the flow in relation to turbulization of curves occurs at $n > -2$, which corresponds to the value $Ta > 0$. For a Kepler accretion disc $n = -3/2$, which gives $Ta = 16/9$. If $n = 0$, we are concerned with solid-state rotation and, in this case, there is no shear flow and consequently, no energy is available for the development of the turbulent flow and this is reflected in the value of the Taylor criterion $Ta = \infty$.

The problem of turbulence is very important for the theory of accretion discs. The observed X-ray irradiation in accretion of matter on a relativistic object in double stellar systems is determined by the rate of accretion: $L = \eta_{\text{conv}} M_{\text{accr}} c^2$ erg/s, where η_{conv} is the efficiency of conversion of gravitational energy, M_{accr} is the rate of accretion of matter on the relativistic object, c is the speed of light. For a neutron star, $\eta_{\text{conv}} \approx 0.1$–$0.2$; for black hole, $\eta_{\text{conv}} \approx 0.06$–$0.4$. The rate of accretion is determined by the mechanism of transport of the angular moment on the disc to the outside. There are several possibilities for the redistribution of the angular moment. Evidently, molecular viscosity is not important in the astrophysical conditions. However, according to Taylor's studies of the laws of rotation with increase of the angular moment to the outside, the instability to turbulization of the flow may be suppressed.

The traditional model of the α-disc links the viscosity with the temperature of the matter of the disc. This is characteristic of the

Kolmogorov spectrum where the small scales of turbulence annihilate into heat. In this model, the efficiency of the transport of the angular moment is characterized by the parameter $\alpha = v_t/c_s + H^2/(4\pi\rho c_s^2)$, where $\rho c_s^2/2 = (3/2) \cdot (\rho k_B T)/m_p + \varepsilon_r$ is the thermal energy of the matter, ε_r is the density of radiation energy, c_s is the velocity of sound, and v_t is turbulent velocity. In this assumption, the tensor of the tangential stresses $\sigma_{r\varphi}$ is written in the following form: $\sigma_{r\varphi} \approx \alpha\rho c_s^2$. Usually, it is assumed that the parameter α is in the range $10^{-15}(M_{accr}/M_{accr}^{(critical)}) < \alpha < 1$. In fact, this parameter should be determined on the basis of modeling the turbulent flow in the investigated conditions. The definition of parameter α also includes the strength of the magnetic field H.

The magnetic field may influence the transport of the angular moment. However, the following problem must be taken into account. If the density of the energy of the magnetic field is higher than the kinetic energy, the structure of the flow is determined by the magnetic field and there is no turbulence. On the other hand, if the density of the energy of the magnetic field is smaller than the kinetic energy, the magnetic field has no effect on the flow. Magnetic viscosity becomes considerable if the values of the kinetic energy and the density of the energy of the magnetic field are of the same order of magnitude. This situation exists in the case of the magnetic dynamo.

(2) The structure of the accretion disc has been studied in a large number of investigations, starting from different analytical and simple modeling constructions up to the calculations of the structure of the accretion disc in the three-dimensional approach. We shall describe the structure of the accretion disc taking into account the study [141].

The accretion discs or gas formations in the vicinity of gravitating bodies represent interesting objects for examination in astrophysics and theoretical physics. These formations are typical of the nuclei of galactic and double stellar systems where the processes in the disc control the evolution of the binary system as a whole. Special attention in these objects is paid to the examination of the mechanisms of the loss of the angular moment by the matter of the disc and subsequent accretion of the matter on the compact gravitating object.

The rate of transfer of matter on the central body from the accretion disc determines the nature of radiation and this is especially important for investigations.

The methods of numerical modeling, based on different models and computation algorithms, are used widely at present for examining the processes in accretion discs, together with analytical methods. In particular, the authors of the papers [142–144] carried out numerical calculations and the results of these calculations show the formation, in an accretion disc, of a double stellar system of spiral shock waves. It has been reported that these formations represent an important mechanism of the loss of matter by the disc; in addition to this, the redistribution of the angular moment is caused by viscosity, the magnetic field, and the formation of large-scale structures. In [145, 148], the authors modeled the formation of an accretion disc and the formation of its structure taking into account the intercalation of matter between the components of the binary system. In these studies, the calculation region contains both components of the binary system and, consequently, the vicinity of the star accretor where the disc forms. The evolution of the gas flow is described by a system of Euler equations disregarding physical viscosity. The presence in the model of only numerical viscosity restricts the possibilities of detailed investigation for different values of viscosity because the minimum value is restricted by the limit of numerical viscosity. Nevertheless, changes in the structure of the flow with the variation of viscosity in wide range may be obtained on the qualitative level. Due to the numerical viscosity dependence (for the selected difference scheme) on the time and spatial resolution, in order to obtain the dependence of the solution on the viscosity, a sequence with a decreasing step of the grid was used.

In [149], it was mentioned that the sources of turbulence in the accretion disc may include Coriolis forces; consequently, it is necessary to solve the problem of the possibility of development of turbulence in the nonlinear regime (as the term taking into account these forces in hydrodynamical equations may be a stabilizing factor). Consequently, it is necessary to solve the problem of numerical viscosity in multidimensional calculations, modeling the development of the

nonlinear phase of turbulence. In traditional experimental hydrodynamics, the presence of viscosity might prevent the development of
turbulence. In other words, the flow in a disc with low Reynolds
numbers remains laminar, whereas the development of turbulence
requires high Reynolds numbers so that the problem of numerical
modeling of turbulence is very difficult to solve because of the high
values of numerical viscosity.

However, a solution was found in the following direction. The
numerical viscosity in solving Euler equations is not identical with
small Reynolds numbers. The point is that Euler equations have their
dynamic instabilities from which turbulence structures may form in
calculations, whereas the applications of the Navier–Stokes equations
could not result in the formation of such structures because of high
numerical viscosity, restricted wavelength (in comparison with the
size of difference grid) and the existing turbulent instability. In [150],
test calculations for comparison of the code of Navier–Stokes equations and the code of Euler equations were carried out. It was shown
that the code for the Euler equation far more efficiently resolves
the fine structure and short waves in comparison with the Navier–
Stokes equations. It was concluded that the Euler equations may be
used for modeling with high Reynolds numbers with much better
resolution.

Recently, the problem of the coherent transport of the angular
moment has been discussed in a number of studies. The problems of
the structure of the accretion disc and the mechanism of the transport of the angular moment has a controlling effect on the structure
of the flow in the disc. The application of high values of the turbulent viscosity or high numerical viscosity leads, as in the model
of α-viscosity, to rapid heating of the matter of the accretion disc.
The problem of the high temperature of the disc has been reflected
in a large number of investigations into the advection of energy in
accretion discs. The redistribution of the angular moment through
large-scale structures does not increase temperature. In this case, the
structure of the disc will be calculated using efficient difference grids
enabling the development of turbulence of large-scale structures in
the calculations.

The authors of [151] investigated the mechanism of coherent transport of the angular moment as a result of the development of large-scale vortices. The initial instability of the shear flow is explained by the development of Rossby vortices [149]. In [151], the development of Rossby vortices as a result of the tangential injection of compressed air resulting in a radial pressure gradient was investigated. In particular, this gradient leads to the formation of heterogeneity in the distribution of vorticity and increases the intensity of the Rossby instability. Thus, it may be assumed that the nonmonotonic form of the distribution of density, entropy and pressure gradient lead to the formation of Rossby vortices. They, in turn, lead to the development of turbulence in accretion star discs.

References

1. McComb W.D., *The Physics of Fluid Turbulence*, Clarendon Press, Oxford, 1992.
2. Hinze J.O., *Turbulence*, 2nd edition, McGraw-Hill, 1975.
3. Landau L.D., Lifshitz E.M., *Fluid Mechanics*, Pergamon, New York, 1968.
4. Ladyzhenskaya O.A., *The Mathematical Theory of Viscous Incompressible Flow*, Gordon and Breach, New York, 1969.
5. Majda A.J., *Compressible Fluid Flow and System of Conservation Laws in Several Space Variables*, Springer-Verlag, Basel, 1986.
6. Segel L.A., *Mathematics Applied to Continuum Mechanics*, Macmillan, London, 1977.
7. Temam R., *Navier–Stokes Equations*, North-Holland, Amsterdam, 1977.
8. Chorin A.J., Marsden J.E., *A Mathematical Introduction to Fluid Mechanics*, Springer-Verlag, New York, 1990.
9. Cercignani C., *Mathematical Problems in Kinetic Theory*, Macmillan, London, 1969.
10. Courant R., Friedrichs K.O., *Supersonic Flow and Shock Waves*, Wiley-Interscience, New York, 1948.
11. Resibois P., De Leener M., *Classical Kinetic Theory of Fluids*, Wiley-Interscience, New York, 1977.
12. Orszag S.A., *Dynamics of Fluid Turbulence*, Princeton University Press, Plasma Physics Laboratory Report PPL-AF-13, 1966; Representation of isotropic turbulence by scalar functions, *Stud. Appl. Math.* **48**, pp. 275–279, 1969; *Lectures on the Statistical Theory of Turbulence*, pp. 239–374, 1972; Analytical theories of turbulence, *J. Fluid Mech.* **41**, pp. 363–386, 1970.
13. Kundu P.K., Cohen I.M., *Fluid Mechanics*, Academic Press, San Diego, 2002.

14. Drazin P.G., *Introduction to Hydrodynamic Stability*, Cambridge University Press, Cambridge, 2002.
15. Faber T.E., *Fluid Dynamics for Physicists*, Cambridge University Press, Cambridge, 1995.
16. Lamb H., *Hydrodynamics*, 6th edition, Cambridge University Press, New York, 1975.
17. Kochin N.E., Kibel I.A., Roze N.V., *Theoretical Hydromechanics*, Gos. Izd. Tekh.-Teor. Lit., Moscow, 1963 (in Russian).
18. Batchelor G., *An Introduction to Fluid Dynamics*, Cambridge University Press, Cambridge, 1967.
19. Tritton D.J., *Physical Fluid Dynamics*, 2nd edition, Clarendon Press, Oxford, 1988.
20. Boon J.P., Yip S., *Molecular Hydrodynamics*, Dover Publications, New York, 1991.
21. Nishida J., Fluid dynamical limit of the nonlinear Boltzmann equation to the level of the compressible Euler equation, *Commun. Math. Phys.* **61**, pp. 119–148, 1978.
22. Caflish R., The fluid dynamic limit of the nonlinear Boltzmann equation, *Commun. Pure Appl. Math.* **33**, pp. 651–666, 1980.
23. Ellis R., Pinsky M., The first and the second fluid approximations to the linearized Boltzmann equation, *J. Math. Pure Appl.* **54**, pp. 157–182, 1975.
24. Schechter M., On the essential spectrum of an arbitrary operator, *J. Math. Anal. Appl.* **13**, pp. 205–215, 1966.
25. Kawashima S., Matsumura A., Nishida J., On the fluid-dynamical approximation to the Boltzmann equation at the level of the Navier–Stokes equation, *Commun. Math. Phys.* **70**, pp. 97–124, 1979.
26. Grad H., On the kinetic theory of rarefied gases, *Comm. Pure Appl. Math.* **2**(4), pp. 331–407, 1949.
27. Nikiforov A.F., Novikov V.G., Uvarov V.B., *Quantum-Statistical Models of Hot Dense Matter and Equations of State*, Birkhäuser, Basel, 2005.
28. Rozhdestvenskii B.L., Yanenko N.N., *Systems of Quasilinear Equations and Its Applications to Gas Dynamics*, Moscow, Nauka, 1978 (in Russian).
29. Grad H., Asymptotic theory of the Boltzmann equation, *Part 2*, in *Rarefied Gas Dynamics*, Vol. 2, Academic Press, New York, 1963.
30. Crandall M.G., Rabinowitz P.H., Bifurcation from simple eigenvalue, *J. Funct. Anal.* **8**, pp. 321–340, 1971.
31. Kato T., *Perturbation Theory for Linear Operators*, Springer-Verlag, Berlin, 1966.

32. Rutkas A.G., The Cauchy problem for the equation $Ax'(t) + Bx(t) = f(t)$, *Differential Equations* **11**, pp. 1996–2010, 1975 (in Russian).
33. Zeidler E., *Nonlinear Functional Analysis and Its Applications*, Vols. 1–4, Springer-Verlag, New York, 1985–1987.
34. Deimling K., *Nonlinear Functional Analysis*, Springer-Verlag, Berlin, 1985.
35. Fimin N.N., Chuyanov V.A., The branching of solutions of the abstract kinetic equations, *Math. Notes* **73**(1), pp. 103–109, 2003.
36. Dorning J. *et al.*, Rarefied gas models with velocity dependent collision frequencies, *J. Quant. Spectr. Radiat. Transfer* **11**(11), pp. 1007–1021.
37. Nicolaenko B., Dispersion laws for plane wave propagation and the Boltzmann equation, in *The Boltzmann Equation*, ed. Grunbaum F.A., Courant Institute of Mathematical Sciences, New York, pp. 125–171, 1972.
38. Dubrovin B., Fomenko A., Novikov S., *Modern Geometry — Methods and Applications*, Vols. 1–2, Springer-Verlag, New York, 1984–1985.
39. Frisch U., *Turbulence. The Legacy of A.N. Kolmogorov*, Cambridge University Press, Melbourne, 1996.
40. Arnold V.I., Khesin B.A., *Topological Methods in Hydrodynamics*, Springer-Verlag, New York, 1998.
41. Tennekes H., Lumley J.L., *A First Course in Turbulence*, NIT Press, Cambridge, 1972.
42. Arnett D., Fryxell B., Muller E., *Astrophys. J. Lett.* **341**, L63, 1989.
43. Buchler J.R., Kondrup H. (eds.), Astrophysical turbulence and convection, *Ann. New York Acad. Sci.* **898**, 2000.
44. Belotserkovskii O.M., *Numerical Experiment on the Turbulence: From Order to Chaos*, Nauka, Moscow, 1997 (in Russian).
45. Bisikalo D.V., Boyarchuk A.A., Kuznetsov O.A., Chechetkin V.M., *Astron. Zh.* **77**(1), pp. 31–41, 2000.
46. Canuto V.M., Turbulence and laminar structures: Can be they co-exist? *Mon. Not. Roy. Astron. Soc.* **317**, pp. 985–988, 2000.
47. Belotserkovskii O.M., *Computing Mechanics: Current Problems and Results*, Nauka, Moscow, 1991 (in Russian).
48. Zel'dovich Ya.B., Friction in liquids between rotating cylinders, Preprint AN SSSR, Moscow, 1979.
49. Molemaker M.J., McWilliams J., Instability and equilibrium of centrifugally stable stratified Taylor–Couette flow, *Phys. Rev. Lett.* **86**(23), pp. 5270–5273, 2000.
50. Belotserkovskii O.M., Oparin A.M., *Numerical Experiment in Turbulence*, Nauka, Moscow, 2000 (in Russian).

51. Belotserkovskii O.M., Oparin A.M., Chechetkin V.M., *Turbulence: New Approaches*, Cambridge International Science Publishing, Cambridge, 2005.

52. Chandrasekhar C., Lebovitz N.P., Non-radial oscillation and convective instability of gaseous masses, *Astrophys. J.* **138**, pp. 185–199, 1963.

53. Lorenz E.N., Deterministic nonperiodic flow, *J. Atmos. Sci.* **20**(2), pp. 112–130, 1963.

54. Gorbatskii V.G., *Gas-dynamical Instabilities in Astrophysical Systems*, Sankt-Petersbourg University Publ., St. Petersbourg, 1999 (in Russian).

55. Peitgen O., *Chaos and Fractals*, Springer-Verlag, Berlin, 1992.

56. Chorin A.J., *Vorticity and Turbulence*, Springer-Verlag, New York, 1994.

57. Benzi R., Biferale L., Paladin G., Vulpiani A., Vergassola M., Multifractality in the statistics of velocity gradients, *Phys. Rev. Lett.* **67**, pp. 2299–2302, 1991.

58. Benzi R., Paladin G., Parisi G., Vulpiani A., On the multifractal nature of fully developed turbulence and chaotic systems, *J. Phys.* **A17**, pp. 3521–3531, 1984.

59. Wilcox D.C., *Turbulence Modeling for CFD*, DCW Industries Inc., La Cañada, 1993.

60. Belotserkovskii O.M., Computational experiment: Direct numerical simulation of complex gas-dynamics flow on the basis of Euler, Navier–Stokes and Boltzmann equation, Karman's Lectures, von Karman Institute for Fluid Dynamics, 15–19 March 1976, in *Numerical Methods in Fluid Dynamics*, eds. Wirz H.J., Smolderen J.J., Hemisphere, Washington, pp. 339–387, 1978.

61. Belotserkovskii O.M., Direct numerical modeling of free induced turbulence, *J. Comput. Math. Math. Phys.* **25**(6), pp. 166–183, 1985.

62. Cantwell B.J., Organized motions in turbulent flows, *Ann. Rev. Fluid Mech.* **13**, pp. 457–515, 1981.

63. Van Dyke M., *Album of Fluid Motion*, Parabolic Press, Palo Alto, CA, 1982.

64. Orszag S.A., *Handbook of Turbulence. Fundamentals and Applications*, Vol. 1, pp. 311–347, eds. Frost W., Moulden T.H., Plenum Press, New York, 1977.

65. Swinney H.L., Gollub J.P. (eds.), *Hydrodynamic Instabilities and the Transition to Turbulence*, Springer-Verlag, Berlin, 1981.

66. Monin A.S., Yaglom A.M., *Statistical Fluid Mechanics: Mechanics of Turbulence*, MIT Press, Cambridge, MA, 1975.

67. Chapmen D.P., Numerical aerodynamics and the prospects for its development, *AIAA J.* **17**(12), pp. 1293–1313, 1979.

68. Townsend A.A., *The Structure of Turbulent Shear Flow*, Emmanuel College, Cambridge, 1956.

69. Goldshtik M.A. (ed.), *Structural Turbulence*, Nauka, Novosibirsk, 1982 (in Russian).

70. Goldshtik M.A., Shtern V.N., *Hydrodynamic Stability and Turbulence*, Nauka, Moscow, 1977 (in Russian).

71. Monin A.S., Coherent structures in turbulent flows, *Russian J. Comput. Mech.* **1**(1), pp. 5–13, 1993.

72. Belotserkovskii O.M., *Numerical Modeling on the Mechanics of Continuous Media*, 2nd edition, Fizmatlit, Moscow, 1994 (in Russian).

73. Belotserkovskii O.M., Numerical modeling of turbulence, in *Etudes About Turbulence*, pp. 137–222, Nauka, Moscow, 1994 (in Russian).

74. Struminskii V.V., *Turbulent Flows*, Nauka, Moscow, 1977 (in Russian).

75. Belotserkovkii O.M., Davydov Yu.M., *Method of Large Particles in Gas Dynamics: Numerical Experiment*, Nauka, Moscow, 1982 (in Russian).

76. Belotserkovkii O.M., Fimin N.N., Chechetkin V.M., Possibility explaining of the existence of vortexlike hydrodynamic structures based on the theory of stationary kinetic equations, *Comp. Math. Math. Phys.* **52**(5), pp. 815–824, 2012.

77. Bandle C., *Isoperimetric Inequalities and Applications*, Pitman, London, 1980.

78. Weston V.H., On the asymptotic solution of a partial differential equation with an exponential nonlinearity, *SIAM J. Math.* **9**(6), pp. 1030–1053, 1978.

79. Jacobsen J. and Schmitt K., The Liouville–Bratu–Gelfand problem for radial operators, *J. Differ. Equations* **184**, 283–298, 2002.

80. Suzuki T., Global analysis for a two-dimensional elliptic eigenvalue problem with the exponential nonlinearity, *Ann. Inst. Henri Poincare* **9**(4), pp. 367–397, 1992.

81. Helms L.L., *Introduction to Potential Theory*, Wiley-Interscience, New York, 1969.

82. Deimling K., *Nonlinear Functional Analysis*, Springer-Verlag, Berlin, 1985.

83. Nicolaenko B., Thurber J.K., Weak shock and bifurcating solutions of the nonlinear Boltzmann equation, *J. Mecanique* **14**, pp. 305–338, 1975.

84. Manela A., Frankel I., On the compressible Taylor–Couette problem, *J. Fluid Mech.* **588**, pp. 59–74, 2007.

85. van de Fliert B.W., van Groesen E., Monopolar vortices as relative equilibria and their dissipative decay, *Nonlinearity*, **5**, 473–495, 1992.

86. van Groesen E., Time–asymptotics and the self–organization hypothesis for 2D Navier–Stokes equations, *Physica A* **148**, pp. 312–330, 1988.

87. Lynden-Bell D., Statistical mechanics of violent relaxation in Stellar systems, *Mon. Not. Roy. Astron. Soc.* **136**, pp. 101–119, 1967.

88. Ellis R.S., Haven K., Turkington B., Nonequivalent statistical equilibrium ensembles and refined stability theorems for most probable flows, *Nonlinearity*, **15**, 239–255, 2002.

89. Naso A., Chavanis P.H., Dubrulle B., Statistical mechanics of two–dimensional Euler flows and minimum enstrophy states, *Eur. Phys. J.* **77**(2), pp. 187–212, 2010.

90. Onsager L., Statistical hydrodynamics, *Il Nuovo Cimento Suppl.* **6**, pp. 279–289, 1949.

91. Bouchet F., Venaille A., Statistical mechanics of two-dimensional and geophysical flows, *Phys. Rep.* **515**, pp. 227–295, 2012.

92. Sano M., Kinetic theory of point vortex systems from the Bogoliubov–Born–Green–Kirkwood–Yvon hierarchy, *Phys. Rev. E* **76**, pp. 046312–046321, 2007.

93. Chavanis P.H., Lemou M., Kinetic theory of point vortices in two dimensions: Analytical results and numerical simulations, *Eur. Phys. J.* **59**, pp. 217–247, 2007.

94. Olver P.J., A nonlinear hamiltonian structure for the Euler equations, *J. Math. Anal. Appl.* **89**, pp. 233–250, 1982.

95. Marchioro C., Pulvirenti M., *Mathematical Theory of Incompressible Fluids*, Springer-Verlag, New York, 1994.

96. Lesieur M., *Turbulence in Fluids*, Kluwer Academic Publishers, Dordrecht, 1997.

97. Fimin N.N., Chechetkin V.M., Hydrodynamic coherence and vortex solutions of the Euler–Helmholtz equation, *Comp. Math. Math. Phys.* **58**(3), pp. 449–460, 2018.

98. Fimin N.N., Orlov Yu.N., Chechetkin V.M., Thermodynamic properties of vortex systems, *Math. Mod. Comp. Sim.* **8**(2), pp. 149–154, 2016.

99. van Saarloos W., Bedeaux D., Mazur P., Hydrodynamics for an ideal fluid: Hamiltonian formalism and Liouville equation, *Physica A* **107**(1), pp. 109–125, 1981.

100. Casetti, L. Pettini M., Cohen E.G.D., Geometric approach to Hamiltonian dynamics and statistical mechanics, *Phys. Rep.* **337**(3), pp. 237–341, 2000.

101. Kambe T., Geometrical theory of two-dimensional hydrodynamics with special reference to a system of point vortices, *Fluid Dynam. Res.* **33**(1–2), pp. 223–249, 2003.

102. Miron R., Hrimiuc D., Shimada H., Sabau S.V., *The Geometry of Hamilton and Lagrange Spaces*, Kluwer Academic Publishers, Dordrecht, 2001.

103. Vasylkevych S., Marsden J.E., The Lie–Poisson structure of the Euler equations of an ideal fluid, *Dyn. Partial Differ. Equ.* **2**(4), pp. 281–300, 2005.

104. Chueshov I.D., *Introduction to the Theory of Infinite-Dimensional Dissipative Systems*, ACTA, Kharkiv, 2002.

105. Vedenyapin V.V., Fimin N.N., The Hamilton–Jacobi method in a non-Hamiltonian situation and hydrodynamic substitution, *Dokl. Math.* **91**(2), pp. 154–157, 2015.

106. Fimin N.N., Chechetkin V.M., The possibility of introducing of metric structure in vortex hydrodynamic systems, *Rus. J. Nonlin. Dyn.* **14**(4), pp. 495–501, 2018.

107. Fimin N.N., Chechetkin V.M., Application of the hydrodynamic substitution for systems of equations with the same principal part, *Rus. J. Nonlin. Dyn.* **14**(1), pp. 53–61, 2018.

108. Chaplygin S.A., *Selected Works* Nauka, Moscow, 1976 (in Russian).

109. Trubnikov B.A., Zhdanov S.K., Unstable quasi-gaseous media, *Phys. Rep.* **155**(3), pp. 137–230, 1987.

110. Karpman V.I., *Non-linear Waves in Dispersive Media*, Pergamon, Oxford, 1975.

111. Richtmyer R.D., Taylor instability in shock acceleration of compressible fluids, *Comm. Pure Appl. Math.* **13**, pp. 297–319, 1960.

112. Meshkov E.E., Instability of the interface of two gases accelerated by a shock wave, *Izv. AN SSSR. Mekhanika Zhidkosti i Gaza* **5**, pp. 151–157, 1969 (in Russian).

113. Henderson L.F., On the refraction of shock waves, *J. Fluid Mech.* **198**, pp. 365–386, 1989.

114. Belotserkovskii O.M., *Turbulence and Instabilities*, MZPress, Moscow, 2003.

115. Rozanov V.B. *et al.*, Experimental studies of gravitational instability and turbulent mixing of stratified flows in an acceleration field in connection with problems of inertial-confinement fusion, Preprint 56 FIAN USSR, Moscow, 1990.

116. Aleshin A.N., Demchenko V.V., Zaitsev S.G., Lasareva E.V., The interaction of shock waves with undulating tangential discontinuity, *Mekh. Zhidk. i Gaza, Izvestia RAN* **5**, pp. 168–174, 1992.

117. Youngs D.L., Three-dimensional numerical simulation of turbulent mixing by Rayleigh–Taylor Instability, *Phys. Fluids A* **3**, pp. 1312–1320, 1991.

118. Town R.P.J., Bell A.R., Three-dimensional simulations of the implosion of inertial confinement fusion targets, *Phys. Rev. Lett.* **67**(14), pp. 1863–1866, 1991.

119. Mikaelian K.O., Numerical simulation of Richtmyer–Meshkov instability in finite-thickness fluid layers, *Phys. Fluids* **8**(5), pp. 1269–1292, 1996.

120. Zhang Q., Sobin S., An analytical nonlinear theory of Richtmyer–Meshkov instability, *Phys. Lett. A* **212**, p. 149, 1996.

121. Cohen R.H. *et al.*, Three-dimensional high-resolution simulations of Richtmyer–Meshkov mixing and shock-turbulence interaction, in *Proceedings of the 6th IWPCTM*, pp. 128–133, Marseille, France, 1997.

122. Ofer D., Shvarts D., Zinamon Z., Orszag S.A., *Phys. Fluids B* **4**, p. 3549, 1992.

123. Yabe T., Hoshino H., Tsuchiya T., Two- and three-dimensional behavior of Rayleigh–Taylor and Kelvin–Helmholtz instabilities, *Phys. Rev. A* **44**(4), pp. 2756–2758, 1991.

124. Zaytsev S. *et al.*, *Proceedings of the 4th IWPCTM*, pp. 291–296, Cambridge, 1993.

125. Dotani T. *et al.*, Discovery of an unusual hard X-ray source in the region supernova in 1987A, *Nature* **330**, pp. 230–231, 1987.

126. Sunyaev R.A. *et al.*, Discovery of hard X-ray emission from supernova 1987A, *Nature* **330**, pp. 227–229, 1987.

127. Van Riper V.K., General relativistic hydrodynamics and the adiabatic collapse of Stellar cores, *Astrophys. J.* **232**, pp. 558–571, 1979.

128. Wilson J.R., Mayle R.W., *Phys. Rep.* **227**, pp. 97–109, 1993.

129. Nadyozhin D.K., Gravitational collapse of iron cores with masses 2 and 10 M, *Astrophys. Space Sci.* **51**, pp. 283–302, 1977.

130. Janka H.-Th., Muller E., Neutrino heating, convection, and the mechanism of type-II supernova explosions, *Astron. Astrophys.* **306**, 1996.

131. Bruenn S.W., Mezzacappa A., Dineva T., *Phys. Reports* **256**, p. 69, 1995.

132. Mezzacappa A. *et al.*, An investigation of neutrino-driven convection and the core collapse supernova mechanism using multigroup neutrino transport, *Astrophys. J.* **495**, pp. 911–930, 1998.

133. Herant M. *et al.*, Inside the supernova: A powerful convective enquine, *Astrophys. J.* **435**, pp. 339–361, 1994.

134. Keil W., Janka H.-Th., Muller E., Ledoux convection in protoneutron stars: A clue to supernova nucleosynthesis, *Astrophys. J. Lett.* **473**, pp. L111–L118, 1996.

135. Chechetkin V.M., Ustyugov S.D., Gorbunov A.A., Polezhaev V.I., *Pis'ma v Astron. Zh.* **64**(12), pp. 34–41, 1997 (in Russian).

136. Zueva N.M., Mikhailova M.S., Solov'ev L.S., Convective instability of a gas sphere, Preprint 65, IPM AN SSSR, Moscow, 1977.

137. Lattimer J.M., Swesty F.D., *Nucl. Phys. A* **535**, p. 331, 1991.

138. Hachisu K.A., Versatile method for obtaining structures of rapidly rotating stars. II: Three-dimensional self-consistent field method, *Astrophys. J. Suppl.* **62**, pp. 461–499, 1986.

139. Belotserkovskii O.M., Chechetkin V.M., Oparin A.M., Physical processes underlying the development of shear turbulence, *J. Experiment. Theor. Phys.* **99**(3), pp. 504–509, 2004.

140. Belotserkovskii O.M., Chechetkin V.M., Fortova S.V., Oparin A.M., Popov Yu.P., Lugovsky A.Yu., Mukhin S.I., The turbulence in free shear flows and in accretion discs, *Astron. Astrophys. Trans.* **25**, pp. 1–16, 2006.

141. Abramovich G.N., *Theory of Turbulent Jets*, Nauka, Moscow, 1984 (in Russian).

142. Matsuda T., Sekino N., Shima E., Sawada K., Spruit H., Mass transfer by tidally induced spiral shocks in an accretion disc, *Astron. Astrophys.* **235**, pp. 211–218, 1990.

143. Molteni D., Belverde G., Lanzfame G., Three-dimensional simulation of polytropic accretion discs, *Mon. Noti. Roy. Astron. Soc.* **249**, pp. 748–754, 1991.

144. Sawada K., Sekino N., Shima E., Instability of accretion discs with spiral shocks, *KEK Progress Report* **89**, pp. 197–206, 1990.

145. Bisikalo D.V., Boyarchuk A.A., Kuznetsov O.A., Popov Yu.P., Chechetkin V.M., *Astron. Zh.* **72**(2), pp. 190–202, 1995.

146. Bisikalo D.V., Boyarchuk A.A., Kuznetsov O.A., Chechetkin V.M., *Astron. Zh.* **74**(6), pp. 889–897, 1997.

147. Bisikalo D.V., Boyarchuk A.A., Kuznetsov O.A., Chechetkin V.M., *Astron. Zh.* **74**(6), pp. 880–888, 1997.

148. Bisikalo D.V., Boyarchuk A.A., Kuznetsov O.A., Chechetkin V.M., *Astron. Zh.* **76**(12), pp. 905–916, 1999.

149. Li H., Finn F.M., Lovelace R.W., Colgate S.A., Rossby wave instability of thin accretion disc, *Astrophys. J.*, 2000.

150. Porter D.H., Woodward P.R., Jacobs M.L., Convection in slab and spherical geometries, *Ann. N. Y. Acad. Sci.* **898**, pp. 1–20, 2000.

151. Colgate S.A., Buchler J.R., Coherent transport of angular moment: The Ranque–Hilsch tube as a paradigm, *Ann. N. Y. Acad. Sci.* **898**, pp. 105–112, 2000.

Index

CPSIA information can be obtained
at www.ICGtesting.com
Printed in the USA
LVHW082013070420
652563LV00003B/6